Good Forestry
From Good Theory
And Good Practices

Essays on Ecological Forestry
& Ecological Design

Second Edition

Alan Wittbecker

Wildlife trees at Altazor Forest (managed by the author in Idaho)

In Appreciation

Very special thanks to Alan Drengson, Editor of *The Trumpeter*, for helping me develop vastly different articles for publication, including two that were published in the journal *Ecoforestry* and the book of the same name. Professor Drengson also worked to help me secure funding and to have me invited to several conferences as a speaker. I am very much in his debt.

Special thanks to Eugene Hargrove, Editor of *Environmental Ethics*, for helping me to develop four articles for publication in that ground-breaking journal, including two later finally rejected by the anonymous reviewers, but published in other journals, with those and further improvements.

And, as always, thanks to Mike Barnes, Norman Bowie, John B. Cobb, Jr., Michael W. Fox, Buckminster Fuller, Twila Jacobsen, Neil & Yoshie Keefe, Devorah Levi, Boyd Martin, Linda Martin, Johanna Metzger, Arne Naess, Eugene Odum, David Parker, David Perry, Theodore Roszak, Paul Shepard, Paolo Soleri, and Precious Woulfe, for their conversations, criticism, suggestions and support, even if I did not avail myself of all of it.

Books by Alan Wittbecker
   Global Emergency Actions
   Topopoetics: Making Good Places
   Eutopias Or Outopias
   Eutopias: Making Good Places Ecologically & Culturally
   REviewing REthinking REturning
   One Earth Many Worlds
   The Poetic Archaeology of the Flesh

Good Forestry
From Good Theory
And Good Practices

Essays on Ecological Forestry
& Ecological Design

Second Edition

Alan Wittbecker

**Cambridge Books &
Urania Science Press
2006**

Published by Cambridge Books & Urania Science Press
  Ecoforestry Institute
  8051 North Tamiami Trail, No. 32
  Sarasota, Florida 34343
  www.ecoforestry.net
  emt@ecoforestry.net

For more information on sites and projects in the text:
  SynGeo ArchiGraph Co.: www.syngeo.net
  Ecoforestry Institute: www.ecoforestry.net
  G. P. Marsh Institute: www.gpmi.us
  Pan Ecology: www.panecology.net
  Rian Garcia Calusa: www.re-design.us
  Eutopian Ecologists: www.eutopias.net

Publisher's Cataloging in Publication Data

Alan Wittbecker 1946

  Good Forestry from Good Theories and Good Practices

Includes Index.
 1. Sustainable Forestry. 2. Forest Design.
 I. Title.

SD398.S87W46 2006

ISBN 0-911-385-21-5

Book Design by Rian Garcia Calusa
Printed in the United States of America

Good Forestry
From **Good Theories**
And **Good Practices**

Contents

Acknowledgments

Ecological Forestry as Relativistic Science, "Process philosophy and ecological forestry," Invited lecture, Claremont College. 11/18/96.

Interactions The *Trumpeter* (www.athabascau/trumpeter) 1999

Gandhian Nonviolence *Pan Ecology* 2(2):1-6. 1987.

Principles 1. *International Journal of Ecoforestry* 12 (3/4):35-40. 1997.

Principles 2. *Ecoforestry* (accepted).

Wild Thinking. Lecture and Notes, Wild Thinking for the 22nd Century Conference, San Francisco, 1998.

Waldgedankenexperiment. *International Journal of Ecoforestry* 12 (3/4):1-4.

Questioning. *International Journal of Ecoforestry* 12 (3/4):7-11. 1997 (developed from an invited lecture at Claremont, 1996)

What is Ecoforestry? Talk at CA SAF meeting, Redding, 2/1998.

Global Planning Flaws (Review of Alan Savory lecture), Invited lecture, 3rd International Ecoforestry Intensive, September, 1996.

No Forestry is Sustainable (Review of Andy Kerr letter), 4th Distance Learning Course, Ecoforestry Institute, October 1997.

Forestry Education (rejected by *Journal of Forestry*, 1999), 3rd Distance Learning Course, Ecoforestry Institute, November 1996.

Why is Old-Growth Cut? Lecture, Ecoforestry Summer Institute, 1995.

Thoughts about Forest Certification *International Journal of Ecoforestry* 11(4):149-157 Electronic version: /www.uidaho.edu/e-journal/ ecoforestry/

Reforestation Goals: Are planning and gambling enough? *International Journal of Ecoforestry* 12(1):1-3.

What is a Forest Worth? *Ecoforestry* 13(1):4-6.

Can Forestry Stand Alone? *Ecoforestry* 13(1):51.

Lies, Dirty Lies, and Dirty Photographs *International Journal of Ecoforestry* 11(2/3): 87-92.

Dying Forests, The Dying of the Trees *International Journal of Ecoforestry* 12 (3/4):53-54.

Trillium Marches on Tierra del Fuego, Editorial, *Seattle Times*,

Fun with Numbers—Forest Financing, Editorial, for Eugene *Register-Guard*, 2/10/98.

Why Forests Cannot be Saved, Lecture, Ecoforestry Meeting, 1996.

Gigatrends. *International Journal of Ecoforestry* 12 (3/4):41-48. 1997.

Ecoforestry Research. *International Journal of Ecoforestry* 11(2/3):69-78. 1995.

Health of Forests. *Ecoforestry* 13(1):18-28. 1998

Forest Practices. *Ecoforestry*. A. Drengson and D. Taylor, eds.
Gabriola Island: New Society Publishers. 1997 (previously in
International Journal of Ecoforestry 10(4):174-183)

Good Forestry. *Ecoforestry Notes* January 2000 (web: www.uidaho.
edu/e-journal/ecoforestry)

Design of Forests. *Pan Ecology* 13(1):1-14. 2001. (originally Distance
Learning Lecture 14, Ecoforestry Institute 1998).

Forest Care Plan. Certification *Report* to Forest Stewardship Council
1998

Forestry as Ecosystem Medicine and Right Action, Closing Address,
Forests for the Future, Vancouver, 1999.

Virtue, Death and Responsibility, Graduation Address, Ecoforestry
Institute, Victoria, 1995.

Preface: Scribbling on Bark

The essays in this collection have been written over a thirty-year period and range from a discussion of the philosophical foundations of preservation and restoration to the design and restoration of forest ecosystems. Most of these essays were published in professional journals, although a few have been taken from lectures or talks at conferences (with minimum changes for the book). In general they are presented in chronological order. This order can also be interpreted as the going from the general to the specific, and from the theoretical to the practical Thus, ideas presented rapidly in the beginning are fleshed out in middle essays and finally applied in specific northwest forests in the final section.

The book is divided into three sections. The first section starts with a presentation of the kinds of interactions in nature, describing normal interactions in nature, such as exploitation and disturbance, and contrasting them with ecosystem-destroying interference, whether natural, e.g., earthquakes, or human, e.g., poisoning with exotic substances. Principles of ecological forestry are derived from philosophical and scientific theories. Metaphysical principles are then applied to radical conservation programs, such as Earth First!, which are designed to protect threatened ecosystems. Conservation and preservation are grounded in a metaphor useful for describing these necessary practices. A larger form of ethics is described, that is mindful of nonhuman beings as well as of ecosystems. Gandhian nonviolence is examined as a model for ethical interactions in using and protecting forests. A series of implications and principles are derived from the sciences of physics and ecology, and then related to ecological field practices in forests. Ecoforestry is shown to benefit from wild thinking, as opposed to domestic thinking, which is dominated by economic motives. Thought experiments are described and recommended for some ecoforestry projects..

The second section questions the philosophy and application of ecoforestry. Questioning is used to examine ideas of global planning, sustainability, education, and certification—the implications of certification are expanded. Through questioning, the worth of a typical forest is addressed, as is preservation and the relation of forestry to other professions and human activities. Clearcutting, new forestry, and forest financing are also questioned.

The third section starts by describing very-long-term trends in forest ecosystem use over the past several thousand years. The health of forests is investigated and defined. Forestry as a science is grounded in ecology and philosophy. A program of forestry research is outlined, using these principles as a foundation. A framework for forest design and restoration is outlined. The practice of ecological forestry is delineated in a series of chapters, using the author's thirty years of experience in the Pacific Northwest. Finally, forestry itself is characterized not only as a biological unified field theory, but as ecosystem medicine and poetic activity, which is the most appropriate kind of activity for working in long-lived, complex ecological systems as a participant.

Readers with a preference for bald assertions and theory can limit their attention to the first two sections. Readers with interest in practices in the field, regardless of the theoretical basis, can start with the third section. Of course, the practice was developed during those years in the field and specifically linked to the sometimes raw or unfinished ideas presented as theories, so readers who can follow the whole trail might enjoy a better understanding of the way things developed.

In any case, these essays are not meant to be perfectly consistent and complete. They cannot be. Nor are they perfectly original. They only extend the ideas of others. Some of the ideas are represented in different contexts over the course of the book. The ideas are meant to stir responses in other people, so that the ideas may be improved and applied, developed and increased, specifically for the purpose of conserving and restoring forest ecosystems at any scale. To discuss any of the ideas, to report errors, or to obtain more information about the Ecoforestry Institute, contact the author at: The Ecoforestry Institute, 8051 North Tamiami Trail, No. 32, Sarasota, Florida 34243 or emt@ecoforestry.net

Introduction: Science Uncertainty Theory Goodness & Practice

This book attempts to place good practices within a framework of good theories, and to stress that good practices require good theories. This is a difficult task, due to the limits of theory as well as to the ambiguity of our ideas of goodness and to the wide spectrum of practices. It is also difficult because many good practices only become evident after generations, and we tend to judge results in months or years rather than in decades or lifetimes. The task is also complicated by the uncertainty of consequences associated with human activities in complex systems. The planet is a complex system, or being, in which we live. We gather information as we live, then use that information to make our lives better. Sometimes we have too much information, and it is contradictory or meaningless. Sometimes our use of information, that is, our practices, result in disasters, which are the opposite of our intentions. So, we develop theories that put the information in a framework that sorts, discards, explains, or predicts the effects of our intentions.

Many people argue that we have too much information; others think not enough. Some people state that we have too many theories; others say not enough. Some people recommend practices that contradict traditional practices. Some modern practices contradict each other. We must create a way to think about these things and determine how theory works with practice, and to see what theories, as well as how many, are necessary.

Science and Limits
In order to make sense of the sheer multiplicity and complexity of their environments, human beings create abstractions. An abstraction is an idea created to refer to all objects that have certain characteristics in common, e.g., all birds. Abstractions can be generalized, e.g., all things that fly, but at each outer level the objects have less in common; thus flying things include insects, mammals, reptiles, seeds, and spores.

Human beings also classify and label their abstractions. The systems of classification are reflexive and pragmatic; that is, they refer to the classifier as well as to the object, and they are guidelines for how to think about, treat, and relate to an object (according to S. I. Hayakawa). As soon as a classification is no longer useful, people stop using it and become open to a better classification.

Science uses abstraction and classification to transcend common sense to describe the fundamental structure of nature. Scientists use classes to limit things to their own mesocosmic scale, from geological epochs to species, although the classes are independent from the things. By comparison, archaic (sometimes called pre-scientific) peoples believe that there is a necessary and intimate connection between the symbol and the object, that the name is attached to the object. This "epistemic naiveté" is dismissed by modern science, which uses its own unique method. The general format of the scientific method:

- Observe phenomena and record facts.
- Analyze the phenomena into components.
- Measure the phenomena (before and after manipulation) allowing the results to be quantified.
- Make guesses and generalizations, using the logic of deduction. A hypothesis is a statement about relationships that can be shown to be untrue.
- Formulate laws from the generalizations about the phenomena. Such laws may describe the behavior of a natural system.
- Describe the laws mathematically, using numbers.
- Develop a theory to predict new phenomena. Theories can lead to new conclusions and sometimes altered perspectives about phenomena. A scientific theory is a statement that postulates ordered relationships among natural phenomena and explains some aspect of the world. It allows one to ask certain kinds of questions, some as specific hypotheses.
- Test hypotheses and theories in a controlled environment. A theory cannot be tested by hypotheses.

The key ideas here are analysis, measurement, guesses, control, and rule-based predictions, and control. The use of these ideas have brought about keen understandings of parts of nature, from the quantum level to the extent of the universe. And, it has lead to amazing changes, from electricity to computers.

But, science, like archaic thought, makes assumptions in the context of specific cultural situations. Some cultural-scientific assumptions have been incorporated into the works of science by scientists. At the time of Bacon, it was assumed that there was: An absolute, immutable, omnipotent God, everything was sorted into a great chain of being, economic subsistence was preferable, and social inequality was unavoidable. Later, Darwin incorporated a different set of assumptions into his theories: Absolute space-time; atoms as discrete units; economic discrimination, and continued social inequality. These assumptions contributed to the misuse of his theories to justify social and economic conditions at the time.

Other assumptions of science are more subtle and more limiting, such as predicate logic, invariance of regularities of nature in time and place, immutable classes (which were also considered eternal), and styles of metaphors. Furthermore, science is concerned with knowledge of kinds, not necessarily knowledge of individuals. The limits at both ends of inquiry, prokaryotes and the whole planet, thus fall out of range of the process. Even scientific facts have been argued to be merely consensuses among socially interacting scientists. A scientific fact is a product of a complex social process beginning with individual observations and terminating with a stylized statement of truth that fits into the scientific knowledge of a society. A short discussion of logic and metaphor can suggest directions for an ecological science to take.

Logic. Most Western science assumes a predicate logic and is constrained by that logic. This Aristotelian logic is deductive and binary (where categories are mutually exclusive, and substance and identity are permanent). In this logic, contradictions are false by definition. Difficulties of this logic are evident in many instances in physics and ecology, e.g., light cannot be a wave *and* a not-wave, or wolves cannot be a keystone species *and* not necessary to an ecosystem.

Other logic systems exist, although they are often ignored. Magorah Maruyama calls Aristotelian logic homogenistic (and notes that it is also the basis of Hindu thought and Western Idealism and science). He also distinguishes three other epistemologies that each have a typical logic based on them: (1) Independent-event, with an inductive, statistical logic (e.g., some kinds of science, and existential thought); (2) Homeostatic, with a complementary logic (e.g., Chinese thought); and (3) Morphogenetic, with a logic incorporating change, harmony, heterogeneity, and nonrepeating and irreversible processes. (e.g., Mandenka, Navajo, and Eskimo thought).

Independent-event, for instance, is the reverse of homogenistic logic. The homogenistic logic can be characterized as classificational quantitative, competitive, uniform, nonreciprocally causal, and hierarchical. Morphogenetic logic is also quite different from homogenistic logic. By contrast, the morphogenetic logic is relational, qualitative, symbiotic, heterogenistic, reciprocally causal, and interactionist. Reciprocal causal processes can increase structure, differentiation and complexity, according to Maruyama. Such a logic is more useful for addressing complex operations in ecosystems. As a many-valued logic, it can address values other than truth and falsehood. It can avoid fallacies embedded in two-valued logic.

There seems to be no superlogic that can combine all these categories, since they are based on different epistemologies and cosmologies. A broader logic, however, such as the morphogenetic, would work at a metalevel and avoid many of the problems of polar thinking.

Metaphors. The entire activity of science is guided by metaphors. Metaphors can allow scientists to deal with complex situations. Metaphors emphasize likenesses between living things and languages or human constructs. Metaphor in the broad sense includes similes, analogies, models, patterns, and paradigms. A metaphor does two things, basically; it:

1. Furnishes a label and
2. Emphasizes similarities.

By doing these things, it not only defines and extends new meanings, but redescribes domains seen already through one metaphoric frame. With good metaphors, logical consistency is no longer required.

A metaphor is composed of two parts, a focus and a frame. The focus of the metaphor is a primary system, which refers to that which is to be understood: for instance, the brain or atom. The frame is a secondary system which provides the pool of names to be used, such as computer or solar system. The interaction of the systems is mutual: the secondary system imposes a reorganization on the concepts of the primary, but the use of the metaphor alters the perception of the secondary system as well as of the primary. Not only are the two concepts altered, but the meaning of the concepts is altered. And this "halo" meaning, in its unique irreducibility, permits things to be said, which could not be otherwise said.

In the secondary and primary systems ('the brain is a computer") opposites interpenetrate and are unified—the metaphor is initially, cognitively perceived as absurd. This interpenetration is a dialectic, where opposites are unified in a metaphorical synthesis, which transcends the initial conjunction of the two systems.

Concept formation in science is a working out of contradictions. Logical opposites (S and not-S) logically combined assert nothing; metaphorical unity transcends that literal absurdity. Certain scientific metaphors exhibit this: Leucippus's "The world is atomic," the Pythagoreans', "The world is mathematical," and Benjamin Franklin's, "Electricity is a fluid." Metaphors in science are termed models, and are used to bestow a more precise meaning to expressions.

Metaphors have far-reaching consequences on behavior as well as on the form of theories. The metaphor of mechanism by Descartes ("the world is a machine") implies that matter is inert and that machines must have a creator, that is, they are not self-making. This metaphor has been used to justify the destruction of species and places (as well as certain kinds of repair and restoration). Although the metaphor is no longer used in physics, ecology is still struggling with the machine, as well as economic, metaphors ("nature is an economy of trade").

The word individual is a metaphor. As such it refers to aggregates (sand), functional units (populations/species), and autonomous, or self-regulating beings (a wolf). The word whole is a metaphor. A 'thing' is a metaphor, though so old it is considered a real term. A thing is that which can be separated from other things, thus intimately related to the idea of boundaries. Where we make boundaries usually depends on past experience. The boundary is Janus-faced. The metaphor has a boundary. The meanings of a metaphor can be ontological (physical things) or structural (quantified or valued).

Most theories are based on old mechanistic metaphors, such as that "nature is a machine" or "time is an arrow." In addition to considering other logic, a good theory could be based on newer, more flexible metaphors.

- Holism (Smuts, 1912) and holons (Koestler, 1968). The whole is a powerful organizing principle inherent in nature. All complex structures and processes of a relatively stable character display hierarchical organization, where the levels of a hierarchy tend to be contained in subwholes that behaves as a whole to its components and as a dependent part in context.
- Self-organization and constructionism (Varela 1982), instead of natural selection. In the autopoietic framework of Varela, every being is embedded in a world and observed by an embedded observer.
- Reciprocally constrained construction (Gray 1988), in place of adaptation. The organism and environment are co-implicative, co-defining, and co-constructing. A process of self-assembly, where the self is the organism/environment system.
- The field concept (Waddington 1962) for development, emphasizing dynamic transformation (form as organized

spatiotemporal domain), in contrast with the particulate concept of an organism. Understood in terms of group dynamics rather than selective advantage or cost/benefit.

- The process view (Whitehead 1920) in which organisms are dynamic structures that are immanent and simultaneous with the process, less than a consequence of natural selection of past random mutations. Like quantum physics, in biology and ecology, the observer is within the theory in a very fundamental way. By the act of observing the observer influences the outcome of a phenomenon, as Wheeler says, taking part in the construction of physical reality.

These metaphors are used in different ways during the discussion of principles is several essays.

Since all science uses logic, Godel's incompleteness theory renders science incomplete. In exploring mathematical reasoning, Godel came to this Incompleteness Theorem; paraphrased, it reads: 'All consistent axiomatic formulations of number theory include undecidable propositions.' What Godel showed was that provability is weaker than truth, in any system. Karl Popper concludes that the very nature of science is to be incomplete. To be complete it would have to give an explanatory account of itself, which is not possible.

Science has a way of working out so that some theories undermine their own positions, and new theories are developed. Quantum mechanics, for instance, eliminated the notion of the neutral observer—the observer participates in the act of measurement and the universe changes thereby. A similar principle of participation can be postulated for ecology, where the observer changes the system with her presence and actions (even those intended to be neutral). The ideas of ecology and physics benefit from cross-fertilization.

Both physics and ecology highlight the chaos and uncertainty we face in dealing with large, complex, long-lived, wild entities like forest ecosystems. Theories are needed because of that uncertainty. But, uncertainty itself is an umbrella term. The different kinds of uncertainty are relevant to our practices and management techniques.

Uncertainty

There are many levels of uncertainty, from the quantum to the human scale. As physicists have found, elementary processes at the quantum level are not subject to a precise description in time and space. Predictions about location and velocity are just statements of probability—this is Heisenberg's uncertainty principle. The effect of this principle on epistemology is that our exact interpretation of things has to be abandoned. There are uncertainties at higher levels of organization as well; for instance, the hysteresis of some magnetic solids determines their subsequent behavior—without knowledge of initial conditions and all past events, it is not possible to predict present or future behavior of those particles or solids. Furthermore, the higher up in levels of organization, the more kinds of uncertainty there are. There can be genetic uncertainty, as well as environmental or social uncertainty.

Uncertainty can also be broken down further (most of these examples are taken from the work of Max Black—all works are listed in the bibliography). There is fundamental uncertainty in the thing/ event/ pattern, as well as in the channels (noise, meaning) of any relationship. At the human level, uncertainty can be distinguished between ignorance and conflicting knowledge. Ignorance itself can be sub-divided into: vagueness (i.e., indeterminate knowledge); probability (i.e. confidence in partial knowledge); incompleteness (i.e., missing elements of knowledge); irrelevance (i.e., the place in the pattern being unknown); and, fuzziness (i.e., overlapping interpretations). Conflicting knowledge can be broken down into: anomaly (i.e., the incongruity of knowledge, or simple error); ambiguity (i.e., alternative interpretations of meaning); inconsistency (i.e., simultaneous untruth); equivocation (i.e., knowledge that is both true and untrue); and, belief (i.e., confidence in subjective knowledge, or taboo). There are probably other components of uncertainty, as well as other ways that they could be combined into a typography.

We have to live with uncertainty, but this means that our decisions are essentially gambles, based on our ability to work within real ecological and physical limits. Gambling is a profession that acknowledges the operation of chance and makes conclusions in the absence of facts—we need to remind ourselves that few people are successful at it. This is an important admission for ecologists and managers, that we do not have facts on which to base our actions, that nature is a stochastic process, and that ecosystems always changing. Furthermore, we do not know for sure what effects our actions will have on ecosystems. Successful gamblers suggest that the proper attitudes for gambling with nature are awareness, humility and courage, rather than arrogance, fear and maximum use. We should adopt a precautionary principle, which asserts that, if harm is threatened, and if there is uncertainty about the seriousness of the harm, then precautionary actions must be taken. Since the 1970s, in fact, this principle has been incorporated into Swedish and German environmental laws.

This principle means that not doing something, "benign neglect," becomes a valid management option. Carl Walters suggests that inaction is an inappropriate alternative to gambling as a result of confusion, but inaction is a very appropriate alternative in the face of confusion—when in doubt about an ecosystem, we should not interfere. As appropriate alternatives, both inaction and action can be suggested by theory.

Theory

Making theories is an intellectual activity, in the large sense of those words. There are three kinds of intellectual activity, according to Aristotle (in the Metaphysics, E.1.1025b25). The first two are theoretical knowledge (*theoretike episteme*) and practical knowledge (*praktike episteme*). These two activities, or sciences, involve different ends. The theoretical sciences involve necessary propositions and have knowledge as their end; the practical sciences subordinate knowledge to action. Philosophy, mathematics, physics, and theology are each a form of theoretical knowledge. The practical sciences, such as agriculture and forestry, derived many propositions from the theoretical. The sciences become less exact as they involve more elements, and thus the practical becomes more dependent on the theoretical.

A theory is an explanation or system of everything. It is an exposition of the abstract principles of a science or a speculation. The dictionary definition states that a theory is a reasoned expectation, as opposed to practice. A scientific theory is a set of general explanatory statements about a natural process; the set is related as a model, which is a human linguistic construction, subject to human perspectives and limitations. Theories may incorporate guesses, critical observations, experimentation, and logical inference, as well as hypotheses and laws. A theory must be specific enough statement, such that experimental results can be assessed as negative or positive. The results of an experiment can confirm or justify a theory, although other experiments could show that the theory is wrong. Good theories tend to have a limited number of generalizations. Theories can also be nested, so that, for instance, the theory of gravitation might be combined with other theories in a unified field theory.

Science works with theories, that is, specific directions of observation, based on previously identified facts, to lead to evidence, or proofs of hypotheses, about how things work, which can be used to predict other actions in the future. Theory is used to guide experimentation. But much experimentation these days occurs without theoretical guidance, or rather the theories used are not capable of addressing the context itself, so that the experiments tend to go out of control. Many experiments are simply conducted improperly (with no control group, no good theory, or no control of scale).

People who know a lot, but do not have theories that bind them to wise actions, tend to ignore ignorance, or to act in an arrogant manner; such behavior can be dangerous. Theory can reduce this kind of danger. Theory can describe the limits of itself, of knowledge and understanding, so that we do not try to that which cannot be done, e.g., perpetual motion machines. That is, theory deals with ignorance, the kinds of ignorance, especially the kinds that cannot be removed. Theory also deals generally with ambiguity and uncertainty.

Theory has to account for those things that cannot be reduced or explained or understood. Some mysteries cannot be reduced, cannot be understood with human understanding. Some things are ineffable; even if they are understood, they cannot be reduced to words. Our theories and practices are human theories and practices; they are limited by real human limits, by the dimensions and terms of times that we humans experience.

Theories can help science to avoid claiming what things, either knowledge or certainty, that it cannot claim. Theory can distinguish between knowledge and theory, between theory and the unknowable. Theory needs to address the incompleteness and limits of science. There will always be limits and errors; these cannot be avoided. Theory needs to accommodate this truth. Fallibility is unavoidable; it is a part of certainty from a limited perspective with a limited knowledge; it is part of being incomplete and using incomplete knowledge. But, this is the limit of all being; this is the limit of finiteness.

What we know will always be incomplete, but we have to act under those circumstances, and therefore we have to act on partial knowledge in the face of partial ignorance. Theory is a guide to acting, under different kinds of unavoidable and irremovable ignorance, based on thinking about previous actions and projecting future actions. Theory is not necessary to get more information. But it is necessary to understand what can be predicted or understood. Theory can help distinguish between local and global problems. Theory can describe the fitness of our conduct within an ecological and cultural system.

Most theories in this century have been dominated by the impoverished philosophies of modern industrial cultures. These theories have no way of dealing with the profound changes of global warming and habitat destruction, extinctions, social decay, new diseases, or wild technology. These changes are modifying the planet and may destroy us, as well as those things we require for adventure or inspiration.

To be really effective, theories need to be rooted in conceptually rich philosophies, such as process philosophy, phenomenology, general systems theory, ecophilosophy, deep ecology, ecofeminism, or radical ecology, that are more fruitful and considerate (of nonhuman beings and systems)—and in fact consider such problems as what is good or what are our human limits. These philosophies contribute to a new attitude that is more appropriate to the unity and interrelatedness of the earth. These philosophies provide a number of fundamental philosophical, historical, scientific, and cosmological principles that have been presented in other contexts by thinkers such as Cobb, Einstein, Fox, Naess, and Whitehead. Very few of these principles are absolute. Nevertheless, they are essential to the understanding of ecosystems. Principles, combined with common sense and good judgment, are necessary as guides in the absence of definite knowledge. They give us a deep foundation and a broad predictive ability that we need for good practices.

Goodness

Science is a useful system for realizing new knowledge, but it has trouble dealing with concepts like good or bad. We can learn about forest processes and try to manage them, but it is problematic to designate good theories or good practices. Normally we do not think of physical or biological events, such as gravity or cell division, as good or bad; they simply exist in a neutral way. These events are valued in the human realm by human intention, which is open to interpretation and ambiguity, and which is described by human symbols and signs.

Signs are arbitrary, as Ferdinand de Saussure noted; that is to say, they have no necessary connection to things. Signs can make many connections to physical and biological events, thus intensifying them; signs can also become dense with human meaning over time. It is the play of signs in human language and human culture that results in judging something good or bad; that is, the manipulation of signs, in a field of surprise (due to many levels of meaning or uncertainty), produces degrees of goodness.

Cultural structures provide rules that limit play and action (so that it stops short of death or destruction). All cultures work that way. Ethics and economics, for instance, are rules of behavior (and politics is the practice of changing the rules as society changes). Signal play without cultural structures can result in problems (especially when science claims to be independent of culture).

Goodness is thus intention and action in the context of the rules of a culture, using ambiguous signs. Goodness is a feature of the path of actions, as is badness (or evil). Of course, many cultural codes divide everything neatly into good or bad, but, everything is not that simple, and goodness cannot be grown and picked, like a ripe tomato. By trying to focus on either extreme, of pure goodness or pure badness, we miss the ambiguity and uncertainty of everyday situations, most of which occur in a mixture. Our ethics and ideologies are not comprehensive enough to help us live in this mixture, with the inevitability of uncertainty, and with the possibility of enantiodromia (that is, creating the opposite of our intentions). Using new logic and metaphors can help us to adjust.

Good theories are just those that incorporate cultural values with ambiguity and uncertainty. Good theories are grounded in place and in communities. They must be used to place our science and technological capabilities in a culture that has goals and limits for our actions. In this sense the theories are tools for adaptation to place, within place, within specific places that we work towards making good.

When we set goals, as for the goodness of our forestry practices or ecological restoration, we base those on the meaning of the symbols, which can have many more than one meaning. A shift in context can change the meaning. Because there is no absolute good in our practices, it is difficult to determine the best practice to use, especially when applied to long-lived ecosystems, such as forests and deserts. Goodness is relative, but we must make decisions regardless. Throughout this book, the term goodness is related to overall harmony (see Chapter 27 and elsewhere for an extended discussion), but it cannot be considered independent of practice.

Practice
Practice is the actual doing, repeating, and acting (from the Greek, *praktikos*, fit for action, from *prassein*, to do; the Greek *poieein*, means to make). Practice is necessary to supplement science, logic and ideas of goodness (as related to our intentions and actions). Traditionally, practice builds on the successes of the past, without consideration of scale, limits or development. Even under conditions where we know the practices to be ineffective, such as building single family homes with chemically-maintained grass lawns over functioning wetlands, we continue the practice.

Many other approaches to practices are possible, however. For example, a Fuzzy formalism is one approach. Fuzzy sets are a representation of ambiguity (partial knowledge), which is related to uncertainty; probability is considered a subtheory of fuzzy set theory. Because of its generalization, fuzzy set theory has a larger scope of applications to any real world processes involving incomplete or uncertain data. The notion of a fuzzy set is completely nonstatistical in nature.

Game theory, as a theoretical network for decision-making, is another approach; a minimax theory shows that real world situations can be ordered in a game theoretic network. For example, the farmers in Jantilla in middle Ghana, expect dry years 58% of the time and wet years 42%; by creating a two-person-five-strategy zero-sum game (after Peter Gould), they can minimize their losses by planting maize (77%) and rice (22%), rather then a mix with yams, cassava, and millet.

Complexity theory (or Chaos theory) models can yield effective predictions, if both model size and model randomness are both minimized. Newness and intrinsic emergence are incorporated, while the observer is a subprocess in the model.

Common sense (or archaic ecocentric epistemology) is a basic approach; we have learned to survive in nature despite inaccurate estimates; furthermore, we reject many possibilities that are inconsequential. In the Amazon, the Paye (shaman) of the Desana Indians determines the fish populations (meenga-siba, etc.) by consulting with the master of fish (Val-mahse); a number is decided on, even if people have to go hungry, and souls are traded for the rights to fish. Uncertainty is reduced by observation (as in any model). Bottoms-up approaches have worked over long periods of human habitation.

Some of these examples may be complementary with each other. Others depend on a deeper knowledge of resources. Ecological planning may use several of these in a framework. Some of these are used in our restoration and management of forest ecosystems at the Ecoforestry Institute.

Contrary to current popular opinion, good practices really need to be based on good theory. Of course, some good practices can be the result of long traditions or of personal feelings, or even of accident. But, the only way that we can guarantee good practices is to base them on fully worked out, comprehensive theories.

Theory and planning are the foundation for good practices, in every kind of ecosystem, especially forest ecosystems, which are being used and abused at an accelerating rate. With practices based on good scientific theories and designs, ecological forestry, for instance, proposes ecologically responsible use of the forest within its limits of productivity and stability— this book presents some examples. Limiting the practices to net community productivity or even a percentage of net primary productivity would help save forests, but it would require the application of better planning and design, even a reevaluation of human needs. This may take more than theoretical and practical knowledge.

There is a third kind of knowledge, according to Aristotle; it is productive knowledge (*poietike episteme*). The productive sciences have a product as their end, not knowledge or action. Productive, or poetic, science derives many propositions from both other sciences, and because it involves more elements, poetic science is the least exact of all. Yet, in a sense, poetic science is more basic than the others, since all sciences produce an end. Poetic knowledge is inseparable from the power to make. Poetic knowledge is a kind of knowledgeable activity with a product for an end. Aristotle defined poetic art at various times as: having to do with creation; using powers; "principles of change in another thing;" and being concerned with the same thing as "chance." Poetic science is a making, a fashioning of random data into a significant statement of universal relevance. The whole is structurally unified into a complex thing. This unity is based on the organic theme in Aristotle. Art is a mimesis of nature, but it does not copy nature's products. It presents unified wholes, in a like manner. Works of art are structured like living things; when things exhibit unity, then they have order, and when things have a definite size and order, then they are beautiful, in art and in nature. In fact, only by being organically structured can art be mimetic; and only by being mimetic to life can art works be organisms. Poetry imitates, not fragmentary reality, but the essential whole.

Our designs of ecosystems need to be poetic productions. Ecological forest design, for instance, is the design of whole communities. We can design places as organic wholes to promote the well-being of individuals and the common good. But, we can only really do so as participants of ecosystems (and this takes us back to the fundamental lessons of physics). Humans need to recognize that they automatically participate in everything. Furthermore, due to the uncertainty in dealing with large, long-lived systems, we have to learn to accept that the system needs most of its own productivity and to limit our use of the systems to well below critical limits, that is, to within the flexibility of the system.

The proper role, ultimately, for theory is to guide us in our choice of tools and strategies of practice for dwelling on the planet. The survival of human societies depends on the consciousness of the local and global systems in their complexity, limits, and connectedness. Like the Buddhist concept of "right action," which recognizes that the individual has to make decisions based on the consciousness of the effects of those decisions, our practices can fit good theories to suitable actions and creative productions.

Theory Science & Uncertainty

Chapter 1
Ecoforestry as Relativistic Science

Ecoforestry has been defined as selection forestry or restoration forestry. It has been defined as a context-based, community forestry that is based on traditional wisdom combined with scientific knowledge. It could be defined as the pragmatic study and use of genealogical actors in ecological roles (in the evolutionary play in the ecological theater)—the activities of forestry certainly make good theater, with protesters trying to save redwoods, the deafening crashing of trees and roaring saws, and the ant-lines of full logging trucks.

Ecoforestry is basically concerned with using a new metaphor for forests and forestry. Although ecoforestry sounds like a qualification of the modern, industrial forestry, it is in fact entirely different—it is based on a different metaphysics, a broader ecology, a more comprehensive economics, and it is sensitive to limits, ethics, aesthetics, and spiritual values. *Its larger perspective incorporates industrial forestry as a special case, much like the theory of relativity incorporated Newtonian dynamics as a special case.*

Ecoforestry addresses forests and human societies. Ecoforestry includes humans and ultrahuman beings as an integral part of the whole system. Ecoforestry addresses both large and short-term scales (in time, size, and design). Many forests, however, are occupied by people with different (and often better adapted) cosmologies and economies. Industrial forestry partially uses physics and ecology. It ignores philosophy, ethics, and equity. These are basic tools for ecoforestry.

To some extent, ecoforestry is a crisis science, much like conservation biology, from which it has taken many ideas and principles. Consider that one to six billion trees are being cut every year; entire forests are being removed or converted, while others are degraded and simplified. Traditional forestry has been defined as "caring for forests," but in the US alone, 6 million acres of forest are cleared annually, 5 million acres are degraded, and 3 billion cubic meters of wood are consumed. Notice that forestry is concerned with numbers and measurements, but rarely do they reflect whole trees and living habitats and individual animals; 3 billion cubic meters of wood probably came from 1-6 billion trees (imagine, 1 to 6 billion trees). And each tree is home to squirrels, birds, beetles, lichen, and fungi. Some beings, such as pileated woodpeckers, are so territorial that they die with the tree, just as hamadryads were once thought to scream and die as their tree was cut.

The dominant form of forestry is based on the agricultural model of simplify, harvest, and replant. This, and the economic problem of short-term debt loads for forest corporations, causes numerous problems with forests: Wild forests are clearcut; diversity is diminished; erosion increases; and aesthetic ruin can be seen from airplanes and highways.

Unlike forestry, ecoforestry includes humans and ultrahuman beings as an integral part of the system. Ecoforestry addresses both large and short-term scales (in time, size, and design). Let me make the contrast explicit by comparing the study of ecoforestry with the curriculum of a forestry school.

Table 1: Ecoforestry Extent

Ecoforestry	Industrial forestry
History of forests	(ignored)
History of forest use	
Cosmology and Myth	(assumed)
Economics	
Ecological, Coop.	
Physics and Energy in Forest	(partially used)
Ecological Science	
Structure/Function	
Succession/Maturity	
Characteristics	(mentioned)
Applied Ecology	
Conservation Biology	
Landscape ecology	
Applied Philosophy	(ignored)
Ecological Phil, Practice	
Ethics	
Applied Education	
Forest Production	Production
Kinds of forestry	
Valuation	
Preservation	Silviculture
Restoration	
Protection	Protection
Mensuration	Mensuration
Monitoring	Surveys
Procedures/Tools	Valuation / finance
Silviculture	Silviculture
Regeneration (Wild/Artificial)	
Regulation	Regulation
Management	
Tree Farm Design	
Administration/Finance	Administration
Forest Utilization	Forest utilization
Wood properties	Wood technology
Harvest Cutting	Harvesting
Beginning/Intermediate	(cutting/transport)
End Techniques	
Ecoforestry Design	
Certification	Forest Economics
Sustainability	Policies
Goals (Local, regional, global)	
Practice as a way of life	

Industrial forestry ignores the history of forests as well as the history of use. It assumes the cosmology of the industrial age. Many forests, however, are occupied by people with different (and often better adapted) cosmologies. The Yaruro, for instance, believe that all beings are equal, not just men, as the industrial cosmology assumes; a culture that sees humanity and nature as existing in a mutual harmony will have a different kind of forestry than the modern version. In believing that imperfection cannot be avoided, the Dahomey, for instance, are less likely to try to create a perfect set of circumstances, unlike the industrial cosmology, which holds that 'men are perfectible.' Neither the Dahomey or Yaruro have a formal kind of forestry, but neither destroy their forests.

Industrial forestry assumes also the current economic system, which is based on winners and losers and on inequality. Economics has not been unsuccessful with its models, for instance of buying behavior, but it has become a highly abstract academic discipline. All its abstractions are applied to the real world without acknowledgment of the high degree of abstraction involved. The philosopher A. N. Whitehead warned that the economic method would triumph if the abstractions were judicious, but even judicious abstractions had limits, and the neglect of those limits lead to disastrous oversights. Considering a fictitious human nature under imaginary circumstances and thinking it is real is the fallacy of "misplaced concreteness" according to Whitehead. Daly and Cobb suggest that the classic instance of the fallacy in economics is "money fetishism," where the characteristics of an abstract symbol, such as limitless growth, are applied to real commodities and values. Misplaced concreteness also occurs in forestry itself. Genetic reductionism is a good example of the fallacy. Genes only function within the organism, which is the creative being; genes are not creative.

Industrial forestry partially uses physics and ecology. It ignores philosophy, ethics, and equity. These are basic tools for ecoforestry. Let me make several general, simple comparisons to distinguish the two.

Table 2: Forestry Opposites

Industrial is	Ecoforestry is
linear	nonlinear
ordered	chaotic
short-term	long-term
material economics	ecological economics
profit	nonprofit
decreasing	increasing
capital use	interest use
infinite	limited
functional replacement	interworking
material	spiritual
large-scale	small-scale
masculine	feminine & masculine
mechanical	organic
single-value-timber	being-value-multiple

simplicity	beauty
partial	holistic
isolated	reciprocal
independent	dependent
solo	in harmony
competitive	cooperative
growth	development
static	evolving form
fast	slow
shallow	deep
unit	pattern
extrinsic	intrinsic
amoral	ethical
symbolic	material/spiritual
perfectible	satisficing
information	wisdom
corporate	community
global	local
homogeneous	heterogeneous
dominate	participate
absentee	presence
stand	watershed
agricultural model	forest model
anthropocentric	ecocentric
machine logical	other logical

Forestry came to be treated as a special form of agriculture. Indeed, forestry has several parallels with agriculture. Like agriculture, forestry uses soil to produce a crop for the purpose of increasing wealth (or perhaps just revenue). Like agriculture, forestry is renewable (unlike mines or oil extraction). Like agriculture, forestry is based on knowledge of many fields, including botany, soil, and meteorology. Like agriculture, forestry deals with vast areas. Unlike agriculture, however, forestry deals with wild plants on wild soils. Furthermore, trees are very long-lived, unlike crops of annuals, and are related in complex cycles, much more so than annuals. Trees are directly responsible for soil fertility and tilling. Despite similarities, forestry cannot be considered a form of agriculture.

As a science, forestry is based on the modern paradigm of science, the study of discrete units, like atoms, in isolation from their context-since the context would increase the complexity of the situation beyond the capacity of the mathematics to describe it. Western science is based on an Aristotelian logic, a Cartesian dualism, and atomism. It is basically concerned with regular, reversible events. This kind of science has never been good with unique, irreversible, long-term, complex, or catastrophic events, the kind that occur in forests, although some new approaches, such as chaos theory, are promising.

Western logic is Aristotelian; it is deductive and polar; it is either/and rather than both/and. Ecoforestry considers the morphogenetic logic of the Mandenka or Navajo. Of course, having a different logic means that ecoforestry does not make lists like the one above.

Ecoforestry supplies the missing dimensions in forestry. European forestry in the mid 19th century was concerned with the entire forest and was relatively holistic. Tourney spent a good part of the book describing the reciprocal influence of trees on their environment; perhaps only a paragraph on end cuts, such as clearcuts. By the early 1900s at Yale, for instance, Hawley's book and later Smith's concentrated on end cuts; with perhaps a paragraph on influence and ecology. For example, compare the subject areas in two books on silviculture (the care and cultivation of forest trees, from the Latin word for forest, *silva*), written by two professors at Yale University, James Tourney (1928) and David Smith (1986).

Table 3: Forestry Text Comparisons

Tourney	Smith
Introduction-Definitions	Silviculture
Environment of Forests	Intermediate Cutting
Influence of Forests on Env.	Regeneration
Forests	Harvesting
Form and Life of Trees	Reproduction
Development of Stands	Clearcutting
Growth and Yield of Stands	Shelter wood
Tolerance	Selection
Vegetational Unit	Class Regeneration
Forest Succession	Management

Tourney was concerned with the definition of silviculture-the development and care of forests-where Smith addressed how to cut, spray, and clean them. Tourney allows one page for a discussion of clearcutting—Smith over 40 pages. Tourney spent over 300 pages discussing the relationships of trees, animals, plants, fungi, and humans to the forest; Smith devoted a few pages to climate and fungi.

What happened? Actually, Smith's book was based on Ralph Hawley, whose book editions ran from 1921 to 1962. Hawley addressed foresters who had their "feet on the ground." (Who knows where their heads were.) Toomey's work is based on the German tradition and is written for foresters, teachers, researchers, and ecologists. Toomey's book was still used in the forties and fifties. Tourney is out of print now. The Hawley/Smith approach has been dominant since the 1950s. Smith's book is still used at Idaho, OSU, and other schools of forestry today.

Philosophy

Ideas and theories in this century, not to mention practice and technology, have been dominated by the impoverished philosophy of our modern industrial culture. A.N. Whitehead argues that the impoverishment of our conceptual universe led to the disaster of physicals and to no metaphysics at all. Philosophy has not addressed the profound changes of global warming and habitat destruction, extinctions, social decay, new diseases, or technology. These changes are modifying the planet; some of them result from our dysfunctional styles of living, which may destroy us as well as those things that we require for adventure or comfort.

To be effective in this age, ideas need to be rooted in conceptually rich philosophies, such as process philosophy (Whitehead), deep ecology (Naess), phenomenology (Merleau-Ponty), general systems theory (von Bertalanffy, Laszlo et al.), ecological resistance (John Rodman), ecophilosophy (Henryk Skolimowski, David Klein et al.), ecofeminism (Daly, Griffin et al.), and radical ecology (Wittbecker, Merchant et al.).

The matter of these areas highlight the uncertainty we face in dealing with large, complex, long-lived, wild entities like forests. Forest managers have to live with uncertainty; this means that management decisions are essentially gambles. Gambling is a profession that acknowledges the operation of chance and makes conclusions in the absence of facts—few people are successful at it. This is an important admission, that we do not have facts to base our actions on, that nature is a stochastic process, and that ecosystems are always changing. Furthermore, we do not know for sure what effects our actions will have on forests, which used to live for so long, in such diversity, in so many places. Successful gambling suggests that the proper attitudes for gambling with nature are awareness, humility and courage, not arrogance, fear and maximum use.

Science

Philosophies, western and eastern, northern and southern, are colored by many underlying myths and logics. Even science is limited by its logic and myths. Industrial forestry is based on modern science, which itself is based on inappropriate metaphors and assumptions that are subtle and limiting. The basic metaphor is the universe as machine. The metaphor of mechanism by Descartes implies that matter is inert and that machines must have a creator, that is, they are not self-making. Although the metaphor is no longer used in physics, ecology and forestry are still struggling with it.

Virtually all of the characteristics of forest science can be deduced from this metaphor. A neutral, bodiless observer predicts the macroscopic behavior of a stable system, which is independent and isolated from other systems. The behavior is often assumed to be global, that is, it can happen anywhere, and it is linear, that is, it progresses through stages. The regularities of nature in time and place are considered invariant. Sense data are discrete. Problems can be analyzed into parts. Everything part can be measured. Resources are unlimited.

Many forestry ideas are derived from the mechanist worldview. These fundamentals, burdened with pre-evolutionary legacies such as economic discrimination and social inequality, are at odds with a nonmechanistic systems philosophy. They ignore physiology, metabolism, and diversity. They fail to describe the reciprocity of a living environment—after all, life does not adapt to a passive prior environment, it produces and modifies its surroundings.

History of Ecoforestry

There have always been voices for alternatives in forestry. Ever since Muir and Pinchot, forestry has been divided. Hundreds or thousands of people have worked out their own programs of forestry. In the 1960s and 1970s, ecological ideas began to be applied to forestry again. Individuals began applying ecological principles to small areas. Merve Wilkinson started selectively logging his forest, Wildwood, in 1945, limiting the yield, even though it meant that the forest could not provide a full-time income. For Merve, what is left behind is more important than what has been removed.

Orville Camp bought bombed out land in 1967 and nursed it back to health with selective thinning and clearing. He developed his own holistic program in forestry based on natural selection. Operating as a forest farm, Orville has taken logs and firewood out of the forest while improving the health. He also teaches and lectures widely on his system.

Herb Hammond evolved his ideas of wholistic and ecologically responsible forestry in the 1970s while working as a profession forester in British Columbia. Michael Pilarski uses Restoration Forestry to cover a movement and discipline that draws on old traditions to heal degraded forests and provide a steady yield of high-value timber. Many others have been developing new kinds of forestry, from community forestry to excellent forestry, social forestry, and sustainable forestry. All of these kinds of forestry, as is ecoforestry, are part of a general movement away from industrial forestry. Industrial forestry itself is renaming its program as adaptive, or new, or stewardship, or sustainable, but without changing the basis or goals of its program.

Alan Drengson bought property on the Olympic peninsula. The Ecoforestry Institute arose out of a meeting between Orville Camp and Alan Drengson in 1990-both with their concerns about destructive harvesting. The Institute organized in 1991 and became a nonprofit corporation in early 1994. The history and ideas of the Institute make interesting readings .

I started planting trees on an old raspberry plantation, then acquired another area with some old trees. I reforested and restored about 75 acres, using ecological principles. This land, Altazor, is now a teaching forest in Idaho.

The mission of the Ecoforestry Institute is to foster ecologically responsible forest use, through education and related programs and services, and:

- To engage in dialogue with interested persons aimed at deeper understanding of ecological forestry (ecoforestry);
- to develop an educational process to facilitate a paradigm shift from industrial forestry to ecoforestry;
- to educate and train interested persons in the knowledge, techniques and arts of ecoforestry so that they may serve as ecoforestry practitioners, consultants, and teachers;
- to develop and monitor demonstration ecoforests on private and public forest lands where the public can see working models of ecologically responsible forestry and fully functioning natural forests from which forest goods are being harvested in a sustainable manner;

- to develop, in cooperation with others, criteria for ecologically responsible forest uses, and standards for certifying ecoforestry practices, practitioners, materials, products and artifacts;
- to research and communicate to a wide audience the deepening knowledge of the multitudes of values and functions of natural forest ecosystems, as reflected in leading edge work in conservation biology, landscape ecology and related disciplines; and
- to respect and cooperate with indigenous peoples, to learn from them the wisdom of the places where they have dwelled for centuries.

Principles of Ecoforestry

Forestry is defined by Webster's New Collegiate Dictionary as 1. forest land, 2a. the science of developing, caring for, or cultivating forests, and 2b. the management of growing timber. These definitions are relatively new, although humanity has been using forests for tens of thousands of years. Ecoforestry is the management of the human use of forests for necessary goods at an appropriate scale for the forest. It puts the forest first and considers what to leave. It is also context-based community (ecosystem) forestry that is based on traditional wisdom combined with scientific knowledge.

Contrast the dictionary definition with Buddha's definition of a forest: "a peculiar organism of unlimited kindness and benevolence that makes no demands for its sustenance and extends generously the products of its life activity; it affords protection to all beings, offering shade even to the axeman who destroys it."

Within general topics, such as history or ecology, a number of characteristics, principles, and standards will be presented that should allow you to address every situation (to some extent). Characteristics are qualities that distinguish unique individuals, systems, or patterns; Gregory Bateson calls them differences that make a difference. Principles are fundamental rules or laws that we can use to create images or models. Standards are models or examples of quality or value established by authority or consent. Sometimes these statements will be made explicit in the summary.

For example, one characteristic of a mature forest is its wildness. The corresponding principle is that forest is self-making and self-ordering without human control and management. Our objective for this forest is to allow the foresting process to continue, whether we take resources from the forest or not (forests can be influenced or interfered with by acid rain, pollution, and other industrial effects). We can set standards that are likely to keep mature forests wild: Limit biomass removal to 2 percent of the total forest; use appropriate techniques, e.g., single tree selection, horse skidding; retain mature structure, e.g., 19 snags per hectare, 23 nurse logs per hectare (in mature Ponderosa pine for instance); preserve surrounding landscape patterns.

The principles of ecoforestry are based on a number of fundamental philosophical, historical, scientific, and cosmological principles that were first presented in other contexts by thinkers such as Whitehead and Einstein. The sample characteristics, principles and standards below can be expanded to form the basis of ecological forestry.

Characteristics

Forestry deals with primarily wild ecosystems.

Archaic peoples lived within the limits of forests.

Agriculture and modern forestry exceed the limits of forests and land.

Changes in scale put pressures on human and natural systems.

We ignore very long-term trends in nature and human history that have very dramatic influences on our activities and management styles.

Our history of use of forests has been to exploit them to collapse and then move on.

We justify our temporarily successful behavior with myths that allow us to continue the behavior without being responsible.

Principles

Change. Everything changes (according to Heraklitus). Individuals changes; patterns change. Neither plantations or old growth can remain unchanging.

Continuity. Forests proceed through distinct continuous steps in relation to past environments and disturbances; that is, a plantation cannot become old growth without developing through intermediate stages of community.

Irreversibility. Forests pass through stages that are never repeated, despite superficial similarities; that is, tree-planting cannot reverse clearcutting (although another old-growth forest may develop in time). The history of land limits or determines its future.

Uniqueness. History creates unique patterns, especially in forests. Each forest is unique in its parts and structure, in its matter, energy, forms, information, and in its dynamics and history.

Morphogenesis. Our species is shaped by forests, as well as by other species and ecosystems. Changing wild forests to plantations, fields or deserts will profoundly change our psychology.

Synergy. We have been most successful working with nature rather than controlling nature, and when we give back to the system.

Standards

We should consider the consequences of all our actions in their contexts.

We should allow wild ecosystems to continue to be self-creating and self-managing.

We should modify our myths to provide appropriate images of place.

We should apply known ecological ideas to our exploitation of forests.

Many of these statements will be reinforced throughout these lectures. It may be that these are also inadequate to address the first situation, in which case more information is necessary.

Summary

With the preceding philosophical principles in place, many specific changes would occur in the human use of other species in forests. These would allow us to:

- Recognize individuality in trees and other beings.
- Recognize the feelings and emotions of animals and the sensitivity of plants.
- Promote a noncommodity, drymoperipheral approach to forestry.
- Minimize the devastation of wild forests and wild ecosystems.
- Emphasize habitat loss as a human ethical issue.

Furthermore, in the near and far future, through its basis in ecological philosophy, ecoforestry could:

- Suggests ways that biosphere cultures can be converted to ecosystem cultures (after Ray Dasmann), characterized by wonder and wildness.
- Promote ecosystem lifestyles, characterized by frugality and joy.
- Relate the success of microorganisms to the ultimate success of living in self-sustaining forests.
- Develop a basis for management techniques for finite ecosystems, especially wild forests.
- Link ecological sustainability with the richness and diversity of ecosystems, as well as with human pleasure.
- Work for immediate solutions to inequity and destruction under worsening and thankless conditions.
- Educate with confidence and energy for ecological enlightenment over the long-term, despite short-term wobble.
- Learn to gamble wisely.

The description of ecoforestry is not a linear path leading to a definite conclusion. As a dialectical spiral, it requires the creative effort of the participants to be put together. By the way, although we may be smart enough to go beyond Aristotle's logic in our actions, we should remember that, for Aristotle, the goal of action was always contemplation, that is, knowing and being, rather than seeking and having.

Chapter 2

The Spectrum of Interactions in Nature:
Disturbance, Exploitation, and Interference

Introduction

This essay examines the parallels between the interactions of processes, of animals and of humans in ecological systems; it concentrates on disturbance and exploitation behavior and contrasts them with the interference behavior that characterizes the nonecological activities of the dominant human, industrial culture. Examples of each will be taken from wild ecosystems, forestry, animal cultures, archaic human cultures, and industrial culture. The word interactions is used, instead of words like 'events' or 'catastrophes,' to describe the feedback and cyclic nature of actions.

Humanity is exploiting nature recklessly, without attention to the minimal health of ecosystems. Many ecologists, such as Eugene Odum, have observed that complex communities have existed for thousands of years in relatively stable environments, even though these environments are characterized by regular disturbance and constant exploitation. These environments are now vulnerable to human interference, which is a different thing from disturbance or exploitation. Disturbance, by definition, is an event that can be caused by climate, biological entities, or other actors. Exploitation is the normal use of a resource or of a species by another species, including the human species (this ecological definition differs from a sociological definition, which means 'selfish or unethical use,' although it may suffer from negative connotations due to the latter); in fact, ecological exploitation has a rejuvenating effect on populations. Exploitation is contrasted with interference, an activity that can degrade, destabilize, or destroy entire ecosystems. Interference is not a form of disturbance, exploitation, or competition; it is destruction without gain to any species; sometimes it is caused by planetary events, but in the case of human interference, it is the destruction of the structures and processes of evolution for large-scale, one-species, short-term gain.

The interactions of living beings in ecological contexts may have positive and negative effects on themselves and other species, as well as constructive and destructive effects on ecosystems and the operation of biogeochemical cycles. Human interactions are also considered. The pandominance of ecosystems by humanity is related to the biological and cultural characteristics of the species. Ignorance and indifference are identified as major reasons for continued interference.

Interactions

Living organisms in a given area interact with the physical environment so that an energy flow leads to the defined trophic structures and material cycles that comprise an ecosystem, according to Eugene Odum. An ecosystem can be analyzed into parts, including organisms, energy circuits, food chains, diverse patterns, nutrient cycles, development and evolution, and control. No organism can exist by itself or without participating in an ecosystem.

Organisms interact in a number of ways. Interactions can be positive, negative or neutral in effect. In 'neutralism,' for instance, neither population is affected; in 'competition' each group adversely affects the other for resource use; in 'parasitism' one benefits and the other is adversely affected; and, in 'mutualism' both benefit in a necessary relationship.

Competition was once considered the basic interaction between individuals and species. The better adapted organisms survived and reproduced; others died. However, cooperation is now known to be as effective a strategy as competition and as necessary. Survival of the fitter is correct only up to a point; beyond that it is survival of the more cooperative, as Konrad Lorenz has pointed out. Competition is necessary but limited. Cooperation is what creates communities. Both old studies (Reinheimer 1910, Kropotkin 1972) and new research (Fox 1974, Lorenz 1952) stress the primary importance of cooperation.

Some commentators feel that cooperation should replace competition as a primary interaction, and that positive interactions like symbiosis should be maximized. Of course, neither interaction can dominate without sad consequences. If all plants were symbiotic and none were competitive, we would have no trees or flowers. If every being was competitive and not symbiotic, multi-cellular beings would probably not have evolved.

Most interactions are not simple, but are complex and paradoxical, involving symbiosis and competition, which makes proper definition difficult. For example, predation is regarded as being where one population adversely effects another, but is dependent on other, like parasitism. This is not so; predation benefits both. Predation is more of a mutualism and less a parasitism, especially with wolf-deer predation, for example. Since competition also regulates populations to some extent, according to Stanley and others, its effects are not entirely adverse. Furthermore, an interaction, like parasitism, that is negative on an individual level may have benefits on a species level. The predator/prey (or parasite/host) are not excluding opposites, but generate a whole unity, an autonomous domain where there is complementarity, stabilization, and survival values for both species. The effect on a species, by disturbance, exploitation, or interference, is also complex.

Exploitation

Animals and plants, algae, bacteria, fungi, live together in ecosystems. Living together involves many kinds of interactions, from competition and conflict to cooperation and mutualism. Interactions may be reciprocal or complementary. They may dominate or control. Interactions are multidimensional. A wolf, for instance, may howl to communicate, or to restore proximity with a mate, or for simple pleasure. Many animals, such as wolves and caribou, develop together over time, adapting to each other's strategies. Paul Ehrlich and Peter Raven refer to this mutual adaptation as coevolution. Coevolving systems never completely adapt.

Every species uses some part of other species or of the environment. This use is termed exploitation. Insects, diseases, and animals, more than being simply agents of mortality, are native components of complex food webs in ecosystems, and they contribute to the selection of species. In a Ponderosa pine forest in the Pacific Northwest, insects exploit trees; they pollinate some trees and overwhelm others, but rarely more than 1 percent of a forest. Diseases exploit trees; they remove stressed trees—also probably a low percentage on the order of 1 percent. Their effect on the long-term health of a forest, however, is positive.

Birds and mammals eat foliage and seeds; they also disseminate seeds. Mammals, the best regulated of more recent species according to Frank Golley, change their habitats to suit themselves by chewing, digging, and burrowing. Rodents can dislodge earth at a tremendous rate. In many cases, these activities improve the conditions for the growth of vegetation. Mammalian grazing promotes vegetative regrowth and the movement of seeds. Bison and prairie dogs were responsible for much of the character of the American plains. Rodent caches may account for a good percentage of pine seedlings; possibly 15 percent of a Ponderosa pine forest rises from such seed caches. Beavers and other rodents create their own microsystems. Wide-ranging caribou and wolves transfer energy between systems.

Predation increases the survivability of a prey species. Predation also increases the diversity of species, according to Steven Stanley, by limiting the most populous prey species. Rarely do predators kill all of the prey. Rarely do animals interfere with the operation of the biogeochemical cycles in the environment. The exploitation of the system by plants, insects, and animals contributes to the health and continuity of the ecosystem. Exploitation is not chaotic; there are limits and rules.

Rules. Animals obey rules of behavior. Many animal communities have codes of behavior that regulate interactions. In birds and less complex mammals, these rules may be very rigid and predictable. With increasing brain complexity, however, learning takes a larger role. For example, young white-tailed deer in Idaho have to learn to cross highways. They appear to use rules of thumb (not the best phrase), finding a proper balance between safety and reasonable progress, between no traffic in sight and bumper to bumper congestion.

Animals like wolves have behavioral inhibitions against killing too many prey or killing their own kind, against coupling with a mated or disinterested female, or against attacking nonprey species. Animals that break such inhibitions are usually sometimes attacked or ostracized. In general, food is shared by all members of a wolf pack. Adults will regurgitate part of their food for adults who stay behind with juveniles. The members of a wolf pack cooperate bringing down an elk, but then compete for the choicest parts of the prey. Rules are not always strict; wolf mates raising pups may consciously deceive one another to get a break from the responsibilities, according to Michael Fox.

The social structure of a wolf pack is most important. Breeding, playing, hunting, feeding and territoriality are tied to social structure. Wolf pups are taught how to behave and how to hunt. Much of the behavior of wolves is directed to keeping the animal's status in the pack or to raising it. Quarrels take place often and the entire pack seems to take an active part.

Actual battles, however, are rare. Ritualized squabbles result in few physical injuries. Wolves do kill each other, however. Wolves that behave strangely, such as epileptic pups or adults crippled in the chase, are sometimes killed by the pack. Disputes over the alpha position may end in death. Foreign wolves may be killed if they do not flee. Prey may be killed in excess during times of denning. Rules of encounter are complex and the outcome depends on numerous circumstances, such as the abundance of prey, the size and health of the pack, and stress; that is, the rules often depend on limits.

Figure 1. Wolf in the Rila Mountains of Bulgaria (where the author conducted wolf monitoring 2000-02).

Limits. Mammalian behavior is controlled and population regulated through the use of space in general. Most mammalian populations, wolves for instance, regulate their density well below the limits of the food supply, often by as much as 50-70 percent. Territoriality limits populations, but populations can also be limited by specificity of prey or plant source, size of prey or plant populations, predators, natural events, or even individual tastes.

Wolves in the Arctic disperse with the migration of the caribou. According to David Klein, they prefer Caribou to other often more easily obtained species, such as mice. This preference reduces their hunting efficiency, however.

The goals of an organism are limited by the life-images of its species. Each animal is a participant in a field of existence. Using its senses, each participant creates an image of nature, or world (umwelt, life-world, is the term used by von Uexkull), from the sensations that are meaningful to it. Each animal fits itself to its unique world as completely as it can—simple animals to simple worlds, complex ones to well-articulated worlds. Each fits its place as well as it can. Konrad Lorenz, Michael W. Fox, and others have elaborated this kind of fitness in more detail.

Each organism is inseparably related to a place; breaking the bond may result in death. Organisms and places shape each other. This is true of archaic human cultures as well.

Traditional Ways of Archaic Societies. Most human cultures are located in a particular territory. This is especially true of the Campa, who live in a tropical forest in Peru, and for the Ituri pygmy, who live in a tropical rainforest in Africa. The features of their cultures are unique to their place. They literally could not live with images of desert or ocean, like the Taureg of the Sahara or the Samoans of the Pacific. The very circumstance that makes each culture unique—being in a unique place—ensures that it can fit a place. This fitness ensures a limitation of exploitation.

Particularly in agricultural societies, cultures are gauged closely to seasons. The culture makes the world manageable by limiting it. A local culture is also tuned to the limits of the local ecology, within the knowledge of interactions—the long-range ecological consequences of drainage, irrigation, or overexploitation can contribute to the success or failure of a culture. Many, but not all, archaic cultures are a form of fitness and limitation. Most archaic groups try for adaptation before domination. For instance, according to Gerardo Reichel-Dolmatoff, the goal of the Desana Indians is the cultural continuity of their society in its place in the rainforest.

The Chipewyan Indians in Northern Canada occupy the same territory as wolves and compete for the caribou, although their niches are not identical. Both social systems are adjusted to the hunting of barren ground caribou. Chipewyan hunters depend on animals other than caribou, which migrate out their Indian grounds. The cultural decision to hunt caribou as the primary item of subsistence, however, has produced many similarities between the two species in their utilization of land and in the formation and distribution of social groups. The cultural decision to hunt caribou results also in a population density lower than what would result through other decisions regarding the utilization of resources.

For hunting, Chipewyan use dog teams, snow shoes, and boats to increase their mobility and rifles and bows to supplement the traditional spears. The strategy of the Chipewyan, according to Henry Sharp, is to kill caribou at any opportunity. They increase their opportunities by walking aimlessly, watching, and driving the caribou, although the Chipewyan expend less energy by watching and ambushing. Wolves follow the more active strategy because of their increased and superior mobility. Both species adopt a pattern of dispersing and concentrating with the caribou. The basic choices regarding subsistence patterns, social organization. demography, terrain usage, and yearly cycles are made on the basis of the internal logic and structural characteristics of the two cultures (wolves do have culture, in the general anthropological sense, according to Sharp and Fox).

Although the two species do not compete directly, both Chipewyan and wolves are predators that put pressure on caribou populations. Sharp suggests that the commitment of both to caribou hunting is ecologically inefficient, since both species could spend more energy on secondary sources of food. For the Chipewyan, a deliberate "underutilization" of moose, rabbit, grouse, birds, and fish, is the result of their cultural values, including their willingness to live below the carrying capacity of the local environment—a characteristic of most hunting/gathering societies—the complex practice of drying caribou meat, and the reciprocity of their kinship system, i.e., caribou is a better basis for future relationships. The cultural decision to hunt caribou as the primary item of subsistence has produced a unique pattern in the utilization of land and in the formation and distribution of social groups. Wolves also underutilize their resources.

Regardless of cultural order or cooperative interactions, part of the process of life is uncoordinated, unfitting, disorderly, unbalanced, and destructive. Therefore, suffering often occurs. Suffering is an unavoidable part of disturbance or exploitation. We cannot intervene in every case, nor can we eliminate the possibility of suffering. We cannot maximize the self-realization of every being, and we cannot make evolution into a perfectly functioning machine; the functionless features of evolution are part of the process, but, we can protect the process. The mode of operation of nature consists of a rhythm of dissolution and reformation. The extravagance and beauty of the natural world features many more species in an ecosystem than would be necessary if exploitation alone were its organizing principle.

Disturbance
Disturbances in nature are regular but unpredictable events. Disturbances are caused by geological events, climatic events, physical processes, and biological agents. Hurricanes, for example cause disturbances, as do volcanic eruptions, windstorms, tidal waves, disease outbreaks, and acid rain.

Disturbance is one thing that causes change in an ecosystem. On a small scale, a single tree falling over is a disturbance. Although an individual dies, species continue. Mortality is a normal part of the life cycle of the forest. The disturbance may be necessary for the ecosystem to continue to mature; for example, according to David Perry, without windthrown spruce that expose mineral soil seedbeds, the northern forest ecosystem would shift to bogs.

Disturbances, if sufficiently regular, become a 'known' feature of the ecosystem. In Florida, some species such as Cypress, need the complete inundation provided by hurricanes to remain healthy. Yet, even catastrophic disturbances like hurricanes rarely damage more than 5 percent of a forest; for instance, the 1938 hurricane in the eastern US blew down less than 4 percent of the trees.

As the frequency of a disturbance increases, the forest becomes adapted to the disturbance; even pine plantations in the southeastern United States that are managed with controlled burns are less damaged by lightning-caused wildfires. Many disturbances in forests, such as insect explosions or fires, kill low percentages of trees.

After long periods without a major disturbance, however, a catastrophic disturbance becomes more likely. Where wind and fire are absent, for example, the probability of insect and disease outbreaks increases. By trying to prevent one kind of mortality, ecosystem managers often establish conditions for another kind.

Fire is regarded as catastrophic. In some ecosystems, for instance tall grass prairies in Illinois, fire is required to suppress competition from trees. In some forests, such as lodgepole pine in Washington state, fire is required for the cones to open and the trees to regenerate. Forest fires rarely damage more than 10 percent of the whole forest. Even the Yellowstone fire of 1988, still regarded by some as a tremendous disaster for "Smoky the Bear" policies of prevention (resulting in dead material forming fire ladders), caused limited damage as it leapfrogged along, leaving healthy untouched stands that became the source for the regeneration that is now being observed.

A typical percentage of death is the normal condition of an ecosystem, necessary for its renewal. The rate of death per year in an old forest is remarkably consistent at about 1-2 percent, even with wind storms, fires, disease outbreaks, and animal damage.

In some cases, a larger percentage of the forest is affected. For instance, high elevation balsam fir forests are subject to bands of dieback that progress up the slopes parallel to the contours of slopes. These "fir waves" seem to be triggered by cold winds striking exposed forest margins. A new stand regenerates where the trees have been killed.

Disturbance may change the direction of maturity in an ecosystem, but it also is stimulating for those species adapted to it. Disturbance may continue succession or it may deflect it, according to Bormann and Likens. Because of the range of scales and intensities with disturbance, it is a complex concept.

Disturbances that are not part of the history of an ecosystem may cause irreversible changes to the system, because the system has not evolved a defense or response mechanism to such a rogue disturbance. A meteor strike would be such a disturbance, especially if the landform was altered by a crater. Human disturbances, in the form of acid rain or clearcutting, are both novel and threatening. If they are small enough or rare enough, however, the ecosystem may rebound.

Very large scale disturbances, such as volcanic eruptions or meteor impacts, can destroy entire ecosystems or disrupt global biogeochemical cycles. However, such very large scale disturbances are rare, and the ecosystems often have thousands or millions of years to become reestablished, although changed.

Interference

Although rare large-scale or novel disturbances can interfere with ecosystem processes, the term 'interference' is reserved for constant large-scale or novel effects. The destruction of ecosystem processes in nature by the action of one or more species is rare; any species that did so would become extirpated or extinct, unless it was not dependent on a single ecosystem, as is the case with wolves. Many commentators have accused mammals, wolves for instance, of overkilling their prey. It is fairly well established now, by David Mech and others, that wolves will take prey in excess of their immediate needs. This behavior has been interpreted as useful in maintaining not only the wolf but also secondary predatory and scavenging populations, for example, foxes and ravens. Indian informants are aware of this aspect of the wolf's excess kill, but they attribute to the wolf sufficient foresight to kill an excess of caribou near the den site in order to have an adequate food supply when the caribou are absent. Regardless of the wolves' intent, excess kills of caribou by wolves seem to be linked to the pup-rearing part of the pack that follows behind, as well as providing some food for the reverse seasonal migration—wolves can eat the remains of kills that are up to a year old.

Like wolves, human beings, as part of the process, interact with the individuals of other species or with entire species. Human beings are mammals who, as George Woodwell puts it, live in a biosphere whose essential qualities are determined by other species. Mammals are bound by biological requirements that must be met if a population is to survive.

Like other mammals, humans change their habitats to suit themselves. Humans have modified animal and plant associations in a different way from other mammals, simplifying patterns of energy and chemical exchange and solidifying themselves at the end of many food chains as a dominant species. (A dominant is a species with greater influence than any other in its biotic community, changing the lives of other species and the character of the habitat.)

Human populations have increased exponentially, with billions in giant urban ecosystems. Agriculture has produced monumental yields, but only at the cost of tremendous erosion and great subsidies of fertilizers and pesticides. Dams have been built all along rivers, and riverine forests have been cut, altering rivers and fishing grounds. Changes have been made without regard to the long-term impact on the ecosystem or on its human population. We dominate entire ecosystems.

Pandominance. By its influence of all ecosystems, humanity has become a pandominant species. As such, humanity reclaims, overgrazes, clears, depletes, and wastes at a scale that interferes with the stability, processes, and existence of many systems. One of the ecological consequences of human activity is the degradation of wild habitats for human developments and the introduction of novel elements into the biosphere—elements that have not been added slowly over time as the result of natural processes.

The biomass of the human species probably exceeds the biomass of any nondomestic mammalian species, and that biomass is supplemented by the tremendous biomass of domestic animals, which is far greater than the human biomass and consumes much of the same food as humans, including milk, fish, and grain. The domination of humanity is related to other characteristics as well, its large annual population increase (over 2 percent), high structural organization (of information and matter), and high energy use (globally 13 times the total of all other mammal equivalents).

This dominance has major effects on ecosystems: Transient perturbations in energy relations (from oil spills and burning); chronic changes/shifts of systems (from dams, irrigation, and chemical wastes); species manipulation (from the import and export of exotics); and interference competition with wild species. None of these effects are exclusive to humans as a species, but they are excessive, rapid, compounded, and very large-scale. Humanity has upset the balance of nature in favor of its own needs. Animals, plants, and habitats are being destroyed because of short-sighted, short-term economic interests.

Human beings have contributed to the extinction of species and to the destruction of ecosystems. Human hunters are hypothesized to have wiped out the most of the large mammalian species of the Pleistocene through overhunting—not for future food, but rather from the style of hunting—by driving herds over a cliff. There are other instances. In the 1880s, soldiers and cowboys slaughtered buffalo as a political strategy to reduce the resources of native peoples. Farmers and loggers destroyed the dense forests of Ohio and other states. Settlers and industrialists in the Amazon are destroying vast tracts of rainforest, as part of a political strategy to move peasants out of cities. Industrial forestry in the Northwest is content to take a high percentage (well over 90 percent) of a forest for wood and pulp, destroying the basis for the continuity of the forest, as well as all beings that depend on the old-growth, fungi, and physical properties of the forest to live.

Human exploitation at the tremendous physical scale that occurs in industrial states is different from exploitation by other species, because it results in the destruction of the entire system, the very basis for renewal of a system that human beings (as well as other species) need for life. Human actions are damaging global biogeochemical cycles, such as the carbon or nitrogen cycle. For instance, deforestation, burning, wetland loss, and industrial processes are releasing massive quantities of carbon dioxide into the atmosphere, which disrupts the carbon cycle. Although the destruction of large species, from whales to frogs, has a dramatic effect on ecosystems, the destruction of microbes, which generate oxygen and recycle nutrients, has a critical impact on the entire food web. These actions are global, like a large volcanic eruption, but, unlike a volcanic eruption, they are constant and hourly. These human activities are best referred to as interference.

Industrial Ways. The cosmologies of archaic cultures have been limited to historical places and by human perception, tradition, and technology. Modern technological cosmology, beyond being another kind of order, more linear and abstract, is wrongly considered the evolutionary successor to traditional cosmologies, and is displacing them rapidly—although we cannot afford to suppress the diversity of thought necessary for adaptation to the diversity of environments or to eliminate ecosystems and the societies adapted to them.

Our modern problems reflect an unbalanced and immature image of the earth, the earth as a machine, for instance. People sometimes constructed their worlds from preconceived notions, and many of these worlds did not survive, because they could not adapt to the environment. Our modern cultures are defective for this reason. The modern attitude toward nature as a resource has resulted in pollution and depletion of resources. It has allowed humans to overpopulate their habitats. Recent productivity studies indicate that the optimum sustainable human population is far below the current world population (Wittbecker, 1983).

Even worse, decisions regarding resources are still made exclusively on short-term economic rationalizations and lead to material shortages and environmental degradation. The crises of environmental degradations are crises of cultures. Monocultures of the industrial kind lead to 'dedifferentiation,' that is, the decomposition and destabilization of complex structures. A species or culture that destabilizes its ecosystem through misbehavior risks its own extinction. Human beings make changes to ecosystems that endanger themselves.

Humanity is calculated by Norman Myers and others to be using over 40% of the ecosystem productivity for the entire earth (56% by 2004, according to Stuart Pimm and others). Humanity influences virtually every ecosystem to some extent, destroying some, interfering directly with many, and exposing the rest to exotic chemicals and materials. Species normally use a percentage of system productivity without disrupting the processes of production. The human species interferes with the processes.

Based on limited scientific and cultural perspectives, humanity fails to value those beings and communities for which no use is known. But, as Aldo Leopold (1949) notes, the majority of the beings in nature have no human uses. Even ecologists cannot think of uses for many large birds and mammals. The real danger is genetic loss, which is frequently grossly underestimated. As wild areas grow smaller, even wild species interbreed. As species are lost, the ecosystems become simpler or start to collapse.

Discussion

The problems of ignorance and inappropriate images are multicultural, ecological, and cosmological, and must be solved on those levels—the entire activity of culture is guided by metaphors. Metaphors emphasize likenesses between living things and languages (or human constructs). A metaphor furnishes a label and emphasizes similarities. It not only defines and extends new meanings, but redescribes domains seen already through one metaphoric frame.

New metaphors already exist for interactions in an ecosystem. These include a process view (A.N. Whitehead, from the 1920s) in which organisms are dynamic structures that are immanent and simultaneous with the process, rather than consequences of natural selection of past random mutations; a field concept (C.H. Waddington, in the 1960s) for development, emphasizing dynamic transformation (form as organized spatiotemporal domain), in contrast with the particulate concept of an organism—understood in terms of group dynamics rather than selective advantage or cost/benefit; self-organization (Franceso Varela, in the 1970s) or autopoiesis, which refers to the dynamic self-producing and self-maintaining activities of living beings, which incorporate materials through physiological processes; and reciprocally constrained construction (R. D. Gray, in the 1980s), according to which the organism and environment are co-implicative, co-defining, and co-constructing in a process of self-assembly, where the self is the organism/environment system.

Combining metaphors, we can see that organisms put together (enfold) structures based on historical patterns, and move (unfold) through a filter of limits, like minnows through a fish net, rather than behaving as interchangeable units competing for resources. These metaphors could form the basis of a new image for humanity, where we are an integral part of food chains and part of an organic cycle of birth and death. We humans need to recognize that we automatically participate in everything, and that we cannot unparticipate by choice. Participation starts at the quantum level and extends through the ecological and cultural. Human nature does not find meaning in an absurd world, but discovers its structure through interaction with the surrounding order. Human identity exists partly in relation to nature; the destruction of ecosystems may lead to the destruction of human identities mediated through cultures.

A culture that fits a local ecology is adapted and more likely to survive. Fitness is a way of reducing negative effects to make cultures more flexible and longer lived. An understanding of ecology, with an emphasis on limits, can lengthen the life of a culture, but ecology is not enough. Good metaphors are necessary, as are good rules of behavior.

New Human Rules. Human beings have no complete guidelines to interacting with other species in an ecological context. Cultural ethics are usually restricted to some members in a local ecosystem; such ethics are assembled inductively, from experiences from living in specific places. Philosophical ethics have been traditionally restricted to the human species and human situations. The areas of concern of ethics are not broad enough; their foundations are not deep enough. Philosophical ethics is the ethics of the human species living alone, without wild animals or plants in modified ecosystems. An ecological ethics has been developing.

Aldo Leopold proposed a land ethic, dealing with human relationships to land, plants and animals. This land ethic was a sense of ecological community between humanity and other species. "When we see land as community to which we belong, we will use it with love and respect." Such an ethic would change the human role from master of earth to plain member of it. Predators are members of the community; and no special interest group has the right to exterminate them for the sake of benefit for itself. This attitude is important for habitat protection. Leopold describes the extension of ethics as "actually a process in ecological evolution. Its sequences may be described in ecological as well as in philosophical terms. An ethic, ecologically, is a limitation on freedom of action in the struggle for existence. An ethic, philosophically, is a differentiation of social from anti-social conduct. These are two different definitions of one thing. The thing has its origin in the tendency of interdependent individuals or groups to evolve modes of cooperation."

The extension of ethics to animals and land is an ecological necessity with human pandomination. This extended ecological ethics defines a social conduct that is a mode of cooperation and, ultimately, symbiosis. Leopold argued that voluntary limitations of freedom are necessary in a complex world of which we remain incredibly ignorant. Extensions of ethics are developed in response to problems that arise from increasing knowledge. Science has phenomenally increased our knowledge of physical and biological processes. It has now become the basis of our moral code, but it cannot very long be a science divorced from feeling and art if that code is to help us survive. To do this science requires aesthetic perception as well as disciplined thinking and feeling. As there is a rational component to ethical judgments, so there is an intuitive and emotional one, also. An ecological ethics suggests that humans avoid tampering with complex evolved systems, not because they are good, but because they are the basis of life at this stage of development. Ecological ethics is situational because ecology is the study of changing systems. It is pluralistic, as Stone notes, because of the variety of entities involved. The morality of the act is determined by the current state of the system. Adaptive modes should conform to ecological patterns. An ecological ethics is based on attributes of ecosystems and human compliance with ecological laws. The aim of an ethic must be harmonious with the whole population of living beings.

An ecological ethics is a set of rules for living together with other beings (in fact, the word 'ethics' is derived from the Greek word *ethos* meaning 'custom,' which itself came from the Sanskrit word *svadha* for one's 'own doing.' Since it was used in the plural, it meant 'doing together'). It is based on ecological knowledge, grounded "in the breadth of being," in Hans Jonas' words, founded on principles discovered in existence. An ethics based on ecological knowledge places human behavior in vital social and biological communities in nature. The frame of reference of ethics is enlarged, as Albert Schweitzer predicted, leading to appropriate behaviors in a larger context, through reverence for life. Skolimowski and Callicott recommend a reverence for life ethic. We must develop specific rules to live with other species, more formal than isolated cultures like the Campa and more comprehensive than modern cultures like the French or German.

Ecological ethics is a series of rules for living together. Most sets of ethics make the rules easy to follow. They emphasize the differences (relativism) or similarities (absolutism) of human beings only; or of the individual or the group; or of good feeling, reason, or desire. But ethics has to confront the individual, embedded in a community, located in a bioregion, on earth. And, the rules really are not as easy as human systems have presented. Schweitzer made them too difficult, with a constant valuing, but neither are they that difficult. An ecological ethics can be detailed only on a local level—even when it uses a global strategy.

An ecological ethics is not distorted by human needs and wants when it argues for the preservation of animals and habitats themselves, because they are, as they are. Because of the uncertainty of human actions, ethics has to encompass the far past and distant future. No one knew that when DDT killed mosquitoes, it would concentrate in the food chain to kill birds. Values are time dependent, and ecological time can be very long indeed. The futures we invent are viable only if compatible with constraints imposed by evolutionary past. An ethics that requires a long-range responsibility also requires a new humility, since technological power exceeds the ability to foresee its consequences. An ecological ethics recognizes the moral obligation to leave the world habitable for future generations.

Rights seem to follow the expansion of the sphere of ethics, as formal statements of intuitive knowledge. But codifying rights is more difficult, especially for philosophers, who tend to limit rights with a series of restrictions. Paul Shepard says the argument is not new, and that its application is ambiguous because "unlimited rights" will conflict with human interest. But, there are two bad assumptions: that human interests are not ambiguous—they are—and that animals will be granted unlimited rights—they will not.

The strongest argument for rights is interrelatedness in communities, which is the basis for assigning rights to nature. Garret Hardin considers interrelatedness, but interprets it narrowly. He considers rights as rules of competition; every right is a ploy in the struggle for existence, and every right implies an obligation to furnish it. This is good as far as it goes. However, life is more than competition; it involves cooperation and play. Rights are formal rules for living together. It is foolish not to assign rights to animals, plants, and the earth because of contractual formalities. The reverence for all beings is concerned with the right functioning and right numbers in the right places, according to standards of health and quality of life.

One problem with the current legal system is that all nonhuman beings are given the status of inferior human beings, legal incompetents, thus keeping humans in a guardian role. A new legal category is needed that would respect the existence, competence, and excellence of natural beings. Christopher Stone recognizes that the judicial system has granted rights to a variety of inanimate holders, trusts, corporations, and nations, for instance. The legal system already operates with fictions, so the extension to natural entities should not present an insurmountable problem.

To be sure, formal rules should to be altered to account for unconscious, interdependent beings. Current legislation on animal experimentation and protection implicitly recognizes the right to life and to a healthy habitat. Laws are needed to protect entire habitats of animals and plants from human interference.

We act by intuition and feelings. Like the inductive creation of cultural ethics, we are building a framework of intuitions, feelings, theories, and principles. The whole is recognized as a valuable end by hunters and actors as well as by scientists and politicians. The framework is supported by principles and theories developed by ecologists and philosophers, by the working rules of conservationists and activists, and by specific instances from cultural traditions as well as from the industrial paradigm (determined to become its own worst enemy as well). Stone considers that these things are only part of the framework. In The Laws, Plato has the Athenian say to a youth that all things are ordered with a view to the preservation of the whole, each portion contributes to the whole, and every other creature is for the sake of the whole. Ethics has expanded in wholes, from the family, to the human community, and to nature, on which everything depends. With Ervin Laszlo, ethics encompasses all systems in the universe.

Observing Ecological Limits. Life involves a vast number of interacting structures. Living consists of complex behaviors whose limits are defined empirically. The earth is suitable for life because of three kinds of limits: (1) solar radiation has stayed within certain limits for 4 billion years; (2) the biogeochemical cycles of oxygen, carbon, nitrogen, phosphorus, sulfur, water have stayed within certain limits; (3) the environment has been constant enough for organic evolution, but variable enough for natural selection to operate.

Animals and plants stay within limits of an ecosystem; for instance, Klein concludes that caribou populations are limited by food supply. Wolves are sometimes limited by stress. Trees are limited by water supply.

Traditional cultures have often stayed within the limits of an ecosystem. A sense of place, with its beings and features, is necessary for information on how to live, get food, and stay dry. The ecological benefits of rootedness are that people will take care of a place if they realize they are going to be there for a long time. Having a place means that the inhabitants have stock in it and participate in its unfolding, through planting and caring. Detailed understanding of the plants in a locale allow gathering of food and medicine. People in place—being in place as used here means in a human scale in unique surroundings—acquire a sense of community, share a set of values and concerns, and reap physical and spiritual benefits.

Practicing Noninterference. Exploitation, in the ecological sense, is necessary and beneficial to biological populations. A machine metaphor approach, with its assumptions of interchangeability and quantity, apparently has difficulty distinguishing between exploitation and interference. An ecological metaphor, that is more receptive and reverential, may be more appropriate to understanding organisms and nature in general. Such an approach would stress noninterfering observation rather than controlling manipulation.

Applied to nature, human intelligence discovers the significance of natural rules of interaction and exploitation. The reverence for beings as they are results in the rule of noninterference (Wittbecker 1984). A rule of noninterference states that human beings ought to avoid behavior that disrupts essential ecological processes or destroys biotic communities. As Paul Taylor states his rule of noninterference, it requires a "hands-off policy" for whole ecosystems and biotic communities; the rule stated here is concerned with limited and sustainable exploitation of ecosystems already shaped to some extent by human activities. Many other ecosystems, perhaps covering 50 percent of the land area of the planet, would be reserved by law for predominately natural ecosystems or adapted first nations. Noninterference also means "letting be" (after Martin Heidegger), or "letting alone" in the words of E. O. Wilson. Noninterference is not indifference, which is diffuse. It is caring. Noninterference will not lead to chaos, poverty, or stagnation. It permits the rational exploitation of resources.

We need to practice the rule of noninterference so that all beings can enhance their lives and habitats. Noninterference can be derived from nonviolence (or taoistic nondoing, a metaphoric expression for the nonbeing of nature) or even from English Common Law, which is well-established in Western law; it includes a precept: "Use what belongs to you in such a way as not to interfere with the interests of others" (*Sic utere tuo ut alienum non laedas*). This rule could be defined by positive laws and by negative restraints on behavior. This attitude would entail using what is necessary, exploiting some ecosystems completely, changing a place to fit human aspirations, and killing plants and animals for sustenance. But, it would also mean limiting humanity and its technological effects, limiting human use to local impacts, and letting other beings live without interference. It is not necessary to dominate or terraform the earth completely. Humanity could contain itself to a small percentage of the planet's surface and ecosystems and only visit or ignore the remainder.

Conclusion

Interference has been a rare phenomenon on earthly ecosystems; it has happened in the past as the result of global catastrophes, such as meteor impacts. Now, interference, as opposed to more limited and predictable disturbances or exploitations, is threatening the stability of all ecosystems. It is dangerous to interfere with the processes of ecosystems because it disrupts the communities on which other species, and ultimately human communities, depend. Furthermore, in the deepest sense, it violates the idea of living together with other species on the planet. The proper relationship of humanity with nature includes competition and exploitation and mutualism, but not interference.

We kill millions of animals in laboratories to insure our safety, we kill billions of plants and animals for food and clothing and products, while indulging in the sentimental preservation of some individuals of other species. Animals do not need to be saved from natural death, a great regulator of life, but from unnecessary suffering, experimentation, and premature extinction. The world would not be a better place without sharks, silverfish, rats, cockroaches, or hyenas. They need their own places. The places, entire ecosystems, need to be saved. If we diminish variety in nature, we debase its stability and wholeness. To save ourselves, we must preserve and promote the variety of nature.

A start has been made. Ethics now considers almost every human being and human interaction. The restriction of ethics to exclusively human modes of existence, however, leads to a troublesome isolation. Human beings are not separate from their social and biological communities and these communities are embedded in ecological contexts with biogeochemical processes. Human communities are one of many communities that make up an ecosystem. Human ethics describes only a small part of the rules for living together in communities, perhaps only the self-conscious part. Ethics must be extended to the entire framework and to the surrounding communities in the framework, without which there would be no human health or wealth. Through our efforts, we understand that communities of other beings have their own values and rules for living together. It remains for us to integrate and codify human rules that recognize the values and rights of other beings to live in healthy ecosystems and that limit human use of those ecosystems.

Chapter 3

Gandhian Nonviolence and Defending the Earth

Some groups of people, concerned with defending vacant lots, wilderness areas, ecosystems, and the earth as an organic body, have advocated using any means necessary. Earth First!, for instance, showed a lot of us arm-chair sympathizers that saving the earth requires more active participation than just letter-writing and circulating academic articles. They were among the first to put their bodies where their ideals were, and their stance has made some profiteers and environmental rapists more cautious. But, Earth First! made its wild reputation by monkeywrenching. Although the group has been effective so far, there is a possibility that its tactics may cause long-range difficulties in the form of violent retribution or the overreactive destruction of wilderness, as well as polarization inside other environmental groups. This article contrasts the tactics of Earth First! with a Gandhian nonviolence.

Dave Foreman of Earth First! has said,[1] "Monkeywrenching is nonviolent resistance to the destruction of natural diversity and wilderness. It is not directed towards harming human beings or other forms of life." Many Earth First! tactics, however, make this statement questionable. Is Earth First! really nonviolent, according to Gandhi's definition and practice of nonviolence?

Mohandas K. Gandhi characterized his ethics of group struggle by the Sanskrit word *ahimsa*, meaning "nonhurting" and "nonviolence." Ahimsa is a closely related set of prescriptions and descriptions. Gandhi said: "Ahimsa means avoiding injury to anything on earth in thought, word, or deed."[2] He adopted a wide interpretation of 'injury.' The subject 'anything' included all living beings and perhaps nonliving things.

Ahimsa is the absence of *himsa*, meaning hurting from the root *hins*, meaning hurt, a form of the root *han*, with a larger number of meanings, such as strike, kill, destroy, or dispel. Gandhi mostly had living beings in mind, but injury to nature or to natural processes, could come under the general principle of ahimsa.

Indeed, the concept of ahimsa is so wide that it could be interpreted to be an act of violence to abstain from efforts to prevent injurious acts, such as the exploitation of wilderness for profit. But, Gandhi referred specifically to destruction as part of sabotage as himsa, even if the things destroyed were not the property of anyone.

Gandhi also said,[3] "Ahimsa really means that you may not offend anybody, you may not harbour an uncharitable thought even in connection with one who may consider himself to be your enemy." Mental forms of injury include hurting people's feelings, their dignity, or their relationships— but the feelings and relationships must be positively valued, that is, it would not be himsa to hurt feelings of hatred, nor to save a victim from wrongdoing. Furthermore, some actions may be in accord with ahimsa if they are performed 'wholly unselfishly', although Gandhi does not accept the postulate of unselfishness as sufficient for the qualification of nonviolence. Most selfless terrorism, however, is still *not* nonviolent.

Mental forms of injury seem to occur in Earth First! campaigns, from name calling and humiliation to suggestive publications. By its verbal and physical stance, Earth First! seems to be violent towards many groups, including loggers and RVers. Foreman tacitly admits this when he makes the distinction between blockades (nonviolent civil disobedience) and monkeywrenching, while recommending that they not be combined in the same campaign.[4]

By its attitudes Earth First! polarizes most people. The Earth First! slogan is, after all, "No compromise in Defense of Mother Earth!" This really polarizes the opposition. One problem with opposition is its either/or character: "if you're not with us, you're against us." Positive action does not have to be bilateral or dualistic; it can transcend simple opposition and be positive without being adversarial. Gandhi was always willing to compromise on nonessentials. He characterized himself as a man of compromise because he was never sure that he was right. Compromise is an essential part of the nonviolent person, satyagraha; *ahimsa,* as unselfish love, demands compromise. There are principles that admit no compromise, and furthermore, if the compromise fails, the *satyagraha* is ready for "battle" (Gandhi's term).

Could Earth First! be truly nonviolent and still be effective? Could Earth First! be more visible and less destructive? Perhaps it could, with the adoption of a Gandhian campaign to defend wilderness. Before this campaign is presented, however, let us consider how violence may be a dangerous tactic for groups interested in preserving and protecting wilderness against an industry-dominated consumer public.

1. The hostage, wilderness, is very large and vulnerable to counter-attacks, as well as to swings in fad (which often determines how people vote to dispose of the object of the fad).
2. The destruction of materials, such as bulldozers, is usually illegal. Property is considered sacred in America and Norway, as well as in many industrial and consumer countries.
3. Violence tends to polarize opponents and some of the undecided against the long-range goals of a group, regardless of how well supported and argued.
4. Wilderness is symbolic of people's right to earn a living—even if the people in this case are loggers, drillers, or roadbuilders, who may destroy it in the process of using it.
5. Violence leads to escalation by opponents to protect their equipment and property. Violence leads to violent conflict as a *style* of opposition.

Most of the thousands of direct actions on behalf of the environment have been nonviolent in the Gandhian sense and some have been effective. The "hug the trees" movement in India, for instance, physically blocked excessive logging in the Himalayas. The Chipko (meaning "hug" or "cling to") movement started out to preserve trees by embracing them before axes could be used and has resulted in a ten-year ban on tree-felling in over 550 square miles of the Uttarakhand in India, a major source of timber and water power. The main goal is the "judicious use of trees," according to Chandi Prasad Bhatt, the founder, and not complete preservation. The movement is pressing for a complete remaking of forest policy; they are also responsible for planting trees (over a million so far). One tenet of the movement is that the erosion of human values follows erosion of the land.

Both ecology and human ecology offer the best support for saving wilderness and humanity (which depends on wilderness). The principles from these sciences can inspire a series of hypotheses and norms (based on the Spinozan forms used by Arne Naess) that are the foundation of a Gandhian nonviolent campaign.[5] This series may also serve to clarify the aims of Earth First!

Hypotheses
The hypotheses are segregated by common themes so that the derivations are more obvious and can be linked more easily. In this instance, hypotheses are presented from three broad levels, so that you can see the process of induction. This procedure is genetic rather than logical.

Third level hypotheses.
- Humans have the same genetic stock and the same basic interests as other species: food, welfare, self-actualization, love of place.
- Human make their worlds of facts from observations and theories colored by needs and wants, within perceptual and imaginative limits.
- Humans live together in human communities within biological communities in geographic places.
- Humans can communicate and do so to build cities and make art.
- Humans are capable of great trust and exhibit this in their use of automobiles, weapons, and money.
- Humans are capable of violence, for many reasons, from self-preservation to misunderstood symbols.
- Living in separate locations with differences in languages and cultures can lead to misunderstandings.
- Humans are autonomous beings and can choose their behavior.
- Humans can change their behavior.
- Nature is self-making and self-maintaining.
- Where nature is left alone, by definition wilderness, animals and plants reach their own balance.
- Animals and plants live in communities in geographic place.
- Animals and plants are autonomous beings.

Second Level Hypotheses.
- All beings (human and ultra) have long-range interests, especially in the continuation of their ecosystems.
- The presentation of facts, as uncolored by needs and wants as possible and within the limits of perception and imagination, increases the probability of understanding.
- Living together in the same community fosters cooperation and understanding on immediate, common goals.
- Working together on common projects increases cooperation.
- Constructive work is more binding; a constructive program is more

meaningful and generates trust and communication.
- Work abstracted from the community can cause its opposite (enantiodromia), that is, destruction instead of good.
- Misunderstandings can be corrected in peaceful confrontation.
- Constructive confrontation concentrates on faulty understanding in a situation and not on the opponent's action or personality.
- The focus on misunderstanding reduces violence.
- Humans are responsible for their own actions.
- Humans enjoy being with animals and plants in general and being in wilderness and using it for living and recreation.
- Human economies are based on wilderness, ultimately.
- Wilderness should be saved for many reasons.

First Level Hypotheses.
- The campaign to save wilderness should be constructive and positive and be addressed to issues and relationships, using unbiased facts. It should be well-advertised and simple, sticking to goals.
- The campaign should be nonviolent and given to the possibility of agreement, including compromise on some aspects.
- Violence against transgressors of wilderness may result in further violence against wilderness or perhaps further violence on the part of defenders to guarantee wilderness.
- The campaign should apply to a local area and address those who are knowledgeable in the local community.
- Responsible persons, acting in a group with an appropriate lead time, concerned with their own community, present an undeniable case.

Norms
From these hypotheses can be derived norms that regulate personal and group behavior in a nonviolent campaign.

Third Level Norms.
- Live in the community with your opponents!
- Formulate the essential, shared interests and try to cooperate on the basis of these interests!
- Refrain from provoking or humiliating your opponent!
- Seek personal contact with your opponent and his group and be available for meetings!
- Trust your opponent!
- Learn about wilderness!

Second Level Norms.
- Act as an autonomous, responsible person!
- Choose attitudes and actions that reduce conditions that lead to violence!
- Find common interests to build on!
- Present unbiased facts!
- Be flexible and ready to compromise!

- Be constructive! Suggest alternatives!
- Trust!

First level Norms.
- Act nonviolently, peacefully, and responsibly!
- Defend wilderness!
- Don't stop!

For Earth First! (a norm perhaps?), the norm "Be constructive!" could take the form of alternative suggestions for loggers, such as private land management or unions or legislation to restrict (by tariffs) lumber sold below the real cost. The Marsh Institute, for instance, uses an educational tactic by offering to performing a consulting service for free on a small scale; the farmer and logger then can compare directly the results and the methods, such as pesticides versus integrated pest management to reduce aphids on a barley crop, or sanitation cuts versus single-tree trimming to eliminate dwarf mistletoe infestations. Then, there is defensive monkeywrenching, such as tree spiking—with signs notifying cedar thieves, and passive resistance, such as not informing hunters or biologists of the whereabouts of a bear. In general, logging and farming prices are unrealistically subsidized by free goods from nature, as well as government subsidies. Prices should be geared to costs, and tariffs and taxes used to balance trade discrepancies. The answer is not to fight over trees in wilderness areas or to give up wilderness as a one-time boost to timber interests. It is to attack the old, unreal ways of cutting, to attack unrealistic economic policies. The long-term strategy is to save most wilderness as completely as possible, but also to include the cost-benefit analyses of alternate plans for some areas. Sawmill workers have been complaining that wilderness removes real jobs from their grasps; this misperception needs to be exposed as the short-coming of a problem economy.

Anti-ecological decisions usually come after years of planning by government or industry bureaucracies, at extravagant costs. Therefore, it is best to seek out and address plans at the earliest stages, before momentum can carry the plan. This is hard to do when the bureaucracies resort to secrecy. One of the goals of Earth First! is exposure of destruction and the increase in public awareness. It might be more effective to team with reporters with cameras and send news flashes on destruction and illegal use than to destroy equipment.

These norms provide a consistent guide to reacting to dishonest or violent opponents in the struggle. For instance, if your opponent uses biased reporting, merely provide a factual presentation, with evidence, without resorting to provocation or name-calling. Although the stupidity or badness of opponents is not an issue, some people are stupid and bad, so it is a factor to be considered. The stupid and bad must be neutralized some way, by simple diversion through relatively harmless assignments.

Nonviolence is easily misunderstood. If your opponent respects violence in defense of property, and you misjudge your opponent and offer nonviolence, which is perceived as weakness, prompting him to violence, what should you do? Stay in the center of the issue and be active, but do not respond violently yourself. Most such opponents eventually recognize perseverance as an indicator of strength.

It is important to formulate one very clear, concrete, easily understandable goal for an action and alert the opponent to that goal as soon as possible. This is very hard to do when it is difficult to know who the opponent is and how to reach them. Any action can be part of a larger campaign, however, and this may be important for psychological reasons, since many actions are unsuccessful in reaching their goals, although that does not reduce their importance. The success of the campaign does not depend on the success of each single action. One side effect of actions is to attract the attention of the public, whom it is assumed would act rightly with knowledge.

A campaign may be part of a larger movement, such as one to eventually put 40 to 50 percent of the landscape (of North America or the planet) into special preserves. Such a movement may take many campaigns and a hundred years. But, the goal is important.

Wilderness campaigns are not really constructive in the sense of architectural or educational campaigns, because they are defensive. Furthermore, what is defended is, first, in the process of change, so it can not be saved as it is; second, it will always be vulnerable; third, it is an ambiguous concept misperceived by opponents; and fourth, it is ultrahuman. What is to be saved is the potential for evolution of uninhibited development of species and ecosystems.

Violence polarizes opposition. A stance of "no compromise" polarizes opposition. To turn your opponent into a supporter, compromise is far more effective than violence or coercion. We have to compromise. We don't know enough not to—we are not sure enough of the consequences of our actions, which often have the opposite effect of the one intended. Furthermore, compromise can be such that it satisfies the opponent's ego without giving up much, because it creates a state of cognitive dissonance, where a little token is enough to convince people that something is owed in return, a much bigger concession in this case, regardless of whether they know about how cognitive dissonance works.

The code of nonviolence, as presented by Gandhi, is not a rigid system. Exceptions are possible and, under some situations, even desirable. Arne Naess suggests that a small piece of a technical installation (a dam, for instance) could be destroyed in order to avoid the greater destruction of an area. Nevertheless, this violence is an exception and not a norm. Earth First!, by compromising on nonessential issues and using violence only as a warning, might increase its effectiveness. Either way, Earth First! already exemplifies an important, and neglected, aspect of Gandhi's philosophy: *you should follow your inner voice whatever the consequences.*

(This article was rejected by *Earth First!* journal without explanation; it was subsequently published in *Pan Ecology*)

Chapter 4

Principles of Ecological Forestry (Part 1)

The agenda of ecoforestry can be presented through a number of principles. Principles are fundamental rules or laws, based on the characteristics of the forest systems, that we can use to create images or models to meet stated objectives, that is, the goals towards which our action is directed, e.g., a healthy forest or strong beautiful lumber. Principles unify our images. The principles are introduced briefly to show the depth and breadth of forestry.

The principles presented are derived from the typical characteristics of forests. Characteristics are qualities that distinguish unique individuals, systems, or patterns; Gregory Bateson refers to characteristics as differences that make a difference. From these principles, standards for our activities in forests can be established. Standards are models or examples of quality or value, established by authority or mutual consent, that can be repeated as procedures. (Standards will be presented in future issues.)

For example, one characteristic of a mature forest is its wildness. The corresponding principle is that a forest is self-making and self-ordering, without human control and management. Our objective for any forest is to allow the foresting process to continue, whether we take resources from the forest or not (forests can be influenced by human effects such as acid rain, pollution, and other industrial effluences). We can set standards that are likely to keep mature forests wild: Limit biomass removal to 2 percent of the total forest; use appropriate techniques, e.g., single tree selection or horse skidding; retain mature forest structure, e.g., leaving a good number of snags and downed trees; and preserve surrounding landscape patterns.

The principles of ecoforestry are based on a number of fundamental philosophical, historical, scientific, and cosmological principles that were first presented in other contexts by thinkers such as Whitehead and Einstein. Very few of these principles are absolute or universal; in fact the further one gets from physical or chemical principles, the more likely there are significant variations or exceptions. Nevertheless, they are essential to the understanding of forests and quite useful in applications in forests. Principles, combined with common sense and good judgment, are necessary as guides in the absence of definite knowledge. They give us a broad predictive ability. For each principle, we have to ask, how will it affect our objectives for that forest? Will the standards vary? Should forest operations be modified as a result of understanding?

I. General, Metaphysical, Metaphoric, Historical, Cosmological
Principles at the most general level apply to the universe and to various facets of the universe. We will call them global principles.

I.A. Being. This is an ontological principle that states that everything has its source in existence. Simply, forests are. When they disappear, they are not. Many archaic peoples living in forests accepted their existence; in fact the forest was often considered another being, as a god or mother.

I.B. Flow or Change. Although botanists recognized that change was inescapable as a principle of the new science, Frederic Clements insisted that change was not an aimless wandering, but a steady flow (he thought it was towards a stable final state, however). Continuous change has been identified recently by ecologists and evolutionary biologists as the context for ecosystems and species; in fact, biodiversity is an expression of continuous change. Individuals changes; patterns change, forests change. Neither tree plantations or old growth forests can remain unchanging.

Nature is in flux. The climate will change, the shorelines will change. Of course, thousands of years earlier, Heraklitus noted that everything changes. And Alfred North Whitehead made change the basis of his metaphysics. In his process view, organisms are dynamic structures immanent and simultaneous with process, rather than a simple consequence of the natural selection of random mutations.

Process is a fundamental feature of reality. Processes that generate form and variation at every level occur before natural selection is said to act; evolution can be understood in terms of this process, more than in terms of maximum fitness (as exemplified in protobiotic evolution and molecular genetics). Form and variation are not arbitrary or random.

A forest is a process. It grows; it is not made like a model or a plastic tree. It smells, it cools, it involves. It is not a form or a web of human words that we can manipulate into endless imaginary variations (as the deconstructionists are intent on doing in their nihilistic ecology).

The principle of separation (by walls barriers or membranes, that is, nonflow): Barriers are necessary to maintain form and integrity of individuals as well as ecosystems. By removing some barriers, such as releasing carbon that has been locked up for eons, we unbalance natural cycles. It is the problem of free flow or division by membranes. At each level of a natural system, from cell to biosphere, the units involved do more exchanging internally than externally with other units at the same level. Flow and division must be in balance.

I.C. Organism Forests are composed of living organisms; the forest itself changes, lives and dies in ways similar to a living entity or organism.

Whitehead regards organism as a universal principle, applicable in every field of reality from metaphysics to ethics. Everything that exists has its place in the order of nature. This does not mean that reality is an organism or that everything is reduced to biological terms. It does mean that every thing resembles a living organism since its essence depends on the pattern in which they occur, and not on its components. The organism is what it does. The organism expresses an order particular to its place and time, within limits. In Whitehead's metaphysics of experience, the world is an ecosystem, an intertwining of all things.

An organism is characterized by wholeness. Wholeness is the organizing principle in nature, according to J. C. Smuts. Wholes are self-making systems composed of subwholes (or holons) in a hierarchical system. All well-unified wholes are organic; all wholes are involved in organic wholes. The notion of part and whole is derived from extensiveness, which is the pervading generic form to which the morphological structures of the world conform, according to Whitehead. In discussing the reciprocity of part and whole, Whitehead uses the terms organism and environment. There is internal relatedness between organisms and environments. Life refers to complexes in which parts are modified according to principles derived from the whole. Organism can refer to molecules and ecosystems, or to any general sense of organic unity. Each individual organism is only a partial, however. They are like Arthur Koestler's concept of the holon; from above each is a whole; from below each is a part.

Figure 2. Slugs on an Austrian Pine, Vets Forest, Bulgaria.

The forest as a whole remains the same, according to W.S. Cooper (1913), "the changes in various parts balancing each other." Partly this is because the whole is a nested system that turns over at a rate much more slowly than the parts.

The principle of self-regulation. Organisms are self-regulating. They have evolved to maintain their stability in the face of many kinds of disturbances from planetesimal impacts to changes in atmospheric composition.

I.D. Field. The forest acts as a field, containing organisms. David Perry notes that any removal, even a single tree, sends ripples through the forest system; this ripple effect may be good or bad for the health of the system, depending on the chain of consequences. Thinning produces a larger ripple effect; because more light reaches the ground level, it stimulates herbs and shrubs, which may compete for moisture and slow tree growth or increase the rate of nutrient cycling and enhance tree growth.

The ideas of field and particle were indispensable to physical inquiry by the end of the 19th century (Faraday developed the concept in the 1860s). Although Whitehead noted that though the two concepts were considered antithetical, they are not logically contradictory. Ordinary matter was considered atomic, whereas electromagnetism was conceived as arising from a continuous field. A general space/time/energy/matter (STEM) field has many characteristics: discretion, participation, connection, consistency, limitation, wholeness, and self-development. No component of this field is ontologically subordinate to another; energy, matter, and pattern all have equal status—or, put another way, process is not more basic than structure or function.

By the time Alexander Gurwitsch (1922) used the term field in biology, gravity was also regarded as a field (then, nuclear interactions were described in terms of fields). The field concept was useful in his investigation of mushrooms where nondifferentiated structural units resulted in highly regular and specific shapes. The source and extent of a field was not confined to an organism, but was the result of geometric properties.

Paul Weiss objected to this concept of field, since it was based on ideal geometric constructions and not on the structural complexity of the organism. His own conception of the field was as a system of organizing factors that proceeded from already organized parts to developing regions, resulting in "typical patterns." Weiss described amphibian tail bud transplantations to the limb area: If the tail bud were transplanted early in the development, it gave rise to a limb in the new area, but if it were transplanted late, it produced a tail. The forces pushing cells into specific forms were a function of development. The system of forces of organizing action was named a field. Fields divided into smaller fields during development until the organism was a system of coordinated patterns. Growth and pattern are emergent field effects.

Weiss saw the field as a symbolic term for the unitary dynamics underlying the ordered behavior of the collective. It denotes properties lost in the process of analysis. In living organisms, the patterned structure of the dynamics of the system as a whole coordinated the activities of the parts. The parts of the organism are not assembled, but integrated. In the operation of a field, every part knows the activities of every other and responds to a collective equilibrium.

Although recent experiments support the existence of some kinds of biological fields, scientific descriptions are still unsatisfactory. Waddington regarded his own concepts of chreods and morphogenetic fields as descriptive "conveniences." The topological, qualitative models of Rene Thom depend on fields, but Thom admits that the use of local models implies nothing about the ultimate nature of reality.

Locality. Fields exist on many scales. Every field is limited by what can happen within a unit of time—its locality; thus, fields are independent, although not unaffected, of other fields. The principle of locality is biological also, that is, each tree or owl interacts primarily with other organisms in its local neighborhood—not with all organisms under all conditions. One thing this means to forestry is that global approaches may not always work.

Global. The problem of global flow and local flow. Local flow is needed for renewal, as is some global flow. But too much global flow (of matter, energy or form) can destroy relationships and diversity. Too little local flow stifles diversity and renewal.

I.E. Patterning. Patterns are the key to understanding the nature of a forest. Nature, for Whitehead, consists of patterns whose movement is essential to their being. These patterns are analyzed into events. Everything that exists has its place in the order of nature. This does not mean that reality is an organism or that everything is reduced to biological terms. It does mean that every thing resembles a living organism since its essence depends on the pattern in which it occurs, and not on its components. In some ways, patterns are prior to things, in helixes, light, fields, and ecology. Ecology attends the overall pattern of relationships. Paul Shepard and others have written that relationships are as real as the objects that result from them.

The genotype determines the physical and chronological pattern of an organism within the limits of an environment and interactions—that is, an organism unfolds as an embedded part of an overfolding environment. The being of a species is the reality of the pattern of its members, that is self-sustaining, self-organizing, reproducing units (compare to the idea of other holons at other levels). Selection operates as a survival filter that passes any structure with the integrity to persist.

The patterns and process have the same ontological status as the individuals, but not by making them into individuals, just by changing the epistemology to emphasize patterns as well as the ultrahuman pieces that make them up. Organisms put together (enfold) structures based on historical patterns, and move (unfold) through a filter of limits like minnows through a fish net.

Richard Hart suggests that the actual substance of which the forest environment is made consists of patterns rather than things or individual species. The forest environment is generated by a patterning of the ecological ebb and flow of energy, substances, individuals and species across a suitable landscape. Successful adaptation to this complex system requires an enormous amount of minute local adaptations by a large number of individual organisms from a large number of species. The distinction between growing and declining patterns is not arbitrary, and can be arrived at objectively, through monitoring.

Based on a broader metaphysical foundation, with more comprehensive values, measuring and monitoring need to address patterns of being in a forest and not just a few commodities dictated by a economics. One challenge to ecoforestry is to set up long-term programs to identify and study patterns and relate them to a healthy sustainable human use of forest ecosystems. But, the tools have to be used in new ways in a new framework, perhaps with topology and holograms as metaphors (topology provides the mathematical model for processes; a hologram provides a model for wholeness).

Monitoring is crucial to understanding forests. Until we understand how forests change and move around the landscapes, we will not know which changes are important and inevitable and which are the unhealthy result of human interference. Until we understand the changes, we will not be able to adjust our needs to the limits of forests.

The principle of patterning has several subalterns that refine its definition.

Limits. All patterns are defined by their limits. Limit gives form to the limitless (Pythagoras). Every motion, energy, or force is subject to some limit. This property is also called discretion. Entities, activities, patterns, and events are uniquely discrete, not averages, means, or samenesses.

Polycentricity. The universe has more than one center; it has multiple frames of reference. Having multiple frames of reference means that things can only be fully described through an idea of complementarity (after Bohr). Any event can be described by different frames of reference that may be mutually exclusive but also may complement one another. Things that in nature may seem contradictory are functions of perspective or the tools we use for examination. Every human culture puts itself at the center of its universe. Every culture has an image of the universe, that is usually anchored in place, in a landmark or a forest, for instance.

I.F. Continuity. Forests proceed through distinct continuous steps in relation to past environments and disturbances; that is, a tree plantation cannot become an old growth forest without developing through intermediate stages of community. Continuity can be qualified in three other ways.

Connectivity. Everything is connected, however weakly, to everything else. In a local system the connections are often strong. In a global system, connections between local systems may be weak or invisible. Reactions often propagate like ripples through the systems, or like a tug on a spider web.

Participation. The human species participates in its environments. Interactions and interrelationships are undeniable. Cobb and Griffin state "The whole of nature participates in us and we in it." Like quantum physics, in biology and ecology, the observer is within the theory in a very fundamental way. By the act of observing the observer influences the outcome of a phenomenon, as Wheeler says, taking part in the construction of reality. And, of course, every individual observes, from fungi to humans.

Complexity. As it continues, the universe becomes more complex; things change, patterns build. In the generation of organic forms, physical and chemical processes provide organizational principles that coordinate detailed biological mechanisms, including viscoelastic changes of a cytoskeleton and the expression of different genes. Evolution increases the levels of complexity through the operation of natural events. Nature may be more complex than we can understand, as G. P. Marsh, Barry Commoner, and Jerry Franklin have all said.

I.G. Historicity. History creates unique patterns, especially in forests. Each forest is unique in its parts and structure, in its matter, energy, forms, information, and in its dynamics and history. An individual entity, according to Whitehead, whose own life history is part within the life history of some, larger, deeper, more complete pattern, is liable to have some aspects of that larger pattern dominating its own being, and to experience modifications of the larger pattern reflected in itself as modifications of its own being.

Irreversibility. Forests pass through stages that are never repeated, despite superficial similarities; that is, tree-planting cannot reverse clearcutting (although another old-growth forest may develop in time). The history of land limits or determines its future. Fundamental particles and chemical compounds, as well as magnetic fields and living forms, exhibit this.

Indeterminacy (after Heisenberg). Some part of nature is always fuzzy.

Heisenberg's uncertainty principle states that predictions about location and velocity are just statements of probability. The effect of this principle on epistemology is that our exact interpretation of forests has to be abandoned. There are uncertainties at higher levels of organization as well; without knowledge of initial conditions and all past events, it is not possible to predict present or future behavior. Furthermore, the higher up in levels of organization, the more kinds of uncertainty there are: Genetic, environmental and social, at least.

I.H. Novelty. As an ordering / disordering process, a forest continually creates new forms and new patterns. Every forest is unique. Forests decay, as well as become more complex.

Boltzmann, Hirth, Brillouin, and Whyte are among those who first recognized that there are two great universal tendencies: One towards dynamical disorder, and the other towards spatial order. The first process has been called entropy; the second ektropy (or negentropy, anti-entropy, morphic order, or syntropy, in Wittbecker 1976).

Entropy, from Clausius, means generating a transformation in an abstract phase space. Ektropy, from Georg Hirth means generating order or form in ordinary space. All visible things have been formed by a combination of processes. Entropy equals shuffling; ektropy equals sorting. Entropy is a measure of infinite individual order. Ektropy is a measure of complex order and of limited individual order. The entropy of order, of energy, of information, is derivative of the entropy of turning. The two tendencies are directly related. When one increases, the other increases, although there is no *rate* of change. The universe is one turning composed of at least two polar processes, entropy (inward turning) and ektropy (outward turning). These processes occur roughly equally for the creation of order and disorder in dynamic tension. No theory can consider only entropy or ektropy. No metaphysics can be based on only one of the processes.

Creativity. The process of nature is not merely rhythmic change, it is a creative advance, producing new forms everywhere. "There is an all-embracing fact which is the advancing history of the one universe," Whitehead states. For Whitehead, the creative advance is from disjunction to conjunction.

Creativity as a fundamental metaphor—the dance of creation; in expressing itself it moves around the floor but has no single direction. Creativity, an ultimate principle in Whitehead, relates the many and the one in a manner productive of the pulsations of process which are the actual entities. Ecoforestry embraces creative complexity, as opposed to the simplification on which industrial forestry is based.

Surprise. The interactions of billions of small actions cause a change in quality, that is, quality emerges from quantitative action. Thus, rare events may shape the entire course and texture of the universe and its systems. Where does rarity go in our understanding? Unusual and exceptional events, such as the origin of life on earth, must be factored into scientific understanding.

Intrinsic Value. Value is mentioned sometimes as if it is just one thing, the economic market value of forest commodities indicated by price, but there are different kinds of value. John B. Cobb, Jr. contends that everything can have instrumental value (in the sense of having consequences for other things or beings) and living beings have an intrinsic value. Every species has some value (Deep Ecology Platform 1): Being unique, creative, historical patterns, all beings have value. These values are independent of the usefulness of the nonhuman world for human purposes. Whitehead has stated that existence is the upholding of value intensity; for a being itself and shared with the universe, from which it cannot be separate. The value (intrinsic worth) each being has for itself is shared by others. Each exists for itself and for others; is a value in itself and for others. Value is achieved through an ongoing process in nature, not a static one.

II. Physical, Energy, Cycles

There are laws that have been identified by physicists and chemists, which form the basis of life. Some laws have been called "impotence principles" by the biologist E.T. Whittacker. They cannot be proved true as laws, but they limit what we can do, i.e., we cannot use all the energy in a system, and we cannot return the system to a previous state. The first two principles here are the same as the first two laws of thermodynamics, which are impotence principles.

Energy. Energy can be transformed but not created or destroyed; energy is not created in the sun, just changed from its state in matter, mostly hydrogen and helium. "You cannot win," one of my professors used to say. A forest, which grows only from solar energy, is just a stage in the transformation of energy into production of "flesh." This principle is why old growth forests have no extra energy in the form of net primary productivity.

Entropy. Energy transformation cannot occur unless some of it is degraded into a dispersed form where it cannot be used again in the same system. "You must lose," the professor would add. The biosphere, and forests, obey the law of entropy only as a general limit. The entire process is exentropic since energy flows from sun to earth and long wave radiation flows from earth to the sink of space. Entropy does appear to force a historical direction to forests and the biosphere. Possibly this is one reason for the maximum of 6-8 trophic levels in a food chain in a forest; each level loses some energy that it cannot use.

Waste. That quantity of energy and material no longer of use to the system is wasted (for that system). Often it goes through another system, where a percentage of it is used (energy is not considered to recycle, although it can be used several times within a system and by several systems), but eventually it is lost to the interplanetary space surrounding the earth.

Cycles. Chemical elements, especially those used by life, circulate in the biosphere in characteristic paths known as biogeochemical cycles. Very little is actually lost to space, but often the elements concentrate in sinks, where they may be unavailable for ecological or geological periods of time. Forests, for instance, act as sinks for carbon; the ocean bottom acts as a sink for phosphorus. The rapid release of sinks can affect other atmospheric or terrestrial cycles.

Limits. Biological order is built on physical and chemical orders. That is why life is limited to such a narrow range of conditions. And that is why the most complex orders are vulnerable to changes in their substrates; energetic radiation can alter and destroy an individual, a small change in climate can destroy forests and civilizations. The earth is suitable for life because of three kinds of limits:

1. solar radiation has stayed within certain limits for 4 billion years;
2. the biogeochemical cycles of oxygen, carbon, nitrogen, phosphorus, sulfur, water have stayed within certain limits;
3. the environment has been constant enough for organic evolution, but variable enough for natural selection to be challenged.

Life is limited by elements and physical factors (light, water, gas, salt); too little of an element limits life(Liebig's law); too much of an element limits life (Shelford's law of tolerance). Regardless of how plentiful nutrients are, for instance, without water a forest cannot exist.

Productivity. Energy is bound into organic material, and is measurable as productivity. This energy can be partially used by living beings or released by disturbances, such as fires. The gross primary productivity is the capital of a forest ecosystem; the net ecosystem productivity is the interest. The kinds and numbers of organisms in a forest are limited in varying degrees by the productivity of the system.

Energy relation. Within a range, biological activities increase with increases in temperature. Metabolism and respiration increase in animals; reproduction and growth increase in plants. Rapid, introduced changes in temperature, however, shorten life cycles and increase micro-organisms. Sudden modification of a forest, such as from clearcutting, can cause changes in temperature and unexpected effects.

Food Chains. The transfer of energy and materials through organisms is referred to as the food chain. It is of various lengths, depending on the system but is rarely more than 7 or 8 layers deep. Mature forests have longer food chains.

Trophic Structure. The interaction of individuals in a food chain in a local physical environment results in the trophic structure of communities (ecological pyramids), which interacts with material cycles. Mature forests generally have more steps on the trophic pyramid.

Maturity. The energy required to maintain an ecosystem is inversely related to complexity; succession decreases the flow of energy per unit biomass until the system reaches maturity (Margalef's concept of maturity). In a mature forest almost 100% of the energy is required to maintain the state of the forest. Any system formed by reproducing and interacting organisms must develop an assemblage in which production of entropy per unit of information is minimized. It is a general property of some systems that acquired information is used to limit further inflow. A mature system needs less information, since it works toward preservation. The limit of maturity allows variability between systems with slight external differences, such as temperature. Ecosystems consist of different prefabricated pieces, species, and since the supply of species is limited, succession becomes asymptotic at a mature state.

Synergy (Fuller's concept). Reactions at the chemical, organism, or ecosystem level, when combined, produce unexpected positive results from the sum of single reactions. The forest ecosystem has emergent properties that are different from the sum of community interactions. They also affect biogeochemical cycles. Health is a dynamic quality of the whole, the result of a harmonious interaction of all the analyzable parts that comprise the whole forest with the surrounding larger environment.

Summary
This outline of principles does not include emergent principles. Part II will present Ecological, Evolutionary, and Forest Principles, as well as Sociological, Political, Economic, Educational, Forestry, Valuation, Planning, Design, and Management Principles.

Most of the principles stated so far have been global principles; some principles may apply to a particular region; others may exist at a very local scale. This is even more true for standards. The global provides the framework for the regional, which provides the framework for the local—the level of detail and participation. Standards are very important for forest practices or certification, to provide consistency of judgment regardless of the owner or certifier; they also form the basis for any appeal of judgments. They should be clear, unbiased, and applicable.

These principles, whether we accept or deny their applicability, influence our interactions with forests, from clearcutting to preservation. They influence our objectives, our standards, and our operations. The interplay of these principles with examples and exceptions will refine our approach to and understanding of forests. This is part of the process of living with and understanding forests.

Chapter 5

Principles of Ecoforestry (Part II)

VII. Political/Human Community (Sociology)

One concern of ecoforestry is to ensure the survival of human communities in place in forest communities, which is, in fact, the goal of politics. Because survival is in nature, politics must rest on an ecological foundation. As a philosophy, ecoforestry investigates the normative aspects of living together, that is, ethics, and the maintenance of the affairs of a community, that is, economics and politics. As a science, ecological forestry describes the interrelationships of organisms and environments, that is, the experience of living together in the biosphere. As a noetic discipline, it provides information on the state of nature, but recognizes that human beings are participants in nature, as part of the food chain, for example, as well as participants in the societies that are trying to survive. It offers a new perspective of humanity in the total field of nature and defines balanced relationships with ultrahuman beings and species.

Ecoforestry principles will often overlap several topics, for instance, irreversibility works on many levels, from quarks to ecosystems. Human fitness (Morphogenesis). Our species is shaped by the richness of forests, as well as by other diverse species and ecosystems. Changing wild forests to plantations, fields or deserts will profoundly change our psychology. Humans have no right to reduce this richness and diversity except to satisfy *vital* needs.

Present human interference with the nonhuman world is excessive, and the situation is rapidly worsening. The flourishing of human life and cultures is compatible with a substantial decrease of the human population. The flourishing of nonhuman life requires such a decrease. The richest countries are expected to reduce their "excessive interference" with nonhuman world. The reduction may take time, but it is extremely serious. Interference (Wittbecker). When an ecological system is stressed beyond its flexibility, its carrying capacity is exceeded, or its processes interfered with, the system usually collapses, reducing the amount of life it can support or materials it can cycle. The amount of interference determine how soon it can come back, if at all. A stressed system is more vulnerable to disturbances.

Policies must therefore be changed. These policies affect basic economic, technological, and ideological structures. The ideological change is mainly that of appreciating *life quality* (dwelling in situations of inherent value) rather than adhering to an increasingly higher standard of living. Quality of life cannot be quantified, but there is no need to do so.

Principles of Human interaction
- Membership. Humans participate in natural systems (stated by Aldo Leopold). Humanity is part of the large order of things, but the proper attitude is citizen rather than conqueror.
- Dependency. Human beings are dependent on nature for inspiration, ideas, resources, and recycling. Wild nature, as the sources, must be

preserved for future inspiration, ideas, resources, and recycling.
- Synergy 2. The forest is effected by human activities, which can compact soil, remove biomass, and interfere with animal and plant patterns—but, it can also improve patterns and structures as well as reintroduce plants and animals.
- Cultural Transference. The family is the fundamental unit of culture. The continuity of families, however defined, assures the survival of societies.
- Compliance with laws of other cultures. Management shall respect all applicable laws of the country as well as all signatory international treaties and agreements.
- Tenure and Use Rights and Responsibilities. Tenure and rights shall be clearly defined, documented, and legally established. (FSC No. 2)
- Indigenous people's rights. The legal and customary rights of indigenous peoples to own and manage their lands and resources shall be recognized and respected. (FSC No. 3)
- Self preservation and self-interest. Be healthy. Contribute to the health of your community. Contribute to the health of natural communities.
- Ethical behavior (suggested by Deep ecology). Recognize the intrinsic value of beings, connections, patterns. Increase self-realization of others through identification. Act responsibly as part of the community. Live simply (voluntary simplicity)
- Principles of diversity and symbiosis (where diversity means live and let live rather than either/or, and symbiosis means coexistence)
- Cooperation. Modern industrial culture places an emphasis on individualism and competition. Cooperation, with an understanding of rights and responsibilities, is based on cultural understanding. This kind of understanding, once prevalent in many cultures, is the reason why the tragedy of the commons (Hardin) did not always occur with common resources.
- Political Power Limitation (from Paul Ekins). Those who wield power must be elected by, accountable to and replaceable by the people over whom power is exercised.
- Separation. Powers must be separated. Religious power, political power, and economic power cannot be combined; one should not acquire the others.
- Social cooperation. Modern industrial culture places an emphasis on individualism and competition. Cooperation, with an understanding of rights and responsibilities, is based on cultural understanding. This kind of understanding, once prevalent in many cultures, is the reason why the tragedy of the commons (Hardin) did not always occur with common resources.
- Rights. Humanity has the right to coexist in healthy diverse, sustainable conditions. The right of nature and humanity to coexist in healthy, diverse, sustainable conditions
- Wild ecosystems have the right to flourish without human interference (that is not the same as moderate exploitation or regular disturbance). In principle, deep ecology proposes a biospherical egalitarianism, that

is the equal right of all beings to live in place.

- Humans have no right to reduce the richness and diversity of wild nature except to satisfy *vital* needs (from the Platform for Deep Ecology). Vital need is left vague to due requirements of different cultures in their locations.
- The flourishing of human life and cultures is compatible with a substantial decrease of the human population. The flourishing of nonhuman life requires such a decrease. Richest countries are not expected to reduce their "excessive interference" with nonhuman world overnight. Stabilization and reduction may take time but is extremely serious.
- Present human interference with the nonhuman world is excessive, and the situation is rapidly worsening. The slogan of "noninterference" does not mean not exploiting species for human use. Wilderness areas should be expanded.
- Policies must therefore be changed (from the Platform for Deep Ecology). These policies affect basic economic, technological, and ideological structures. The resulting state of affairs will be deeply different from the present. Continued economic growth is incompatible with principles 1-5. Current growth is not sustainable. Things are valued in the current economic system because they are scarce or have a commodity value. Waste and consumption become sources of prestige. The implementation of deep changes requires global action as well as local. Government interference, especially developed world, accomplishes nothing. Appropriate technology can advance cultural diversity and independence.
- Rules and laws should be used for the transition to sustainable continuity.
- Community Relations and Worker's Rights. Management operations shall maintain or enhance the long-term social and economic well-being of forest workers and local communities. (FSC No. 4)
- The ideological change is mainly that of appreciating *life quality* (dwelling in situations of inherent value) rather than adhering to an increasingly higher standard of living (from the Platform for Deep Ecology). There will be a profound awareness of the difference between big and great. Quality of life is considered vague because it is nonquantitative. It cannot be quantified, but there is no need to do so
- Nonviolence. Like seems to beget like, especially violence and ignorance. But this is also true with friendship and resource development.

V. Economic

Economics has many of its own principles, which seem trivial in application for forests and communities. Modern economics has a few principles that are still considered important, such as Quantification (the evaluation process is quantified in monetary units and accounted for)—although quantification is only applicable to part of the human system in nature, the principle is applied to the whole by economics—or Diminishing marginal product (at some level of output, total output increases by smaller amounts as variable resources are put into service—this is basically a principle of limits). Other principles, such as Plenitude (the earth is unlimited), have been allowed to fade away. Economics finds that the unifying principle in markets is the interaction of supply and demand, not the invariable nature of the resulting value.

An alternative ecological economics is being developed that promises consideration of all aspects of human economic interactions, from equity to environmental services.

- Real Plenitude (or Enoughness). There are adequate resources (flow of goods and services) for locally-adapted human populations to sustain their needs indefinitely.
- Conservation (after G. Pinchot). The regenerative capacity of a renewable resource, such as a forest, should not be damaged or destroyed. Ditto for the biosphere.
- The Locus of Value. Economic value lies in the use of goods and services, more than in the things themselves, which exhibit an intrinsic value.
- Self-interest (from Garrett Hardin). Never ask a person to act against her own self-interest. Of course, problems of scale apply to the self as well; one may love one's self, family, friends, home; maybe less so for a city, bioregion, continent, earth. Under certain conditions (starving in W.W.II concentration camps, in some cases), the self becomes locked in the skin and all morality is lost. Classical economics taught that all economics in sane individuals are grounded in the "desire for wealth" (following two laws: a greater gain is preferred to smaller; the greatest wealth with the least labor or self-denial).
- Economic Power Limitation (after Paul Ekins). Everyone should possess sufficient autonomous economic power to guarantee a decent subsistence. That is, people are entitled to health and opportunity.
- Ecological Limits to Economics. Limits are necessary for sustaining economic activity. The land surface area is limited; most resources are limited. Nonrenewable resources should be reused and recycled as much as possible. Economies are limited by a combination of a natural resource base, information, labor, and technology.
- Limited good, that desired things exist in limited quantities.
- Working with Nature (F. Bacon, R. Carson and others). As an inseparable part, sharing and requiring basic processes, human beings must work with nature, within natural limits.
- Economic Context. A healthy environment is a requirement for a healthy economy.
- Local Advantage. Cultures that have adapted to their forests have unique understanding of the forests and their use of them. The attempt to modernize and industrialize local cultures is unwise, since the associated interference, waste, and deterioration may destroy the forests and the local cultures.
- Increasing costs (after (R. J. Vogl). Attempts to increase the production of systems near maximum productivity, depletion, or carrying capacity

are more costly in energy, effort and finance than in healthy systems.
- Energy Advantage (after Merve Wilkinson). The energy cost of obtaining forest resources should not exceed the value derived from the resource (at the risk of economic or personal failure).
- Theft. High, wasteful lifestyles in some cultures can only be maintained through drawdown and inequity.
- Because of our reliance on ecological systems, the costs to those systems must be internalized in our economic systems.
- The best economic use should be determined based on ecological considerations; all possibilities should be considered.

VIII. Education
Education is a process leading from ignorance to knowledge. There are many kinds of knowledge, including wisdom. The process involves play, liberation, and community, to result in equity and wholeness.
- The principle of play. Play is the method of learning for most juvenile animals and a means of enjoyment for many adult animals. For humans, play is imaginative experience, entered into freely. Much human activity is play, in place in a community. Even science is a forms of play, attempting to solve the puzzles of existence, through an open, self-referential, self-correcting system capable of using analytical and synthetic methods. Play may be defined as an activity from "overflowing energy" as Schiller and later Bolwig considered it. It is also an activity with ``no immediate objective'' (Hall 1968); an experimental dialogue with the environment (Eibl-Eibesfeldt 1970); rehearsals performed in a nunfunctional context, of the serious activities of searching, hunting, fighting, mating (Wilson 1971); and behavior that functions to develop, practice, or maintain physical or cognitive abilities and social relationships by repeating or recombining sequences of behavior outside their primary context (Fagen 1985). For Schiller, play is activity for its own sake, where the drives of emotion and reason are harmonized. The state of play is whole and simultaneous. It unifies permanence and transition, chaos and order, duty and selfishness. The object of the senses is life, the object of reason is form, and the object of play is living form—called beauty in the widest sense. Aesthetic play, like physical play, requires order and control. The principle of error is a corollary of play. Error permits learning. Error permits diversity. So many things are thrown at the flow of life. There is not just a little error. Half of everything seems to be error: billions of pollen grains, billions of eggs, millions of species. Transmission is not flawless or efficient.
- Community. Community is formed by beings living together in place. The human community is only one of many. Human relationships embrace other beings as well. This interrelatedness makes human beings less discrete and less alone. The biological term for ``living together'' is symbiosis. Living together, beings do together. The Sanskrit word for ``doing together'' eventually became the word ``ethics'' in English. Ethics (and rights) are simply rules for living

together. Human ethics describe only a small, self-conscious part of the rules.

- Liberation. The rich flowering of human nature is possible only when the constraints of need are replaced by leisure and abundance. Liberation requires a larger perception and larger concept of rationality. For humanity, liberation means an end to prejudice or discrimination based on arbitrary characteristics. The liberation of nature and ultrahuman beings is inseparable from human liberation. Nature and ultrahuman beings are to be free from the obligation to be human (so pervasively presented from Plato to Kant) and from the status of human resource or human artifact. In Schiller's scheme beauty frees us, because we decide how an act is performed, without being restricted by blind necessity.

- Wholeness. As Plotinus and Novalis recognized, education has an outward, social and civil, aspect as well as an inward, personal and self-revealing, aspect. Education has at least four ends: (1) the appreciation of the richness of nature, (2) the comprehension of human existence, (3) the understanding of the nature of human society, and (4) the training for a position in human society. Education has become more universal, but its goal, the well-rounded individual, has been distorted by its fourth aim, training for the economy. To produce wealth for the state and livelihood for the individual, education has become money obsessed. Ethics, in the second and third aims, has been neglected, since it might limit or contradict its economic obsession. In fact, the first three aims are restrictive to a growing, industrial economy. Education, as practiced by public schools, produces unprovocative individuals, adjusted to an unbalanced society.

- Equity. The distribution of symbolic and real wealth is very inequitable as the result of historical trends, old economic rules, and cultural confusion. How can the idea of equity be given determinate content and political force? Society should be organized on the basis of functions, not rights. Ecological rights (customs) could be based on functions. It is foolish not to assign rights to animals, plants and earth because of contractual formalities.

  The issue of intergenerational equity is relevant here. Present need must be weighed against future need. Stable, good societies must pass the social institutions on a material base to a new generation, but without sacrificing everything for posterity. The traditional object of economics is the administration of scarce resources; this needs to be extended to more than one generation. There is the question of the quality of life from one generation to the next; possibly a the distribution of a dowry to all succeeding generations.

  Equity is the promotion of opportunities and rewards for all people, as well as some redistribution of goods due to historical patterns.

- Continuous learning. Because the universe changes, and considering novelty, surprise, and uncertainty, knowledge can never be complete.

X. Forestry Practice

Forests are the foundation of forestry. An ecological forestry must observe the proper scales of forests, especially in terms of size, age, and patterns.

- Soil is a major part of the forest and key pathway of various cycles.
- Natural reproduction sorts tree genetics to microsite requirements.
- Pathogens determine tree mortality, which drives forest and landscape patterns.
- The human relationship to forests should be symbiotic rather than parasitic or cancerous.
- Preserve all components, structures, and functions
- Protect and maintain diversity / preserve the patterns
- Ecological health depends on the maintenance of ecological processes.
- Utilize natural processes for regeneration and protection
- Forestry practice should be holistic since it affects the whole forest. You should harvest in the context of planning, measuring, monitoring, protection, and restoration. Emphasize interconnectedness over separate structures or operations. Sustain the forest before any yield.
- Retain the structures and processes (which produce the complexity and diversity) of the forest, including legacies and special areas. Soil is an important structure. Complexity of interrelationships. Not only is there a wide variety of plant and animal interrelationships, but there are soil, water, atmospheric, and geological interrelationships. Each directly influences the others. Unique regional climates are related on a global scale. Any change produces a chain of effects, many of which are unpredictable or unwanted.
- Design your harvests to maintain connectivity, thereby minimizing fragmentation.
- Do not interfere with the health and stability of the forest.
- Encourage natural regeneration.
- Respect traditional practices in place-based cultures.
- Broaden the base of use of materials from the forest; avoid concentrating on timber only.
- Be flexible—keep your options open.
- Manage for preservation, reservation, protection, and use.
- Scale management to forest size. Approach the forest as in a partnership; do things to benefit the forest as well as yourself. Give back to the system.
- Keep acquiring understanding and knowledge.
- Attempt to set up a transgenerational land tenure system to accommodate forest time, which spans human generations.

X.A. Valuation

The well-being and flourishing of human and nonhuman life; all beings have value in themselves (synonyms: intrinsic value, inherent value, from the Platform of Deep Ecology). These values are independent of the usefulness of the nonhuman world for human purposes. Life includes individuals, species, populations, habitats, and all human and nonhuman cultures. Deep concern and respect for cultures. Ecological processes should remain intact.

- Satisficium. In a system of ethics, if we decide to have a value, should we maximize value? If so, whose? Then when? Should we maximize it for the present or future for humans or for ecosystems? Cobb suggests a few possible principles:
- Selfishness: Act to maximize value for yourself in the present
- Prudence: Act to maximize value for yourself for the future
- Utilitarian: Act to maximize value for all humans for the indefinite future (the greatest good for the greatest number)
- Process: Act to maximize value in general (every entity with intrinsic value) Cobb points out that the first three are unstable and unacceptable. That leaves the final principle; perhaps though, we should aim for an optimum.
- Minimum value: every species has some value
- Limited good—desired things exist in limited quantities.

X.B. Planning
At the spatial scale, planning and carrying out activities at the scale of small landscapes (small watersheds of 200 hectares/500 acres and larger) is a minimum standard. Ecologically responsible forest use requires the development of forest landscape level plans for as large a landscape as is practical, given political and ownership constraints.

Ecologically responsible forest managers must attempt to carry out large landscape level plans in order to qualify for certification. Educating government and industrial managers of forest ecosystems about the need for landscape level plans is a part of the task for PCC and other agencies advocating ecologically responsible forest use.

Ecological limits are physical and biological factors which indicate that various human uses may result in unacceptable levels of modification or degradation of forest ecosystem functioning. Common ecological limits include: shallow soils (less than 30 cm/12 inches deep); very dry or very wet sites; very steep slopes (greater than 60% slope gradient); broken slopes (abrupt slope gradient changes occur regularly across a small landscape); very dry climates (less than 25 cm/10 inches of precipitation annually); cold soils that limit biological activity, particularly soil nutrient cycling; snow-dominated forests characterized by open, canopied forest stands (i.e. park land forest ecosystems); and, riparian ecosystems.

Forest ecosystem types are relatively homogeneous forest areas delineated by their biological and physical characteristics, and by their ecological limits or lack of ecological limits. Stands or patches frequently contain several ecosystem types.
- Create a practical plan before doing anything on the ground. Set goals. Know what you want to do.
- Ensure that all plans and activities protect forest functioning at all scales (time and space) and define ecological limits of various ecosystems to human disturbance (Hammond Principle No. 3).
- To maintain forest ecosystem functioning, forest plans must be designed on temporal (time) scales of 500 years and beyond, as opposed to conventional logging development plans of 1 to 20 years.

In other words, forest plans are developed, as opposed to logging plans.

- Describing a forest ecosystem type as having an ecological limit to human activities does not mean that such an ecosystem type will not grow trees following a human-induced disturbance such as logging. However, the existence of an ecological limit means that sustainable timber crops that have economically viable timber volumes and timber quality cannot be grown in reasonable periods of time. As well, both physical and biological problems, such as landslides and poor regeneration of trees, may result if ecological limits are not respected. If forest users attempt to ignore ecological limits, unacceptable levels of forest degradation will occur in both the short and long terms. (Hammond Principle No. 3)
- Apply the concept of landscape to the forest organism or process under consideration.
- Different forest organisms or forest processes operate at vastly different scales. What is a landscape to a salamander is only a small patch or small stand to a bear. Similarly, the landscape that results from a single tree falling over due to root decay and wind is much smaller than the landscape patterns created by a large fire. A forest landscape can exist at virtually any scale, depending on the organism or forest process used as the point of reference. Applying the concept of a forest landscape, as much as possible, to all scales— from large landscape to small stand or patch level plans—is an essential step in ensuring the maintenance and/or restoration of fully functioning forests. (Hammond Principle No. 8)
- Plan and carry out diverse activities to encourage ecological, social, and economic well-being .In planning for a diversity of human activities in a forest landscape or forest stand, we can use as a model the natural diversity that occurs in forest composition, structures, and functioning, from the smallest forest patch to the largest forest landscape. Diversity in forest composition, structures, and functioning maintains the integrity and resilience of forests. Diversity provides for both flexibility and stability in forest ecosystem functioning. Large natural disturbances, such as fire and insect attacks, that can dramatically alter the forest are simply processes of maintaining and restoring natural diversity, and, therefore, healthy functioning in the forest.

    Because natural forests depend upon diversity, a diversity of ecologically responsible human activities is most likely to maintain natural forest diversity, and, therefore, to maintain fully functioning forests. At the same time, diverse human activities best meet the needs of all interests in human society, and provide for the most stable, sustainable human economies.

- Cut sustainably. Currently, in most forests around the world, the most aggressive and consumptive forest uses are expanding, namely logging and the manufacture of a few wood products such as pulp and 2x4s. Continuing this growth of consumption is not sustainable, either biologically or economically. Ecologically responsible forest use will not replace conventional timber cutting levels with the same level of ecologically responsible timber cutting. Instead, ecologically responsible timber management means reducing timber cutting levels, and, therefore, reducing the overall use of wood. A high priority is placed upon developing and marketing recycled wood products, including paper, 2x4s, siding, paneling, windows, and doors. Local and regional certifiers associated with the Pacific Certification Council will work to educate consumers and suppliers, and to develop sources of, and markets for, a variety of recycled wood products. Reducing consumption and recycling *all* wood products is a high priority for ecologically responsible forest use.

- As well as ensuring that timber cutting and the manufacture of wood products stays within ecological limits, ecologically responsible forest use will also limit the number and scale of activities such as tourism and ranching within forest landscapes in order to maintain fully functioning forests. (Hammond Principle No. 9)
- Plan for complexity, as opposed to simplification, on which modern forestry practices are based.
- Plan for heterogeneity, as opposed to homogeneity. Heterogeneity is important on the genetic level, which in wild forests reflects differences in microsites.
- Apply the precautionary principle to all plans and activities. The precautionary principle means that plans and activities must err on the side of protecting ecosystem functioning, as opposed to erring on the side of protecting short-term monetary profits or annual timber cutting quotas. In other words, if you are not sure that an activity will protect, maintain, or restore ecosystem functioning, do not do it. (Hammond Principle No. 2)

X.C. Design

In a way, it is silly to think of designing a forest, yet we are modifying forests all the time. Therefore, we should design our modifications to maintain the spirit of the forest. According to the Forestry Commission, forest landscape design depends upon an appreciation of six key design principles: shape, scale, diversity, sensory, unity, and 'spirit of the place'.

These elements need to be expanded and related for ecological design (the following set draws from the Commission, Forman and Godron, and Mollison, but adds a few new ones).

There are basic geometric elements of any design, from the 3-dimensional (volume) to 2 (plane), 1 (line), and 0 (point) dimensions. These elements can vary in numerous ways, by number, position, direction, size, shape, interval, texture, color, and temporal. Furthermore, the elements can be organized into groups by nearness, similarity, and difference (diversity), into structures by rhythm, tension, balance, and scale, and finally into a whole with sensory force and a spirit of place (genius loci). All of the elements interact in complex and unpredictable ways. The spirit of the place is the most important principle to be conserved or enhanced.

Sensory Force. All the elements of design can be combined in an image. Every organism creates an image of its place from what is meaningful to it. This image is what fits the organism to its place. Suckers and caddisworms have simple images; coyotes and humans have more complex ones. Boulding notes that the image as a cognitive construct of the world has several aspects: spatial, temporal, personal, relational, value, and affectional (emotional) for each individual. Cognition is an active relationship that is creatively shaped by the participants. Participation is not an option by the way—every scientist or inhabitant becomes part of the system she observes. The total sum of individual images is a world. Some of the images we impose on nature result from idealized notions of pastoralism or technological futures. Thus landscapes abound in nostalgic or consumptive trends on many levels of explication—some are iconic, some invisible. We originally perceive the landscape symbolically, but the landscape has other functional dimensions that increase according to use.

- Irregular shapes are more pleasing
- The eye follows diagonal lines
- The scale of a forest should reflect the landscape
- Optimum diversity in a landscape is valued
- Landscape is unified when scale and diversity are optimum
- Any one sense, even smell or touch, can dominate a landscape
- Design should follow sensory force
- Design should enhance the spirit of place
- Application of design principles to forest standards
- Maintain size and completeness of forest
- Strengthen natural shapes and margins
- Keep density and openness in natural balance
- Preserve the interior/protect riparian zones
- Plan paths to avoid sensitive areas and emphasize pleasing perspectives
- Bundle paths (power lines, utilities) to minimize intrusion
- Preserve the character and all aspects of the forest
- Allow for human participation
- Encourage responsibility
- Conserve continuity
- Work with the forest. Succession can be assisted, slowed, or speeded up, but not skipped or ignored.
- Forest is designed by its limits, time, scale, complexity
- Only details and exceptions form the forest; only generalities are important—so you have to live with contradictions
- Make the smallest number of changes
- Adapt process of change to the site (you fit the forest, not other way)
- Seek best use for products; everything in the forest is a resource for something; many can be directed to human use.
- Extend the life of things through cycling, then return to forest; things can be recycled indefinitely, as in an old growth cycle.
- Preserve the components, structure, and function

- Understand the patterns and connections
- The interdependence of design on nature; distant effect must be acknowledged
- Respect every aspect of human dwelling, from the spiritual to the economic.
- Responsibility for all the consequences of design, from human well-being to nature.
- The creation of long-term value in human artifacts, from letter-openers to buildings and plantations, by avoiding low standards, careless work, or cheap processes.
- Derive operations predominantly from natural energy flows, from gravity to the sun. Use this energy efficiently and safely.
- Incorporate waste into a full life-cycle of products and processes. Waste as a concept only applies to one system; in surrounding systems, it is a resource.
- Acknowledge limitations in human designs. Natural designs make good models, as we know from egg cartons and velcro.
- Productive ecosystems, such as valleys and wetlands, should be reserved (beyond preservation areas) for wildlife, forests, crops, fish, and livestock, instead of being used for city structures and roads. Conservation biology suggests some general principles of reserve design:
- Species that are well distributed across their native range are less susceptible to extinction than those confined to small areas.
- Large blocks of habitat, containing large populations of each species, are superior to small blocks with small populations
- Blocks of habitat close together are better than blocks spaced far apart.
- Habitat in contiguous blocks is better than fragmented habitat.
- Interconnected blocks are better than isolated blocks.
- Corridors can function to make small blocks function as large blocks, although some corridors may be too narrow for many species.
- Roadless blocks are better than accessible roaded blocks.
- Human disturbances that are similar in scale and timing to natural disturbances are less likely to threaten species than those disturbances that are radically different from a natural regime.

X.X. Management
Management, whether it is archaic, traditional, industrial, or ecological, should adhere to basic principles.
- Holistic Practice. Forestry practice should be holistic since it affects the whole forest. You should harvest in the context of planning, measuring, monitoring, protection, and restoration. Manage for preservation, reservation, protection, and use. Create a practical plan, based on the forest (especially riparian areas). Emphasize interconnectedness over separate structures or operations. Sustain the forest before any yield.
- Retention. Retain the structures and processes (which produce the complexity and diversity) of the forest, including legacies and special areas. Soil is an important structure. Preserve the diversity of

microsites.

- Connectivity. Design your harvests to maintain connectivity, thereby minimizing fragmentation.
- Noninterference. Do not interfere with the health and stability of the forest. Do not block flows of air or water. Water Flow. As the medium for life, water is used and transported by organisms and systems. Water is necessary in any organic energy conversion. Air flow. As the medium for gaseous and water cycles, air is essential for most life. Do not block flows of nutrients, materials, or animals and plants.
- Least Effort (also a principle of economics and cybernetics). Encourage natural regeneration; let the forest do as much work as possible—it has millions of years of practice in some cases. Make the fewest cuts, go the least distance.
- Respect. Respect traditional practices in place-based cultures.
- Broad Use. Broaden the base of use of materials from the forest; avoid concentrating on timber only.
- Managerial Flexibility. Be flexible—keep your options open. Principles must be flexible to mirror the flexibility of open systems; flexibility is provided by diversity in fact.
- Flexibility. Small is more flexible (and beautiful as Schumacher says); small mills, small operations, small businesses are easier to run and easier to change with conditions or trends.
- Work with the forest. Succession can be assisted, slowed, or speeded up, but not skipped or ignored.
- Only details and exceptions form the forest; only generalities are important—so you have to live with contradictions
- Make the smallest number of changes
- Adapt process of change to the site (you fit the forest, not other way)
- Seek best use for products; everything in the forest is a resource for something; many can be directed to human use.
- Appropriate Scale. Scale management to forest size. Approach the forest as in a partnership; do things to benefit the forest as well as yourself. Give back to the system.
- Appropriate Time. Work with forest time; do not try to do everything in the industrial schedule. Be as slow as you want or need to be. Attempt to set up a transgenerational land tenure system to accommodate forest time, which spans human generations.
- Openness. Keep acquiring understanding and knowledge.
- Participation. Manage personally. Probably no one will know that forest as well as you do (especially if you live in it)
- Legal Stability. Find the best combination of ownership, trusts, easements, and plans to ensure that the forest will be cared for in the long-term.
- Profitability. Make your business profitable (by profit, I mean simply the excess of returns over expenditures—thus your health and happiness might be considered a return from caring for the forest). Adjust your input to expected goals. If your goal is to restore the forest to health and enjoy it, then the profit may be spiritual.

- Keep forests wild by minimizing control and interference
- Preserve the structures and functions of a forest
- Preserve minimum viable populations of species
- Create standards of forest classifications

Ecological managers attempt to recognize interconnected systems instead of focusing on one crop. Important principles for forestry include:
- sloppy clearcuts
- limited use of heavy equipment to reduce soil degradation
- reduction of slash burning
- retention of large "woody debris"
- distribution of standing large trees
- reliance of natural succession after logging
- protecting riparian areas

The Wholistic Timber Management (WTM) developed by Herb Hammond is guided by 3 major principles:
- Maintain and protect biological diversity in the stand and across the landscape; and to ensure that uncommon or declining species are not sacrificed for common or desired species.
- Maintain and protect natural structures and functions, including soils and water quality.
- Utilize natural processes

Forest Management Principles
- Preserve all components, structures, and functions
- Protect and maintain diversity/preserve the patterns
- Utilize natural processes for regeneration and protection
- Approximation. Management practices that work with natural processes will probably be more successful in the long-term.
- Self-making and self-managing. Species can take care of themselves in their natural habitats.
- Extraction Limits. Humans can harvest low numbers of a species and it will continue to be self-making. Resources removed from the system must be minimized and their costs (to the system and eventually us) recognized; this is to avoid imbalances in species, interference in processes, and destruction of community patterns. Waste materials should be returned to the system.
- Sustainability. To be sustainably used, resources can only be harvested at rates below their productivity or recovery levels (for the unit of time).
- Uncertainty.
- Manage personally. Probably no one will know that forest as well as you do (especially if you live in it)
- Find the best combination of ownership, trusts, easements, and plans to ensure that the forest will be cared for in the long-term.
- Make your business profitable (by profit, I mean simply the excess of returns over expenditures—thus your health and happiness might be

considered a return from caring for the forest). Adjust your input to expected goals. If your goal is to restore the forest to health and enjoy it, then the profit may be spiritual.

- Accept Management Limits. Management is limited, by costs mostly, to regulating animal and plant populations in a system, rather than climate, geology, or water, soil, mineral cycles. Manipulating plants and animals to the environment is more likely to succeed than the opposite. Management limits are determined by the ecological characteristics of the resources, before any economic, technological, or political limits are applied. The use must be limited by the ecology before monetary considerations.
- Management Plan. A management plan, appropriate to the scale and intensity of operations, shall be written, implemented and renewed. The long-term objectives and means shall be clearly stated. (FSC No. 7)
- Benefits from the Forest. Management operations shall encourage the efficient use of multiple products and services to ensure economic viability and a wide range of environmental and social benefits. (FSC No. 5)
- Environmental Impact. Management shall conserve biological diversity and its values—water, soils, unique and fragile ecosystems— so maintaining the ecological functions and integrity of the forest. (FSC No. 6)
- Selection. (after R. J. Vogl). Putting pressure on aggressive species with high tolerance, those very species that we tend to regard as pests or weeds, helps to select resistance or immunity in them—thus, screwing up any attempt to manage them in wild or domestics forests.
- Maintenance (after R.J. Vogl). When human activities remove or destroy native vegetation, the new vegetation can only be maintained at great effort and cost, unless the region is colonized by aggressive (and usually unwanted) weed species.
- Resource Finitude. Almost every resource has limits: energy, light, minerals, soil, air, water, land, and living space.
- Resource Renewability. Resources are renewed at varying rates. Overuse or interference with the renewal process can renew such resources nonrenewable.
- Respect and maintain natural disturbance regimes through time and space in order to maintain forest landscape patterns. Natural disturbances, from the death of individual trees to large fires or windstorms, are responsible for critical composition, structures, and ecosystem functioning necessary to maintain fully functioning forests. For example, the death of an individual tree sets off a process of change, beginning with a standing snag that provides habitat for cavity-nesting birds and ends with a fully decayed fallen tree that serves as Nature's water storage and filtration system. At a landscape level, natural disturbances, large and small, are responsible for diversifying habitat patterns and, therefore, maintaining a natural diversity of plants and animals. Natural disturbance regimes are also critical to the maintenance of soil nutrient cycling and

adequate levels of soil nutrients. Protecting, maintaining, and, where necessary, restoring natural disturbance regimes provides for natural composition, structures, and functioning at the forest landscape and stand levels. (Hammond Principle No. 5)

- Ensure that all forest use activities respect, protect, and maintain Indigenous culture, both traditional and current. Abiding by this principle means that Indigenous people must be fully and meaningfully involved in all forest use planning and all forest use activities that take place on or that affect Indigenous traditional territories. (Hammond Principle No. 10)

X.X. Preservation

Focus on what to leave, not on what to take. Ecologically responsible forest use leaves fully functioning forests at all spatial scales through time. In other words, ecologically responsible timber managers identify the parts of a forest stand and forest landscape that must be protected to maintain short- and long-term forest functioning, and then determine what can be safely removed for wood products and other uses. Evaluating historical Indigenous forest management is an important part of determining what needs to be protected and what can be removed. (Hammond Principle No. 1)

Base Wilderness. Wild areas, undominated by human processes, are required to keep bioregional and global cycles operating, as well as for homes for ultrahuman species. This preservation of natural processes must transcend any economic, spiritual, or aesthetic needs. Preservation is essential for survival and self-preservation. Principles of preservation and protection include:

- Save an optimum amount of habitat for the largest indicator species
- Use many species
- Limit the use of each kind
- Respect, use, and recycle materials

X.D. Protection

Ensure that all plans and activities protect, maintain, and, where necessary, restore biological diversity (i.e. genetic, species, and community diversity). Maintenance and, where necessary, restoration of all types of biological diversity is necessary to sustain life as we know it in forest ecosystems.

- Protect and maintain diversity. Maintaining genetic diversity means ensuring that viable natural gene pools, including the gene pools of trees logged from a site, remain on the site or, in the case of previously degraded forests, are restored to the site following human use. Maintaining species diversity means that viable natural populations of plants, animals, and microorganisms are maintained or restored, in previously degraded areas, throughout the various successional phases for each ecosystem type within a forest landscape. Maintaining community diversity means maintaining or restoring, in previously degraded areas, the variety of forest ecosystem types that result from natural disturbances at a variety of scales through short and long time frames in a forest landscape. Biological diversity must not be viewed as a frill or luxury. Ecologically responsible forest users understand that maintaining natural biological diversity is an absolute requirement to ensure maintaining fully functioning forests through time. (Hammond Principle No. 4)

- Protect, maintain, and, where necessary, restore composition, structures, and functions at the patch or stand level in all plans and activities. Natural composition and structures must be maintained in order to maintain fully functioning forests. Many compositions, structures, and functions are beneath the surface of the ground, within the soil where human beings cannot see while planning and carrying out forest use. Ecologically responsible forest use hopes that by maintaining the forest composition, structures, and functions that we can see, we will also maintain the composition, structures, and functions that we cannot see. For example, avoiding soil compaction during logging activities is absolutely necessary to protect soil structures and composition. (Hammond Principle No. 6)
- Protect, maintain, and, where necessary, restore forest ecosystem connectivity at all scales during planning and carrying out ecologically responsible forest use. Connectivity in forest ecosystems is maintained, in large part, by ensuring the protection of water movement patterns. This includes microscopic water movement patterns in the forest soil and in riparian ecosystems, from ephemeral streams and small wetlands to large river systems and wetland complexes. Connectivity is also maintained in forest ecosystems by protecting and, where necessary, restoring the full range of composition and structures from the large landscape level to the smallest stand or patch. (Hammond Principle No. 7)

X.E. Mensuration
- The health of ecological systems and human institutions should be measured with a holistic index.
- The measurement needs to be regular and long-term, to detect changes at as many times frames as possible.

X.F. Monitoring
Evaluate the success of all forest use activities at meeting the requirements of ecological responsibility. Important questions to ask during an evaluation include:
- Are natural landscape patterns maintained or restored?
- Are natural stand or patch composition and structures maintained or restored?
- Are water quality, quantity, and timing of flow, at all scales, unaltered from the standpoint of protecting forest functioning?
- Are soil structures and soil processes unaltered from the standpoint of protecting forest functioning?
- Have natural disturbance regimes, from the landscape to the stand or patch level, been protected and/or restored?
- Do all ecologically responsible forest users, both human and non-

human, have a fair and protected landbase? (Hammond Principle No. 11)

- Natural landscape patterns and disturbance regimes will be estimated from as much information, both formal and anecdotal, as is available. This is not an easy, straight forward job, but reasonable estimates of the types, frequencies, and extent of natural disturbance are vital to describing natural landscape patterns that are necessary to develop landscape and stand plans. Estimating the historical range of variation for natural disturbance regimes and landscape patterns is a useful tool.

- Evaluation—asking how we did—is an absolutely essential part of ecologically responsible forest use. By evaluating our plans and activities, we learn and are able to improve our relationship with forests and with each other. The questions posed above are inclusive of all aspects of forest functioning as it relates to a variety of human uses. However, more detailed questions need to be developed by local certifiers and forest users in their processes of evaluation for planning forest use and considering the results of forest use activities.

- The Pacific Certification Council believes that much can be learned about appropriate and inappropriate forest uses through evaluation of past activities. In other words, all people and organizations involved in forest use need to spend more time asking such questions as "Did it work?" and "Did our activities protect the forest?" Therefore, members of the PCC are committed to frequent evaluation of our activities and the activities that we certify to determine whether these activities are meeting the principles and standards of ecologically responsible forest use described here.

- Monitoring and Assessment. Appropriate monitoring shall be conducted to assess the conditions of the forest, product yields, chain of custody, management activities, and their social and environmental impacts. (FSC No. 8)

X.G. Production/Tools
- Use appropriate tools
- Use them carefully

X.H. Silviculture
Ivan Sims et al. have said that the general principles by which silviculture should be guided are two:
- to maintain full stands, and
- to direct their development along natural ecological succession (as mixtures).

X.I. Regeneration
- Natural regeneration, in most places is more efficient and appropriate
- Natural controls will probably do a better job of protection of regeneration than fertilizers or biocides.

X.L. Harvesting Kinds
- Limits. Practice appropriate cuts, depending on the limits (such as disturbance regimes, tree tolerances, climatic variations) of the forest. If in doubt, don't cut. Adjust your harvest style to the limits of the forest, as well as to your goals.
- Selection. Selecting trees is an art and can be redone and redone; the more you know about trees the better you can select.
- Minimal Change; use fewer roads, make fewer intrusions, remove less biomass.
- Practice appropriate cuts, depending on the limits (such as disturbance regimes, tree tolerances, climatic variations) of the forest. If in doubt, don't cut. Adjust your harvest style to the limits of the forest, as well as to your goals.
- Preserve the diversity of microsites
- Maintenance of Natural Forests. Primary forests, mature secondary forests and sites of cultural significance shall be conserved—not replaced by tree plantations or other land uses. (FSC No. 9)
- Plantations. Plantations shall be planned and managed in accordance with all Principles. Plantations should complement the management of, reduce pressures on, and promote the restoration and conservation of natural forests. They should not be placed on soils in areas that cannot support forests. (similar to FSC No. 10)

X.M. Harvesting Techniques
- Minimize intrusions into the forest
- Use equipment that has less damaging "footprints," e.g., balloon-tired tractors or blimps.

X.N. Wood Structure
- Aesthetic Union. Emphasize the aesthetics of wood in your plans. Wood has its own unique history and beauty. Often the location of a tree will encourage it to react in ways that we find attractive. Suppressed cedars have very tight growth rings. Redwoods twist the same direction as their parent trees. Scots pine planted in rocky terrain and poor soils get twisted and gnarled. Jeffrey pine twists with prevailing winds. Drought and heat cause spiraling in slash pines. Research at the University of Georgia showed that heat stress caused predominantly right-hand spirals. Red pine develops fluted stems in Connecticut (Yale foresters study that). Old folk-knowledge says that right-handed owls watching the moon travel east while sitting on Douglas-firs cause right-handed spirals. Douglas-fir sometimes develops elliptical indentations (similar to birds-eye) called "bear scratches." Maple can produce birds-eye, quilt or fiddleback figures.
- Use special structures in a sustainable way

X.P. Restoration
As with other forms of ecologically responsible forest use, the process of restoration involves solving problems with finesse and ingenuity, according

to Herb Hammond, rather than with force. Soft approaches that protect all parts of the forest must replace aggressive approaches that label some parts as valuable and other parts as worthless or harmful. Some important restoration principles include:

- Treat the causes of forest degradation, not just the symptoms. In other words, forest restoration involves reducing timber cutting rates to ecologically responsible levels, as well as restoring forest composition and structure degraded by excessively high timber cutting rates in the past.
- Restore whole watersheds and large landscapes. For example, degraded stream channels and fish habitat are the results of degrading timber management activities throughout a watershed. A stream channel cannot be restored simply by replacing missing structures in the stream channel. Instead, all parts of the landscape that contribute to the loss of structures in a stream channel must be restored along with the stream channel.
- Mimic natural ecological processes, particularly in time. People need to recognize that natural processes of ecosystem change, or succession, are the way that natural forest functioning is maintained and the way that forests develop following disturbances. We need to exercise patience in forest restoration activities, and recognize that, just as forests required hundreds and thousands of years to develop, restoration will require hundreds and thousands of years to be completely effective.
- Restore forest composition, structures, and functions at the stand level, while making sure that these activities rebuild landscape composition, structures, and functions. Forest restoration activities must be designed to ensure that landscape connectivity is reestablished.
- Plan and carry out restoration activities with local people, ideally those who inhabit the forest. People experienced in agricultural restoration have found that degrading land use activities have often been designed by specialists and accomplished by powerful technology, such as large machines and pesticides. In contrast, effective restoration requires all kinds of people with all kinds of skills. People with shovels are as important as people with machines. Restoration must be more than a swift afterthought or hopeful resolution to a single problem, no matter how commendable the impulse. People who live and work in local communities are most likely to have both the commitment and the patience for effective forest restoration.

The processes of the forest need to be preserved or restored. The forest needs to be made as or more diverse as it was, not less, with monocultural tree plantations.

X.Q. Certification

- Current knowledge of the forest is necessary
- Acceptance of forest limits is necessary
- Reduction of human demands is necessary
- Forest use cannot be independent of human equity

X.R. Goals
- Make realistic goals
- Make unrealistic goals

X.S. Summary
Most of the principles stated so far have been global principles; some principles may apply to a particular region; others may exist at a very local scale. This is even more true for standards. The global provides the framework for the regional, which provides the framework for the local—the level of detail and participation. Standards are very important for certification, to provide consistency of judgment regardless of the certifier; they also form the basis for any appeal of judgments. They should be clear, unbiased, and applicable.

We can distill a few general principles that you can keep in mind as you work; please feel free to add to these, since they are not complete:

- Learn as much as you can about the history and ecology of the forest you are working in, especially about archaic practices and the previous use
- Be humble and cautious about cutting trees; when you take a tree use everything you take and make sure the rest gets returned to the floor
- Minimize your impacts on the structure of the forest; if you cut over 10% of the gross primary productivity of a forest you are competing with the trees; over 10% of the net primary productivity and you are competing with the animals and decomposers; over 50% of the net community productivity and you are risking losing sustainability (this is what the conference participants in Moscow last week could not understand—they thought all the productivity was ours to take)
- Consider the watershed context of the forest
- Protect the soil and protect the water
- Emphasize labor-intensive, community-building practices
- Monitor, monitor, monitor
- Keep your forest and human community healthy
- Keep yourself healthy—not just the core self but the larger, extended self
- Be frugal
- Play
- Act as if you were wise

These principles are not a final presentation of a limited number. They are meant to be questioned, discarded, or expanded. You can help with that.

Chapter 6

Wild Thinking

For this conference on *Wild Thinking for the 22nd Century*, we the participants, identified as 50 of the wildest thinkers, have been asked to consider wild thinking and its application to ecology, economics, education, and politics.

So, I started writing questions and notes. Are there large-scale forms of thought? Is Philosophy one such form? Religion? Ecology? What is wild? (from the German *wild*, perhaps *Wald*, meaning forest) Not domesticated? Uncontrolled or unmanaged? Not cultivated? Untamed? Savage? Wasteful? In a state of nature? Lawless? Wild as a word has ambiguity and reflexion.

What is thinking (from the German *denken*)? To revolve ideas in the mind? To design, to imagine, to judge? Is ecological thinking intrinsically wild? Is ecological knowledge? Will Wright suggests that knowledge becomes wild when it is critically reflexive and committed to critical access rather than to a version of absolute reality. Thus such wild knowledge cannot be "domesticated" by one particular social institution. It is accessible by individuals.

Otherwise, what is the proper language for humans? What is the proper diet for all humans? The proper mode of expression? These questions in a way are too presumptuous to even be asked.

Ecological thinking is wild because it has a nonhuman component. By contrast, scientific or religious thinking is domesticated or tamed because it is limited by a a human definition of a true reality, or a set of rules for observing a true reality. Science is defined in opposition to religion, with a commitment to neutral observation rather than a moral commitment to tradition. Wild ecological thinking combines technique with moral and ecological concerns. The fundamental emotion of wild thinking is astonishment, literally being struck by lightning.

Of course, there are other forms of thinking. Religious or scientific, which we consider tame. Tame ideas are remarkably persistent—e.g., "more is better." George Orwell referred to these obsolete ideas as "wrong-think." "Double-think" is used traditionally to keep some ideas tame. Gordon R. Taylor suggests that "Non-think" (the failure of good ideas to be recognized or used) is equally obstructive—thus the idea "protect the ecological basis of life" is never considered.

Wild thinking is appropriate for "system breaks," the social discontinuities identified by K. Boulding. We have started to identify the forces setting up the next big break, but we have not defined the forming patterns very well. We need to be rethinking (a form of wild thinking) the basic assumptions of the spheres of civilization, from our economic and political to industrial, religious and scientific.

Talking about wild thinking, perhaps there is another side. Too much anarchy is dreaming; too much feral thinking is noncultural and perhaps dangerous. Too much is unrelated to the important mode of learning by doing.

Ecology is a study of relations. There are too few relations now between honor and community or between work and reward. Ecological thinking could unearth those relations.

Wildness

Is wildness just the nonhuman part of the spectrum? Does it overlap in humans? Is it just difference or craziness? We love and celebrate the wild; also, we fear and suppress the wild. The wild is a quality of being just beyond rules or outside of walls.

Paul Shepard reminds us that wildness occurs in many places, in any species whose sexual assortment and genealogy are not controlled by human beings. Darwin reminds us that humanity is wild, also. The wild, savage mind. Can such a mind be just wild? Not necessarily—we can domesticate our ideas.

It seems easy to talk of wild thinking. Is it meaningful to talk of a wild culture, one that intermeshes with the wild of nature? Are archaic cultures wild? A model of a civilization without walls, without resources, maxima, or weeds?

A Wild Agenda?

What kind of agenda can we make? All we need is new economic, educational, political, cultural, social, psychological, and ecological frameworks. We could try to frame an agenda. We could:

1. Try to rebalance our individual lives
2. Try to rebalance social spins of the patriarchal and matriarchal, between the human and ambihuman
3. Try for individual self-reliance and health
4. Try for community self-reliance and health
5. Try for ecological community health
6. Strengthen community and cultural identity (against globalization)
7. Limit the centralization of power and authority, in style as well as trade
8. Reform corporations to act responsibly as public service organizations (not imaginary individuals)
9. Direct technology in appropriate ways
10. Set up a global commonwealth for global relations.

How wild humans can live on a wild earth is the subject of a Eutopian framework, which is my focus.

Five Major areas are suggested for Proposals and Research Projects. If we were really wild, would we let ourselves be limited to just these?
1. Environmental Values. The disjunction between expression and practice; lack of understanding of connections or ethical models. Response: A problem for cultures and education Proposal: Learn the diversity of values
2. Competing Views about the World. Competing views of what is important; relative conceptions of reality; vital needs versus necessary styles; can sensitive views be taught? Response: This is a problem with cosmologies. And with a prevailing industrial cosmology that is replacing place

cosmologies. Proposal: Create a holocosmological framework

3. Ecological Sustainability and Environmental Policy. Address conflicts between human and nonhuman; social justice and ecological sustainability; resource consumption and policy. Response: This is a problem for ecological planning, which incorporates economic and political planning (See Palouse plan and North Slope plan). Proposal: Balance net productivity with human wants.

4. Population. Growth versus racism or elitism; an ecological goal for population; democratic means or other. Response: A problem for each culture, with limits suggested by an empowered international body. Proposal: Limit populations to overall productivity

5. Education. Fitting education to all else; transforming institutions; ecological thinking and the new paradigm. Response: Create small community schools that can be funded with public school funds; infiltrate public and private schools by offering courses and presentations, as well as practical field experience and projects. Proposal: Incorporate Organic Dialectics of Goethe and the aesthetic education of Schiller; add in Novalis and Wordsworth in a deep ecological framework (see Aesthetic Education). Summary: How can all of these things be put together? In a Eutopian framework.

Notes from the Meetings
Wednesday Evening Introduction
Talk begins with carrying capacity. The speaker (I do not know all the names yet) addressed population only and not resource use or cultural carrying capacity. He did say that humanity has exceeded its carrying capacity. I agree, but would it be so if resource use was drastically reduced? Probably. DT mentioned visions and their realization, and in response to a question, suggested that our audience should be the environmental community, governments and ourselves. Afterwards, AN (only initials are used to protect the participants from my possible misinterpretations of their ideas) wrestles with David Brower. When I tell him to act his age (86), he playfully starts boxing with me—I tell him that I boxed in college, but so has he. We spar while I try to talk, but he is fully involved in the play. He was my professor twenty years before, when he was even less serious.

Thursday morning, Environmental Values
The topic of the hour: Environmental values. KL starts with an introduction of what is valuable in nature for people. He repeats, from his article on nine forms of human limits. His suggestion: Make connections at the community/ individual level.

WI talks about beliefs as models and values as guiding principles. He addresses why values may not translate into action, and broaches the topic of inconsequential values. A short discussion by me of cultural models: limited world, interdependent, unpredictable systems.

CO talks about the "epic of evolution" and evolution as the new human "myth" and a new cosmology, but I mention that Jeremy Rifkin did that and I had argued against it then, as well as now, as a useful cosmology; it is flawed and limited; she ignored me and presented a universe story and vision of the future.

Unfortunately, DT asked me what I thought. I stuttered out that humans needed to link their values to the values of other beings, otherwise they would be unanchored as well as noncontextual. Furthermore, this whole meeting is going to express concern with economic globalization yet we are making the same kind of mistake with cultural globalization by suggesting a global cosmology based on evolution or on the Euro-American competitive cosmology that thinks it deserves to replace the adaptive cosmologies of archaic cultures. What we really should be promoting, I said, is a framework for all cosmologies, a holocosmology which in fact is my proposal for a framework for all adaptive values, that will protect and allow cultural (and nonhuman) diversity—I finally stopped stuttering.

AN points out that values have to lead to norms, such as in his ecosophy, and commitments; norms are implicit in how people act. I contrast violence with Gandhian nonviolence. DF states that his movement is nonviolent, but I suggest examples where it has not been.

HN urges that we become proactive and keep sight of the "big picture." Consumer culture has to be to rejected. I suggest that this can happen when local cultures are strengthened.

AD relates values to feelings, and stresses the importance of narratives. I agree with this idea of big stories, but I am hesitant to hurt his feelings by suggesting that the stories be more scientific or poetic.

DF notes that until the environmental community learns how to talk to rednecks it will be less effective. He also mentions problems that we have with our own images as elitists, leftists, rednecks, characters, or drunks. I think he might have added preservationists or user addicts.

WJ wonders if we can work out a language. In the evolution vein, he suggests an Evolution Protection Agency.

MS states he is getting a cognitive dissonance with the presentations. Ideas versus us are poor choices. He wonders how to make the crises real. I suggest a catastrophic psychology approach to present the crises to people. Reports on destructions and species losses, on hunger and deaths, on long, slow, wide catastrophes rather than short, brief, local ones. Massive changes, massive deaths. I also think that we have problems with anthropological studies, ala psychological studies, when we are not aware of the impact of the studies.

WR notes that people eliminate species. Yes, at a hundred times the natural extinction rate.

AR suggests considering cognition as "something happening in something." I think yes, as in, foresting knowing weathering. I also wonder what do forests value? What do large composite beings like forests, ecosystems or planets value, the sum of all contradictory smaller values? Perhaps most importantly continuity.

DT thinks our talking is disjointed. Of course it is; too many academic reports, ego trips, stories, and responses to each other. Too free form perhaps. But, then I want people to respond to my ideas first!

DR says that options are important; knowing options changes our values. He asks how people change. Again the problem of anthropology: People tell stories, sometimes to please or mislead other people, especially scientists; both psychology and anthropology have to be reflexive and aware of the social uncertainty principle (similar to Heisenberg).

CT notes that values being connected to evolutionary biology is a problematic idea. Language itself leads to a moral code. We need to reform education. SK relates how he works in nature, tending his place.

RO asks how we can persuade people to an ecocentric position. How can morals prevail? By force of argument? How do we act out values? I think part of the problem is centricity, whether eco-, anthropo- or econocentric. Perhaps if we were ecoperipheral, approaching everything sideways from a respectful distance, like big-brained crabs. Then I missed part of WI's talk.

WI asks why do people destroy the environment when they say they value it. I point out that they do not. They destroy a tiny part at a time, not the whole web. Would they, if they could? No.

Now I suggest that people can often face disjunction quite bravely in normal catastrophes, such as earthquakes and fires. We just need to implement a catastrophic psychology. After all, we are inside one: a 50-year collapse, extinction spasm, ecosystem disintegration, and cultural wobble. We need to implement it immediately. We can do so on four levels: With communication of studies in major media; reorganization on an emergency footing; special activities, from planting trees to reeducation; and explicit goals to reverse the turning.

TC suggests a complete accounting, monetary or not, of all resources. RO describes stakeholder capitalism. MS wonders if there is an avenue for green values, perhaps bioregional landscaping workshops. SE adds demonstration projects for various communities. She states that it is better possibly to change behavior before values. DR asks if there is there a feasible revolution? I say we have a model for a Jeffersonian political revolution. We are not quite responding to each other, not directly anyway.

WJ states that agriculture saves wilderness. We should ask questions about agriculture. He urges that we focus on biomass. True, I think, but the patterns of biomass are as important as just the total mass. Are they self-ordering, self-renewing patterns like wilderness or artificial annual (and unsustainable) ones like industrial agriculture?

JL. contributes "life is good" as a mantra. That we should always use biocentric language. I point out that biocentric language is the same Aristotelian logic that got us into our ecological linguistic predicament. We need a better logic not so oppositional, like morphogenetic, like Mandenkan. There are many centers not just one, so we need an ecoperipheral language.

MA. says we need to get in tune with the Third World. Too many white people making decisions for dark-skinned people. Yes, but the solution long-range is education, not color-coding. Besides, three worlds is old hat.

DF suggests that we make proposals with economic incentives, alternatives to market economies. What? We don't go deep enough. We should continue some things at depth.

MI suggests that we sponsor nature camps so that children can identify with nature. Show them the power of this vision. Also, write a new book sponsored by the foundation. I suspect we would all want to be the author.

Then we each make specific proposals, which are written down on large sheets of paper—hopefully to be provided to us later, as I didn't write them down.

Competing Views of the World

I'm just thinking before the session begins. Competing views of the earth or world—they are different, as a world is a human image of part of the earth; each culture has one. Are these views personal, group, or culture? Are we considered to share the same "world"? No, just the same earth, we inhabit different worlds.

A culture needs many things and does many things for its participants: Make a common language; order experience; personalize a place; adapt and preserve aspects of a place; justify human behavior; provide identity and security. Still, most cultures are incomplete (as they have to be since places change), inflexible, and often indifferent to other cultures or individuals. Cultures should not compete or try to be the right or final culture.

DT starts, as moderator. Social systems are incompatible with complex life, he says. We need an ecocentric vision, but there is a schism—which is right? I think this might be a bad question. More centric talk.

GE alludes to the clash between social justice and the environmental movement, as well as the rise of environmental justice. In the 1960s we had three movements: peace, justice and environmental. Now we have deconstructionism with no objective truth, only power struggles for feeble grants.

HN asks for an approach to the issue that does not escalate the government and corporate growth agenda. We need to connect our well-being to the planet. But, I suppose we do not due to our detachment. Her strategy is a more experiential exchange.

DF concludes that the takeover of the environmental movement by the left is destructive. There are three political spectra: conservative, left, and liberal. But none have any recognition of limits I say. Where is this left? Academia? I never know since I seem to be left out.

DT suggests a forest example. I say that the FSC has a real example just as he described: It is with the Hoopa Indians in California. AN says we need to "work with them," and "maximize contact." I say that we are doing that in meetings and visits now through the Ecoforestry Institute, which he helped found.

CE says that GE is too academic in his descriptions of the movements. That the Green Party has its own agenda.

GL wants to bring environmental criminals to justice. Good idea, if the laws specify that kind of criminality; otherwise, we need to change laws first or risk vigilante justice.

SM mentions "political correctness." Then, MM suggests multiscalar legislation. "No one has the right story" (Or perhaps everyone's story is right for their place).

RO identifies the green thread in all thought; also distinguishes between the green and environmental movements.

I make the distinction between competing views of the world and competing world views. For example, ecoforestry is not anti-forestry. It used to be where forestry is ugly, ecoforestry is beautiful, forestry is capital-intensive and ecoforestry is labor-intensive, forestry is anthropocentric and ecoforestry is ecocentric, and forestry uses Aristotelian logic and ecoforestry uses morphogenetic logic—the kind that does not make lists comparing two things. Rather than compete, ecoforestry makes industrial forestry a special case of forestry, appropriate in some circumstances (rare), much like Einstein's special theory of relativity allowed Newtonian dynamics to remain a special case (flat gravity field). Ecoforestry is explicitly based on deep ecology and considers social ecology and ecofeminism. Every cosmology can be characterized by a set of statements about the world. Industrial cosmology is an extension of the European: everything is made of interchangeable units, man is master, man is perfectible, etc. Other cosmologies: everything is unique, gods rule, nothing is perfect. This is how to ground values, in cultures in place.

HN describes how corporate capitalism has changed the world from the old small-potatoes capitalism to global threats. MI suggests a tactical advocacy and specialization. MA dittos that.

AR reminds us to remember that things are connected. GJ suggests that we look for common ground in the zero-cut campaign. And HE dittos that. TR disagrees with SM. SM is unhappy with labels (tags)—suggests a gas tax to restore carbon sinks.

SK states that it is important to be complete. Gandhi recommended complete contact, but with insights from conflict. I wonder what Gandhi would do today; the circumstances are so different—even Gandhi mentioned that his tactics would not have worked with the Germans like they did with the British. Still, there are things to be learned, a different shape of nonviolence.

DT says that we cannot justify logging by corporations. Furthermore, that we must move past dichotomies (didn't I say that an hour ago?). AD notes that we must endorse an ecological response.

The time is late and we will not get to the brainstorming session on competing world views, so I write mine on the blackboard during break: Use new terms to help communication, in fact develop a better language; and of course put cultural thought in a holocosmological frame.

Later, WJ states that as options narrow, diversity increases. I'm not sure I understand that in biological or cultural terms. Perhaps I can ask him later.

TC asks why are we here? Because we care, of course, he answers. We must work on management levels (1-4) before finding answers to questions.

DT states that we must maintain ecological integrity, that human social justice cannot allow extinctions. DT then puts HE in the spotlight and asks him to contribute. HE has not spoken; actually JS has barely spoken either. HE launches into a (well-rehearsed) spiel: We must dismantle corporations; television is making needs in its audience; we must somehow make bridges to those people.

Afterwards, before dinner, I talked with HE. I wondered if we could redo corporations rather than dismantle them (corporate jujitsu or something); if television made needs in people that they really needed to balance their wobbly lives in the void of the corporate world, then maybe television could be used to realize different needs: nature, less consumption, family values (Dan Quayle hey)—in other words we could make the bridge through television (reaching addicts through their addiction). HE rejected these ideas abruptly quoting McLuhan that the medium is the massage—another rehearsed response I think; perhaps inside is a computer and I cannot tell. When he walked off I did not pursue him, choosing to reflect that even massages can improve muscle tone or ruin it.

We gathered around a station wagon outside the sleeping quarters for wine before dinner. In hearing DF say that so much of this was mental masturbation, I overheard JS say that he didn't want to participate because it seemed an academic (talking heads) exercise with no feeling. Too many turgid images. Still, it is a meeting on wild thinking, which may require the head and some self-stimulation to be creative. On the other hand, much talk is regurgitation without sufficient digestion; not a talking heads problem so much as an academic regurgitation—I classify myself as an independent scientist outside of academia, so I say something to that effect.

That evening DT showed slides of his farm. They seem to live modestly by some accounts, despite their millions, and are certainly trying to integrate agricultural ideas with local populaces. I did not see any hint or mention of ecoforestry.

I end up with a group talking about evolution. During this discussion I am thinking about talking heads delivering one-minute sound-bites of one-liners about the environment. Is this enough? Do we need to start a dialogue that can weave an agenda from the bites? It reminds me of the typical problem of getting acquainted: Here is who I am, here is what I do, here is what I think about these topics. Then, if we like each other, we can go to the next level of discussion. Perhaps I have been alone too long. If this is life, I will probably be alone much longer. So, I talk to AN for a while about his projects in Canada and thank him again, for inviting me here.

Day 2, Topic 1: Population

These things are hard to get started; people drift in and out drinking coffee and talking about their families or work. As usual I am first to the table and sit doodling. What is the opposite of wild thinking? Domestic thinking, tame thinking. Is wrong-think (Wells?) wild or tame? Double-think must be tame. Nonthink is very popular. I have decided to suggest that the Foundation establish a think tank to carry on the dialogue. This could be in the image of the Wuppertal Institute in northern Europe—visionaries and activists doing research and monitoring of indicators of "sustainability" and presenting them directly to government (intermeshing if possible).

Ah, it begins. SM repeats her talk about her decision to not have children. She takes Paul Ehrlich's book (population bomb) personally. She makes a good point about how personal decisions lead to employability, but does not consider the obverse side (nonemployability). Here is one person who has become famous, I think, just by writing how she personally is affected by the environmental movements. She asks how many of us do not have children—surprisingly it is almost half (I am in the half who have none); this sets the stage for defensive arguments by those who did have 3-4 children. Much discussion about personal decisions and their long-term consequences. I suggest that we might poll to see how many have also ORVs, boats, snowmobiles, high salaries, and large houses—I am justifiably ignored—perhaps I just wanted to be seen to have the least, with a modest nonprofit salary rather than envy the large academic salaries.

TB urges humor and mockery as strategies to combat corporate lies. I think that might be effective if we target *ideas rather than people*, but then he mentions specific persons; I think that might backfire. He does suggest a nonpronatalist tax structure, and mentions a FOE tax policy project (I need to find out more about that). He also mentions addressing cultural norms as part of the population puzzle. He mentions not having letters behind his name and wonders whether those who do have different attitudes. MI answers that he needs those letters, as a result of his childhood and decision to do research. A short discussion ensues about schooling. I announce that I am self-educated—I went to the University of Oregon (small laughs). I also suggest that there are other ways of being close to children besides having them; I prefer the wolf strategy of being an uncle to my friends' children, and sometimes sponsoring children through Save-the-Children Federation, which I recommend as an exemplary program that works with communities.

MI asks what is the problem with population? Poverty, injustice, economic growth? I think that the main problem is that we never recognize the scale effect and change in patterns that result; ecological planning could go a ways to addressing that. He then concentrates on habitat destruction. But mentions concerns with different problems from income disparity to distribution. "50% of habitat loss from population."

HA suggests that there is a conflict between values and priorities. I think that since population is a human issue, there must be conflict, certainly between cultural values. Perhaps we could agree on something, a series of statements that can be put in a physical agenda. HA continues talking about a mixed discussion.

HN points out that ethnic groups breed for competition. I think immediately of the Sinhalese and Tamil in Sri Lanka. XA mentions that in Sri Lanka they seem to be at peace—everyone smiles. MA states that Sri Lanka is a bad example because the Sinhalese have won, but there is fighting by rebel groups and many have fled the country—furthermore, he is Tamil and knows. Meanwhile HN wonders if the cure is decentralization. I think no, ecological planning, one of my three buzz words that I emphasize repeatedly (no invitation to the farm for me I guess).

DT talks about the fragmenting of cultures and wonders what changes are needed for population stability. RO described the historical causes of inequity and population growth; she proposes that we monitor activities in Geneva. I think yes, but from a perspective of ecological planning; changes at all levels from local to global need to be coordinated.

CT suggests a dialogue with other cultures, with other groups. I think Eutopias is a dialogue more important than separation or over self-reliance (as if).

CE mentions nations' self-determination and International Law. DF questions what we should do after the collapse of nations. MS says: "Norms! Postpone! Refuse! Adopt! MI asks where can Foundation do the most good? Not by distributing condoms or competing with other groups on population. I think "anticipatory" (regarding DF's die-off curve).

XA mentions the destruction of poor people. I ask if the Foundation could coordinate ecological planning through an Institute? HA states that we are not here to decide on actions for the Foundation. DT states that the Foundation's mission includes population, which is symptom and cause. Also, wilderness is a personal interest of his. I think perhaps we should relate ecoforestry to wilderness as a complement to the Wild Lands project, along the nation, whole continent, and perhaps the planet. The idea is met with silence.

HE notes that we are a community of thinkers. I think planning is participating. TC makes statements about children (with a tremor in his voice and a tear in his eye) and offers generalities about what people do.

WJ suggests that we are trapped by myths, such as substitution. I think this is the result of our cosmologies.

WJ proposes a strategy to spend all new money from the foundation, that we should turn away from technology. I think this is extreme (but it seems to be what DT and he want to hear)—rather we should choose technology wisely. Otherwise we have to get rid of anything beyond our fingers.

At lunch I talk with about his ecological footprint idea. I mention that we used to call it ghost acreage , as it has been called, or even an ecological buttprint, since it has depth as well as area. We discuss Holland as an example. I point out that reducing the population one-half percent per year should solve many problems in two generations.

Ecological Sustainability and Policy
HA states that we must take other issues into account. WJ describes his breakthrough with perennial crops and breakthroughs in science in general. He mentions slow knowledge, which need not be improved, and urges development of less ephemeral products. I agree with that but point out that our love affair with speed of any kind might make it hard to sell. RO mentions that sustainability cannot happen without security. I think this is just the opposite of ecological security, which would come from individual nations under a new UN. She suggests bringing democracy. But, I think, democracy is limited to nations with strong middle classes and probably doomed to fail elsewhere, especially Africa and the Middle East.

AN notes that nature does not stand still—it is a process. Humans are replacing diversity with monotones. But nature is refuting Cartesian thinking. Of course, I think, nature refutes all human models. He continues: Nature is more like a jazz band than an orchestra with a score and metronome. Yes, I think, but scores are useful in understanding patterns either way. In fact, I think, we have many problems with the scale of tools. When do they interfere with nature, or do they always impose some structure on nature.

HN addresses ecological security and the link between globalization and urbanization. I argue that security can only be arrived at through a revamped UN, that would call for: Immediate disarmament; recognition of many independent cultures as new nations; the distribution of wealth and education; new rules for global limits, especially regarding transnational corporations; self-governing of nations, not necessarily democracies; and the establishment of global commons and wilds.

MS presents six variables: Ecological integrity, appropriate technology, economic sufficiency, human dignity, decision-making liberty, and ... missed it—I should have recorded all this! He suggests that we build a big coalition of groups and mentions democracy, health, and trade. I suggest that the coalition has to have a philosophical and political theoretical basis. A UN global perspective could address an authentic concept of humanity, economic limits, national/corporation responsibilities, respect for all cultures, and a holistic education beyond cultures. At the same time it has to be local, with small political and economic changes.

MJ states that better information makes no difference. I think neither does better technology in a way.

SK argues for the idea of health. I think we could consider forest ecosystem health as well as human. AD suggests that ecological forestry is an ecologically responsible business. We are on the same wavelength again.

MR talks about problems with trade. WW wonders if consumption is correlated directly with income. HE asks who is the enemy? And answers industrialization and globalization. I think these are all myths, metaphors and movements; perhaps the enemy is us (after Pogo) or the products of our big brains expressed in a big scale.

MI says that we must plan for catastrophe; we must reach mid-America (duh, was I ignored? Will I ever get credit for any idea I say?). DR mentions the New American Dream (and I think of Soleri's article on the old one, with McMansion houses and yards of exotic grasses). DR wonders if there is opportunity in feeling.

JS misses the ritual of a circle. AR describes his New York community, that made lists of visions, developed markets, and held development rights for farms.

SM thinks that artists balance the shrinking spirit (I think they just reflect it). She offers art as a repository for wild thinking (some maybe, but I wonder). GJ suggests we cultivate the human spirit by visioning.

TT suggests writing a sequel to ecotopias (and I think Eutopias of course). Recommends a campaign like Ad-busters to reverse the industrial campaign. He believes that slogans are needed. I agree but they have to be crafted well.

At the break RO tells me about LaMont Hempel and global political ecology, and about the Commission for Global Government, which is making recommendations to the UN. I must write to them.

We try to think of good slogans for deep ecology movement: Earth, Our Self; Be simple to be rich; Save forests to save ourselves.

Sat with AD and others at dinner. Then sat with GJ, who asked a lot of questions; she has good social skills, I should learn that myself, so I ask her the same kind of questions to draw her out, but she does not play this game in reverse I guess.

Education (Saturday morning)
CT starts by describing dimensions of reform. Education now carries forth the ideas of the industrial revolution. There are high status forms of knowledge (e.g., economic) and low forms. New buzzwords: education product, just-in-time education. He asks how we can counter the webization of education. I think this is just the continuation of abstraction. He brings up root metaphors again. States that we do not analyze our technology first (I think that we do not learn the language of patterns either). How do we introduce change or rethink?

SK talks about deep ecology types of programs: outdoor/nature, gardening, science, self-esteem, and a broad spectrum of environmental education. She mentions campus stewardships and environmental councils. The content is critiquing anthropocentrism, activism and bioregionalism. The methods include shifting perceptions, compassion, collaborative skills, acts of celebration, and moral responses to nature.

DF tells how the wild lands projects are experiential. How a decline in mountain lions results in a decline of songbirds due to the explosion of mid-range carnivores like raccoons who can prey on birds eggs. Focal species need nonfragmented lands. With AN and many of us, he wants to encourage humans to maturity. He mentions wildness as awareness of the way things are, Shepard's works, and La Chappelle's work on sexual play.

MA: we need to experience going out. CT returns with computer literacy; how we cannot communicate elder knowledge through computers—the computer hides patterns and commodifies knowledge.

I say that it true, as far as it goes, but like quantum mechanics, the computer bears the seeds for the destruction of its old world view. Perhaps the good can undermine the bad. Computers do much good. They do large scale calculations better than anything; we need them for large scale projects, for higher order problems, for CAD representation of wild lands. Furthermore, computers connect people at some level who might not be connected, from mental patients to fathers and sons. All tools (I'm really on a roll now) have positive and negative implications and effects, depending on their use. The choice is not to either let them take over or destroy all of them; it is to use them as tools in a mature way, understanding their limits and scales. Maybe we do NOT need them at all, but maybe they can be used to save cultures and wild lands.

AD wonders how we can make changes. JS suggests better rituals and ceremonies to be effective. DT recommends that we purge industrial societies. No, I think, it is better to encapsulate them; understand the limits of industrialization and use it within the limits.

CG notes that the computer is a medium. AN describes life quality as an equation: $g2/s+b$, where g=good and s, b are the pains of body and soul. I whisper loudly to AN that we could put that on the computer; AN wisely ignores my attempt at humor.

WJ announces that we are basically ignorant. CO reminds us of the role of story-telling; the magic of the deep time of dinosaurs. I wonder if it is resonance with morphic shapes—could we do that with imaginary creatures?

DR applauds the Northwest Environmental Watch. TJ relates how we are expanding technological problems. He wants to include system analysis of technologies and draw conclusions about impacts. Computers may give us empowerment, but it gives greater corporate empowerment (I wonder if they could be used to jujitsu corporate moves).

At the break, we start to brainstorm; I write: "Use appropriate technology in mature ways; apply three levels of educational effort from local to global; reeducate the old and graduates (of any age)."

HE continues: systematic analysis of technology, then judgment. AD mentions the study of the practice of technology (described in his first book). TC mentions industry influence on education. MI suggests that our task should be to design a university (I think not a university but a center for reeducation for all). AR wants to relate universities to society; educational reform, mimicking a corporate strategy. GE notes the commercialization of campuses, the take-over of schools by corporations who fund what they want and who they want to hire. I think there are different literacies—is it necessary to function in literate ways?

HN continues with globalization, but does not mention how to fight it. How do we fight it? Trap it to participate locally?

CT suggests a deep understanding of metaphor; computers are not just tools. I think, then say, true, nothing is just a tool. He gets angry; I wonder if his anger is a result of his not listening very well (to me?) or due to some other frustration?

DT wants to attack the growth monster, with an arrow in the heart. I ask him later if we are the monster? Or just a part or a cancer? MS makes a kind comment about me; I think he is the only one who matched my words correctly with my name or face. Then, DR makes a kind comment. Oddly, I am the least accomplished of the wild thinkers, but since two of my old professors are here, two of my friends are, and the Foundation gave me grants the past two years, I suspect I got here as an afterthought.

Summary

After lunch, MS starts the summary of the meeting. He asks if we have been wild and answers yes and no (I think mostly no). He notes that deep ecology is at work now, but wonders if a better name could be used to reach out to communities. He offers five steps: Stop the hemorrhaging; create alternate possibilities; empower people; change the story; get the message out (that life is sacred, or people are responsible). He claims that we are theory rich. I think no, we are theory poor, good theory anyway.

HA suggests possible outcomes of the meeting: Papers, a book. AR wonders if the right wing changed our culture or if they just got what they wanted. J. S. says that we must leverage each other.

I suggest that we need four things right away: Creating a think tank to advise and monitor; develop a common nontechnical new language for these ideas; create a big framework with big theory (not less MS but more, to form a real foundation for change rather than piecemeal junk we have now), big ecological plans, principles, and norms; and, take many actions on many levels.

SM suggests that we start a technology fast; set up a task force committee for each of several topics; and create a national organization of environmental education. I wonder if the Foundation could be a better coordinator of environmental groups.

SK mentions specific committees. I worry that we do not consider the enantiodromia with our actions. What are the implications, what are the opposite effects?

MM states that radio is an anarchic guerrilla (I think it interesting that everyone approves of the technology of radio but not computers).

RO suggests a subtitle for deep ecology: Life quality. Suggests having shadow summits, and making technology a verb.

WJ mentions a forestry article by Schultz, ecosystems in forestry (thus, I think, using a physical/mechanical metaphor of system). He recommends ecological community accounting.

DT mentions the need for humility. This has been rare of course at a conference of self-aware, well-rewarded luminaries. He mentions Odum's work, and asks how well we could each live on 1.5 ha with other species. Good idea. He thinks we should penetrate the system and disable our opponents (but there are no opponents, just bad ideas I want to say).

JD calls for global wilderness recovery. Many of us are working on that and readily agree.

The meeting closes and we truck over to San Francisco for a meeting with the press. We are being lured with the promise of luxurious rooms in a great hotel and large quantities of rich food. Because I am at the last of the alphabet, and no standard rooms remain, I get a four-room suite, with a living room and garden. So, I am bought for a day's luxury.

Chapter 7

Waldgedankenexperiment (Forest Thought Experiments)

Humanity is engaged in a great experiment with the planet's forests; unfortunately the experiment is not only bad science (no control planet), it is ill-considered. We are replacing large old complex forests with young simple fragments, in which fires are suppressed, large predators are removed, large herbivore populations are encouraged, exotic species are introduced, soil is compacted, and excessive biomass is removed. Our actions are experiments, whether we want them to be or not. Ignorance, denial, or cupidity cannot unmake this experimental course, which may be global and irreversible.

There must be a way to refine the experiments, to minimize our impacts, to be less reckless, and to anticipate the outcome of our experiments before we finish performing them. Einstein and Infield suggest that knowledge of laws can be gained by the contemplation of idealized experiments created by thought, that is, *Gedankenexperiment*. For example, to address the equality of inertial and gravitational masses, that is, how to show that the problem of general relativity is connected with gravitation, Einstein imagined an elevator at the top of an incredibly high building, and then imagined what research would be done in this local environment. Such experiments might seem "fantastic" he decided, but they might help us to understand what we are trying to understand.

Although forestry is orders of magnitude more complex than simple physical systems, perhaps we could imagine and use such experiments to help us understand what is happening in and to forests. The thought experiments presented below are incomplete, but suggestive of what we could do.

Experiments
1. What if the US Forest Service reopened all sales with new rules. Interested parties would make proposals rather than bids. USFS would sell timber rights for 100 years. The buyer would pay the USFS fair market value for the timber. The size of the sales would be determined by habitat type (so there would be an equal number of small sales). The buyer could then log (within ecological limits) or preserve the area (depending on whether the buyer might be Georgia Pacific or Greenpeace). The buyer would be responsible for building roads, if any, to USFS specifications at buyer cost. The buyer would only pay taxes on detrimental (to the ecosystem and human ecosystems) environmental impacts, such as erosion, loss of fish streams, or aesthetic loss—if any; no taxes would be paid on production or profits. Would this work?
2. Simplify forests by removing species. What would happen if we removed species one at a time, tracing all its connections, instead of removing the entire superstructure? Would we identify keystone species (or keystone mutualists or key linkages or taxa or processes) as things collapsed? Ronald Lanner asks if Whitebark pine is a keystone in some Boreal forests, providing pine nuts to nutcrackers and red squirrels, as

well as fat for bears. Other possible keystones: Fig trees in Amazonian rainforests, beavers in North America, termites and elephants on the African savanna; sea otters, flies, snails, millipedes? It used to be that top predators were considered keystone species, but the systems continued (although diminished) after most of them were eliminated. Would each species be found to have a unique function in the community? How much redundancy would there be? Redundancy is necessary for diversity—it can only come from existing pools of genetic and biological diversity, according to Chris Maser. Would it always be predictable? In clearcutting, we have been simplifying the system by removing biomass, and expecting forests to pop up again ad infinitum. What if we removed different amounts of biomass in similar forests on each rotation—at what point would the forests fail to return?

3. Model the forest by totaling all the behaviors of individual beings—some computer modeling has started to do this with just trees, and although that is one way of getting some information, a forest is not just trees or even a collection of all individual beings. If we could add up all the activities of individuals (with enough people and enough time), we would have a lot of specific information about that forest. Very little could be applied to other forests, in other places, with other components, unless we were able to make general conclusions. Would the general conclusions be applicable to other forests?

4. Replace forest functions with mechanical devices. The machine metaphor dominates modern forestry. This metaphor, and the agricultural model, result in tree plantations, in which many of the functions of the wild forest have to be taken over by human ingenuity. On some tree plantations, many of the functions of a wild forest have to be duplicated. For instance, shade cards are used to protect young shade-intolerant species; plastic sheaths are used to protect bark from predators; fertilizer is used on young trees; and some trees are doped with mycorrhizal fungi. Extending this trend, modern forestry eventually may try to create an artificial forest. By taking this to a ridiculous extreme (the argument known as *reductio ad absurdum*), we might try to create an artificial forest with just one living organism, Douglas-fir trees. For example, we could replace the functions of nurse logs with gigantic nylon sponges. What artifacts or tools would we invent to replace the functions of woodpeckers, bats, insects, fungi, shrubs, or snags in a mature forest?

5. Costing out replacement of services. Suppose that we had to replace every function the forest provides with a human service. Could we afford it? The following sentence uses very approximate numbers. Pure water can be bought in supermarkets for $0.50-$2.00 per liter; canisters of clean air can be bought (in Tokyo for instance) for about $10 a liter, although air can be cleaned for much less; flood control can cost $11,000-$90,000 per linear meter; wind protection increases the costs of buildings; solar protection can cost $300 per square meter; fertilizer and pesticides can cost $40 a gallon; waste recycling costs thousands a day; climate moderation, i.e., cloud seeding, can cost $44,000 per day; recreation can cost $25,000-$25,000,000 per park; genetic modifications can run into millions. As you

can guess, for an area of 50,000 people, the costs of replacing basic forest services could be billions of dollars per year. Many of these functions, like clean water, seem affordable to replace, but many such as climate moderation or soil creation are not affordable in any practical way.

6. What if we got tired of all that work and decided to save half the planet in wilderness? Imagine 6 billion square kilometers of forests set aside, with another 9 billion given over to the control of archaic cultures who reside in them. What changes would we really have to make? Substitution? We are good at that. Higher density living in cities? That trend has been evident for over a century; why not just plan it properly?

7. Imagining a planet without trees. Let us imagine that forestry and conservation both have failed, and the earth is a planet without forests and without trees in general—except for a few artifacts kept in arboretums. For the first time in over a billion years, the planet is not sheltered, the climate is not moderated, and other plants, animals, fungi, and bacteria are not protected. Humans have made shelter for themselves, but we do not want to share it with mosquitoes or grizzlies. Have the ice caps melted? What is the shape of the global system? Are all human crops grown inside? Is the reduction of biodiversity causing unalterable (and yet unknown) changes? Can the forest ever be replanted? What kind of forests could live without fungi and bats and centipedes? Imagine what changes would occur in human psychology. (Have you ever seen the movie *Silent Running* with Bruce Dern? Good movie; bad future.) Would planting trees be required by law in our treeless world?

8. Forests have been changing for millions of years. A macrochronoscope in low orbit over the planet for the last several billion years would show forests moving like shadows over the landscape, with climate changes, glaciers, and shifts in moisture. The forests are connected to the land air and water; everything is constantly changing in a gigantic intricate web. The immense patterns are easy to detect with this imaginary device. On the ground, changes seem chaotic—forests remember, are chaotic systems. A tree is embedded in a vast interconnected process that is creating an intricate and implicate order (physicist David Bohm's term meaning internally related). The patterns are persistent configurations of processes. Order and chaos seem contradictory because of our linear thinking. Linear thinking can be illustrated by industrial forestry: if something works well, such as cutting old growth and planting Douglas-fir, then more is better. Nature, however, is nonlinear. Maybe we should change the way we think, modify our logic and our behavior.

Summary

One thing philosophy has shown is that there is no single "right" answer to a question. The best response to a question is a hypothesis, a thought experiment. Through that, you can create explanations and discover answers in a dialog with others.

Thought experiments can give us clues about what can happen ("And then what?" as Garrett Hardin always asks) and what is the likelihood of it happening. Unlike medical doctors or scientists, we cannot either wait or directly experiment (within a realistic time frame or scale). We cannot experiment at all in a traditional sense, where we hold most variables fixed, while changing one or two in experimental runs. Forests operate over very long time spans; furthermore, their historical nature means that they cannot be restarted.

Large-scale, long-term experiments are expensive and relatively few. Most experiments are short-range, small-scale, isolated, and detail dense. They do not make the hypotheses required for management of forests. Forestry management, because of uncertainties, lack of controls, age, uniqueness, is an uncontrolled, large-scale experiment. Thought experiments can refine the design of our larger experiments by suggesting better hypotheses.

Thought experiments can help us avoid being overwhelmed by details. Thought experiments can help formulate goals and interpret information appropriate to scale. The idea of science is to manage our experiences with generalities. Once the thought experiments are started they can be refined with conceptual or mathematical models, which can simulate the changes and evolution of changes. Computer-based models can permit complex explorations, as well as suggest patterns and further hypotheses. Through thought experiments and models, many of the dangers and expenses of our activities can be avoided.

Thought experiments are vital to understanding the complexity of forests. In practice, erring on the side of preservation—the prudent and conservative course—means minimizing the influence of human activities on the land. It means experimenting cautiously with new approaches to forestry and being properly skeptical about claims for sustainability. It means drastically reducing our demand for wood products, through conservation, reuse, recycling, and human population control, so that the greatest possible amount of natural forest can be left wild and degraded forest lands have time to be restored to health.

Chapter 8

The Philosophy of Ecological Forestry through Questioning

Although ecoforestry sounds like a qualification of the modern, industrial forestry, it is in fact entirely different—it is based on a different philosophy, a broader ecology, a more comprehensive economics, new metaphors, and it is sensitive to limits, ethics, aesthetics, and spiritual values. Its larger perspective incorporates industrial forestry as a special case, much like the Theory of Relativity incorporated Newtonian dynamics as a special case.

Ecoforestry addresses forests and human societies. Ecoforestry includes humans and ultrahuman beings as an integral part of the whole system. Ecoforestry addresses both large and short-term scales (in time, size, and design). Many forests, however, are occupied by people with different (and often better adapted) cosmologies and economies.

Industrial forestry assumes the current economic system, which is based on winners and losers—and on inequality. Economics has not been unsuccessful with its models, for instance of buying behavior, but it has become a highly abstract academic discipline. All its abstractions are applied to the real world without acknowledgment of the high degree of abstraction involved. The philosopher A. N. Whitehead warned that the economic method would triumph if the abstractions were judicious, but even judicious abstractions have limits, and the neglect of those limits leads to disastrous oversights. Considering a fictitious human nature under imaginary circumstances and thinking it is real is the fallacy of "misplaced concreteness" according to Whitehead. Daly and Cobb suggest that the classic instance of the fallacy in economics is "money fetishism," where the characteristics of an abstract symbol, such as limitless growth, are applied to real commodities and values. Misplaced concreteness also occurs in forestry itself. Genetic reductionism is a good example of the fallacy. Genes are not independent creative beings; they only function within the organism, which is the creative being.

Industrial forestry partially uses physics and ecology. It ignores philosophy, ethics, and equity. These are basic tools for ecoforestry, which has a philosophical foundation.

Philosophical Pillars of Ecoforestry
Ecoforestry is different in its science, philosophy, art, and practice. These are four pillars of ecoforestry. Pillars have long been used to represent wisdom (possibly because of their verticality). The emperor Asoka had his edicts on morality and philosophy engraved on pillars throughout India (30 still remain). The pillars of wisdom were identified by Hebrew mystics with the days of the week (and of creation) and with seven trees (terebinth-oak, broom, kerm-oak, almond, quince, pomegranate, willow). The most sacred grove in Ireland, appended in verse to Irish Law, had seven sacred trees (apple, birch, oak, hazel, alder, willow, holly); these represented the seven-

pillared house that wisdom built for herself.

Not being wise, I was only able to think of four; but, four is a good number. It is the number of intuitive human dimensions, the number of corners of an ideal house (in Keresan Pueblo Indian cosmology, for instance—human order was square, while cosmic order was round, as it was also for the Assyrians, Egyptians, and Greeks), and the number of sacred dimensions in many cosmologies. The Oglala Sioux, for instance, identified each cardinal direction (east, north, west, south) with a color and an important action: north was the white cleansing wind, while east was red wisdom (south: yellow, growth; west: black, rain).

The philosophical pillars of ecoforestry are deep ecology (or ecophilosophy), process philosophy, ecofeminism, and general systems. But, then I could not stop at one set of four. And there are other ways to present the platform (see table below; each row is a separate set).

Table 4. Sets of Pillars

Matter	Death	Life	Spirit
Chaos	Ambiguity	Uncertainty	Complexity
Community	Freedom	Play	Understanding
Place	Participation	Cocreation	Interdependency
Cosmos	Wilderness	Garden	City
Forest	Home	Others	Humans
Ecosystems	Communities	Products	Practices
Characteristics	Principles	Standards	Operations

Maybe four is not enough. And, more than increasing their number to seven pillars, or nine, or ninety, the pillars may more resemble a Hindu temple in their profusion. Rather than describe each set of four or all of them, let me just talk about philosophy as a sort of the base of these pillars (the stylobate), the grounding of useful abstractions, and a platform for their stability.

What is philosophy?
Philosophy is a form of systematic inquiry, from the Greek term *philosophos*, meaning "lover of wisdom." Actually, Pythagoras insisted on being called *philosophos*, and forbid people to call him *sophos*, arguing that he was not wise. Even *sophos* is theoretical knowledge, however. The love of practical knowledge, *philophronesis*, never took off—perhaps, due to the Greek cultural character. In a way, philosophy addresses all of the human fields of inquiry, all academic disciplines, all areas of human experience, from language and society to value and science, all knowledge.

What is knowledge?
Already, two kinds of knowledge seem evident, theoretical and practical. Probably there are many kinds:
  • immune-system, genetic
  • personal, intuitive, emotional, tacit, negative
  • cultural, historical, local

- creative, metaphoric, meta-knowledge

Of course, some of these overlap. The important thing to remember for the moment is that both philosophy and science concentrate on theoretical knowledge almost exclusively. How is knowledge different from data, facts, information, imagination, and wisdom?

- Information is the act of imparting form, as in in-form-ation.
- Knowledge is the act of impressing an 'object" into the mind to make it the 'form' of knowledge.
- Data is an abstraction of knowledge. Imagination is a transfer of knowledge to another frame of reference.
- Facts are data transformed (as Goethe first noted) by emotion, theory, and imagination.
- Wisdom is the art of disciplined use of imagination in respect to knowledge of alternatives, exercised at the right time and in the right measure (according to Jonas Salk).

Problems of knowledge are not correctly separated from problems of living, or from ignorance.

What is ignorance?

The complement of knowledge is ignorance. Ignorance has its own diversity. Ignorance has at least two major components, conflicting knowledge and uncertainty. Bertrand Russell suggested that philosophy can teach us how to live without certainty. Uncertainty can further be divided into kinds:

- vagueness (indeterminate knowledge)
- probability (confidence in partial knowledge)
- incompleteness (of knowledge, missing elements)
- irrelevance (place in pattern unknown)
- fuzziness (overlapping interpretations)

There are also levels of uncertainty, from the quantum to the human. Elementary processes at the quantum level are not subject to a precise description in time and space. Predictions about location and velocity are just statements of probability—this is Heisenberg's uncertainty principle. The effect of this principle on epistemology is that our exact interpretation of things has to be abandoned. There are uncertainties at higher levels of organization as well; for instance, the hysteresis of some magnetic solids determines their subsequent behavior—without knowledge of their initial conditions and all past events, it is not possible to predict present or future behavior. Furthermore, the higher up in levels of organization, the more kinds of uncertainty there are. There is genetic uncertainty, as well as environmental and social. There is fundamental uncertainty in the thing/event/pattern, as well as in the channels (noise, meaning) of relationship.

Conflicting knowledge also can be broken down into kinds :

- anomaly (incongruity of knowledge, simple error)
- ambiguity (alternative interpretations of meaning)
- inconsistency (simultaneous untruth)
- equivocation (knowledge is constant and inconstant, true and untrue)
- belief (confidence in subjective knowledge, taboo)

There are probably other components of ignorance, as well as other ways that they can be combined into a formal typography. The main thing to remember is that there are many different kinds of ignorance, maybe more than kinds of knowledge, and that these determine our approach to the practice of forestry. If we think of knowledge as an expanding sphere in a space of ignorance, then as the sphere grows, the surface area in contact with ignorance also grows.

In a way, ignorance shapes knowledge. The tools of philosophy are words (especially definitions), questions, metaphors (images), and the big picture (the framework for everything). Greek philosophy started with questions. What is the ultimate thing: air, water, fire, atoms? Socrates, the gadfly, perfected the art of questioning. He constantly asked everyone about what they really knew and annoyed everyone to death with his questions. Since then, philosophy has continued to be an annoying discipline. Ludwig Wittgenstein, a particularly annoying thinker, decided that a philosophical problem has the form: "I don't know my way about." Asking questions is one way to find one's way. Philosophy is often distinguished by the kind of questions it asks. The branches of philosophy have to do with these types of questions and problems. For instance,

What can we know? How do we know anything? (Epistemology, the theory of knowledge that is concerned with what we can know.)

Why is there not just nothing? What is the nature of reality? (Metaphysics, where the answers that emerge from epistemology structure the study of being as being.)

What is our human relation to reality? (Cosmology describes the place of a culture in reality.)

What is good, beautiful, and of value? (Aesthetics, and the following areas are shaped by the conclusions about what we know, what is real, and what is out place in it all.)

How should we behave? What ought we to do? (Ethics)

What is just? (Political philosophy)

What is the ultimate being? What can we hope for? (Religious philosophy)

As philosophy evolved over thousands of years, it shifted from describing a metaphysical order of being (metaphysics) to the order of knowing—rational consciousness (epistemology). This subjective focus, Richard Rorty suggests, was followed by a historical turn with Hegel and Nietzsche, and then a linguistic turn with Wittgenstein. Each focus had the effect of isolating philosophy. Each focus also kept humanity at the center of the universe; even after Copernicus argued that the sun rather than the earth was at the center, Kant argued that both were categories of the human mind, the true center. Philosophies from German idealism to linguistic analysis and deconstructionism continue this anthropocentric bias. (Idealism, remember, considers that the universe does not exist—only human perceptions or a superhuman consciousness does.)

The rationalists (including Descartes, Spinoza, and Leibnitz) and also the empiricists (Locke and Hume) came to the conclusion that the sciences are independent of traditional metaphysical interpretations and should just try to describe and codify objective observations on their own.

After the divorce, philosophy was left to play with values and methods in its intellectual vacuum. Philosophy had more and more trouble addressing real-world events. Whole schools took pride in being obscure and abstract. Each school forwarded its philosophy as the "correct" philosophy and rejected all others. Philosophical systems were kind of like empty suits of armor on display. The system was designed to be irrefutable, but it enclosed nothing real. Deconstruction took apart the armor. Martin Heidegger and his inheritors, for instance Jacques Derrida, argue that no system can represent the true nature of things; no first principles can define every detail of the universe. But, in taking apart the armor, they also rejected what was supposed to be enclosed inside of it.

Philosophy once offered "lovers of knowledge" not only ideals to live for and principles to live by, but the tools to analyze bogus ideals and principles. As an academic subject philosophy quickly focused on aims rather than ideals and on rules rather than principles, letting students either draw their own inferences or simply reject philosophy as a game of trivia.

Ideas and theories in this century have been dominated by the impoverished philosophy of modern industrial culture. Whitehead argues that the impoverishment of our conceptual universe led to the disaster of physicalism and to having no metaphysics at all. Philosophy has not addressed the profound changes of global warming and habitat destruction, extinctions, social decay, new diseases, or technology. These changes are modifying the planet; some of them result from our dysfunctional styles of living, which may destroy us, as well as from those things we require for adventure or comfort, for instance refrigerators and airplanes.

To be effective in this age, ideas need to be rooted in conceptually rich philosophies, such as deep ecology or process philosophy or phenomenology. There are other philosophies, such as general systems theory (von Bertalanffy, Laszlo et al.), ecological resistance (John Rodman), ecophilosophy (Henryk Skolimowski, David Klein et al.), ecofeminism (Daly, Griffin et al.), and radical ecology (Wittbecker, Merchant et al.), that are also quite fruitful.

These philosophies contribute to a new metaphor that is more appropriate to the unity and interrelatedness of the earth; and it may be more personally fulfilling. To some extent these philosophies are the most recent flowering of a line of thought from Goethe and G. P. Marsh to Paul Shepard and Gary Snyder.

Goethe (circa 1800), for instance, in his organic dialectics, argued for contemplative nonintervention and the primacy of the qualitative as the best methods to reach the deep down "realm of the Mothers." Marsh (1860s) documented the human impact on the environment, especially forests, that could result in extinctions. In his philosophy, Whitehead emphasized the primacy of feeling. Whitehead's ideas support many of the principles of ecoforestry. Albert Schweitzer tried to extend ethics with a reverence for life. Lorenz developed an evolutionary epistemology. Aldo Leopold (1940s) formulated a land ethic. Rachel Carson (1960s) warned of the biological damage from technological quick fixes. Paul Shepard (1960s) noted ecological contributions to the image of the world. Gary Snyder (1960s) suggested poetry as an ecological survival technique.

In the early 1970s, several ecological philosophies and movements were formed. Murray Bookchin promoted a free society based on ecological principles, through social ecology and eco-anarchism. John B. Cobb, Jr. tried to incorporate humanity as part of nature, using the process philosophy of A. N. Whitehead. Theodore Roszak created the framework for a sacramental ecology and an economics of permanence. David Klein described an ecological philosophy. Sigmund Kvaloy also wrote on an Ecophilosophy. Daniel Kozlovsky outlined an ecological and evolutionary ethic. Henryk Skolimowski began to promote an ecological humanism. John Rodman wrote on the liberation of nature and later on ecological resistance. Arne Naess differentiated between a shallow and deep ecology.

Ecological philosophy acquired further definition in the following twenty years as a result of the labors of: Bill Devall, Mary Daly, Susan Griffin, Dolores LaChapelle, Neil Evernden, George Sessions, Alan Drengson, Carolyn Merchant, J. B. Callicott, Paul Taylor, David Foreman, and Eugene Hargrove. Many of these concentrated on defining ecofeminism and deep ecology.

What is Deep Ecology?
The deep ecology movement, as formulated by Arne Naess, is contrasted with the shallow ecology movement, whose main goal is the health of people in affluent countries through reducing pollution and resource depletion. Deep ecology is a movement that goes beyond a concern with pollution and resource use, with its man-in-the-environment image, to consider humanity in a relational, total-field image. The movement promotes human equality, conservation, and local autonomy. In principle, it proposes a biospherical egalitarianism, the equal right of all beings to live in place.

Deep ecology takes its inspiration from the science of ecology. Ecology can be considered an amphibious discipline, with the authority of science and the force of moral knowledge. Studied through its components and relationships, ecology is a way of seeing, a perspective of the human situation in its interconnection. It is a "subversive" (Paul Shepard's term) subject, normative and sensible, offering a "sacramental" (Theodore Roszak's term) vision of nature. Deep ecology is inspired and fortified by ecological knowledge derived from experience, not logic. Its tenets are normative. Its value system is only partly based on scientific research.

The platform of deep ecology addresses the well-being and flourishing of human and nonhuman life, as well as the richness and diversity of life forms, which humans have no right to reduce except to satisfy *vital* needs. The platform also considers that human interference with the nonhuman world is excessive, and that the flourishing of human life and cultures is compatible with a substantial decrease of the human population. Policies must be changed to reduce economic growth and to appreciate "life quality" (which cannot be quantified)—this implies that we have an obligation to try to implement necessary changes. The platform addresses not only ethical concerns, but scientific, cosmological, economic, and political ones as well.

What's the Difference?
These people, whose ideas form the eclectic basis of ecoforestry, asked different kinds of questions from traditional philosophy.

Traditional philosophy asks:
- How can conservation be deconstructed?
- Under what conditions is a preemptive nuclear strike ethical?
- What does linguistic analysis tell us about unreal states?
- Is this chair real?
- Does a tree falling in the forest make a sound?

Ecological philosophy asks:
- How does everything fit together?
- How can ecological philosophy principles subvert the dominant industrial way of life?
- How can the levels of the self be integrated? How can personal actions affect social structures?
- How can natural rights be integrated into the human legal system?
- What actions would reverse the catastrophic destruction of species and habitats?
- How can an optimum human population be calculated and achieved?
- How ought the human species exist on the planet?
- How can the vital needs of the population be satisfied indefinitely, without unnecessary destruction or conversion of ecosystems?
- How can mixed communities be designed to be challenging, safe, and continuing?
- What are the ecological responsibilities of corporations?
- How can wealth be distributed fairly?
- What kinds of technology are appropriate for local conditions and traditions?
- How can spirituality and commitment be encouraged? What is sacred?
- What other belief and knowledge systems are compatible with ecological philosophy?
- How can we make the forest the basis of local culture rather than the source of profit or support for a global economy?

Ecological philosophy also asks different questions specifically about forests and forestry. For example:
- When does clearcutting increase the health of a forest, if at all?
- What is a minimum, optimum or maximum forest cover for various watersheds? Science might identify minima or maxima but philosophy can aim at optima or satisfactions.
- How much wood can be taken from a forest? How much biomass can be taken before the system collapses?
- What kinds of stability should we encourage or maintain: persistence, resilience?
- What is the character of nature undisturbed by human beings (asked by Dan Botkin)? Or what is the quality of nature with limited

disturbances? (These are bad questions—there is no single state anyway. There is no nature undisturbed, but the kinds and rates of disturbances have changed)
- What is the difference between exploitation in a forest and disturbance or interference?
- How much disturbance can a forest take? Is clearcutting a natural-enough kind of disturbance?
- How can we manage with unpredictability? (after Chris Maser) Science can deal with unpredictability to some extent, but philosophy can provide other analytical and synthetic tools.
- At what level should management take place? Individual? County? National? Global?
- Why do forest workers and managers exhaust forests when they know that they are destroying the basis of their own livelihood?
- How will this activity affect our community health and wealth? (suggested by Wendell Berry)
- How can we design forests for the long-term and high diversity? What are the limits of biodiversity? How much of the forest can we design?
- How can we design forests for neutral elements in interrelated processes in a landscape?
- What are we restoring when we restore an ecosystem without all the parts or good knowledge about the ones we have?
- Can we limit our take so the forest is still self-sustaining and self-repairing? Philosophy can provide images of rich frugality.
- Why do we need old growth? We can support it ecologically in terms of nutrient flow and diversity, but we should also recognize what part it plays in human psychology and myth.
- How can we apply ethics to the forest in our current socio-cultural situation that emphasizes short-term profits?
- What kind of certification should we promote? Third-party only? What are the implications of certification?

These questions highlight the uncertainty we face in dealing with large, complex, long-lived, wild entities like forests. Forest managers have to live with uncertainty; this means that management decisions are essentially gambles (gambling is a profession that acknowledges the operation of chance and makes conclusions in the absence of facts—do you know any successful gamblers?). This is an important admission, that we do not have facts to base our actions on, that nature is a stochastic process, and that ecosystems are always changing. Furthermore, we do not know for sure what effects our actions will have on forests, which used to live for so long, in such diversity, in so many places. Successful gambling suggests that the proper attitudes for gambling with nature are awareness, humility and courage, not arrogance, fear and maximum use; it also suggests not gambling with all of our resources at once.

Philosophy is concerned not only with good questions, but with the good answers. Thinking is inspired by questions. The way of knowing is gotten to through questions more than through answers. One thing philosophy has inculcated in people is that "there is no right answer." The best response to a question is a hypothesis, a thought experiment. Through that, you can create explanations and discover answers in a dialog with others.

The business of philosophy is not just to annoy people (the Socratic method) or to fashion "paper flowers in the midst of smiling garden" (the modern method, as characterized by George Santayana). Philosophy has to be positive and constructive, to ask the questions that highlight patterns and rules for living, to contribute to healthier images and myths, to rethink the terms of debates, for instance, of tree farms versus wilderness, so that as many connections as possible are made visible and meaningful, and to provide good metaphors. Ecological metaphors, such as gardening or tending a forest, are more fruitful than images of machines or war.

Summary
With the preceding philosophical understanding in place, many specific changes would occur in the human use of other species in forests. These would allow us to:
- Recognize individuality in trees and other beings.
- Recognize the feelings and emotions of animals and the sensitivity of plants.
- Promote a noncommodity, drymoperipheral approach to forestry.
- Minimize the devastation of wild forests and wild ecosystems.
- Emphasize habitat loss as a human ethical issue.

Furthermore, in the near and far future, through its basis in ecological philosophy, ecoforestry could:
- Suggests ways that biosphere cultures can be converted to ecosystem cultures (after Ray Dasmann), characterized by wonder and wildness.
- Promote ecosystem lifestyles, characterized by frugality and joy.
- Relate the success of microorganisms to the ultimate success of living in self-sustaining forests.
- Develop a basis for management techniques for finite ecosystems, especially wild forests.
- Link ecological sustainability with the richness and diversity of ecosystems, as well as with human pleasure.
- Work for immediate solutions to inequity and destruction under worsening and thankless conditions.
- Educate with confidence and energy for ecological enlightenment over the long-term, despite short-term wobble.
- Learn to gamble wisely.

Ecoforestry is not a linear path leading to a definite conclusion. As a dialectical spiral, it requires the creative effort of the participants—you and me—to be put together. If it asks more questions than it answers, it is for the purpose of stimulating even more questions. If it diverges too much from the area of traditional philosophy or science, it is because it is, as Whitehead expected his philosophy to be, an adventure of ideas.

Chapter 9

What is Ecological Forestry? Art Philosophy Science Practice?
(Talk at a California Society of American Foresters Meeting, 1998)

Maybe I should start by saying what ecoforestry is not. It is not a bunch of long-haired, tree-huggers promoting some hippy fantasy of a perfect wilderness with friendly, talking trees and plants. But, neither is it a knee-jerk reaction to the indifference greed and waste of profit-driven clearcut forestry. Hell, I like driving big yellow machines and dropping trees on the head of a pin as much as any of you. It's just that sometimes I drive big green machines with soft tires; sometimes I ride in and pull out a stem by horse; and sometimes I walk in and walk out without anything wooden between my legs. Yeah, I know, you think clearcutting is woodness, but then you're all ithyphallic, too. I myself would look up those words to make sure they don't mean insane or obscene or something.

What I an trying to say is that perhaps sometimes clearcutting might be the most appropriate form of forestry, although the science is a bit old, weak, unfinished, and sketchy; seriously, the one experiment I know of in the 1800s was never finished. But, I can see where clearcutting exotic species might be the best way to restart a forest, although sunburned soil would still be a problem and patch-cuts might offer better protection.

Modern forestry has been very successful at providing timber efficiently in an economic sense. Certainly clearcutting is the best and safest use of human labor. And, if forests were like grassland species, or domestic fields, it might be okay all the time. Forestry has been treated as a special form of agriculture and is based on the agricultural model of simplify, harvest, and replant.

Forestry has several parallels with agriculture. Like agriculture, forestry uses soil to produce a crop for the purpose of increasing wealth (or perhaps just revenue). Like agriculture, forestry is renewable (unlike mines or oil reserves). Like agriculture, forestry is based on knowledge of many fields, including botany, soil, and meteorology. Like agriculture, forestry deals with vast areas. Unlike agriculture, however, forestry deals with wild plants on wild soils. Furthermore, trees are very long-lived, unlike crops of annuals, and are related in complex cycles, much more so than annuals. Trees are directly responsible for soil fertility and tilling. Despite the similarities, forestry cannot be considered a form of agriculture. Simplification and the economic problem of short-term debt loads for forest corporations, cause numerous problems with forests: Wild forests are clearcut; diversity is diminished; erosion increases; and aesthetic ruin can be seen from airplanes and highways.

Modern, scientific, and industrial forestry has a history of hundreds of years of trying to understand and manage forests, using the best knowledge of the time. Forestry, however, has been limited by its paradigm and the limits of its beliefs and knowledge. Albert Einstein distinguished two kinds of knowledge: The lifeless knowledge in books and the living knowledge in the consciousness of human beings. When the former ignores the latter, it can be very destructive.

Industrial forestry ignores the history of forests as well as the history of use. It assumes the cosmology of the industrial age. The machine metaphor, applied to forests and social conditions, as H. G. Wells noted, results in us making machines out of trees and people. Many forests, however, are occupied by people with different, and often better adapted, cosmologies. The Yaruro, for instance, believe that all beings are equal, not just men, as the industrial cosmology assumes; a culture that sees humanity and nature as equals existing in a mutual harmony will have a different kind of forestry than the modern version.

Industrial forestry assumes also the current economic system, which is based on winners and losers and on inequality. Economics has not been unsuccessful with its models, for instance of buying behavior, but it has become a highly abstract, lifeless, academic discipline. All its abstractions are applied to the real world without acknowledgment of the high degree of abstraction involved. One fallacy of economics is "money fetishism," where the characteristics of an abstract symbol, such as limitless growth, are applied to real commodities and values.

Industrial forestry partially uses physics and ecology. It ignores philosophy, ethics, and equity. Industrial forestry fails in at least five ways: Failure of perception, the failure to grasp the complex operation of a forest ecosystem or to recognize its own efficiency and value; the failure of intelligence, that could tell us not to apply the same approach to every forest on every kind of soil and weather; the failure of integrity, as it is corrupted by greed and stupidity; the failure of will, that is, the inability to try different things, or to be less greedy; and the failure of imagination, where we could picture restored healthy forests represented by an organic metaphor rather than the machine.

To understand what ecoforestry is, we could start with a dictionary definition of forestry as the science of developing, caring for, or cultivating forests. This kind of definition is relatively new, although humanity has been using forests for tens of thousands of years. Ecoforestry is the management of the human use of forests for necessary goods at an appropriate scale for the forest. It puts the forest first and considers what to leave.

Ecoforestry is not just a reaction to industrial forestry; rather, industrial forestry is considered to be one small perspective in the framework of ecoforestry, although industrial forestry has been distorted by excess profits and economic fears. Industrial forestry focuses on wood products, where ecoforestry is concerned with the whole forest community. Although ecoforestry sounds like a qualification of the modern, industrial forestry, it is in fact entirely different—it is based on a different metaphysics, a broader ecology, a more comprehensive economics, and it is sensitive to limits, ethics, aesthetics, and spiritual values. Ecoforestry addresses forests and human societies; industrial forestry addresses logs and profits. Ecoforestry is really not in contrast or contradiction to industrial forestry. Industrial forestry is simply a special case of ecoforestry, much as Newtonian dynamics became a special case of Einstein's theory of relativity.

Ecoforestry supplies the missing dimensions in forestry. Ecoforestry addresses forests and human societies. Ecoforestry includes humans and ultrahuman beings as an integral part of the whole system. Ecoforestry addresses both large and short-term scales (in time, size, and design). Industrial forestry partially uses physics and ecology. It ignores philosophy, ethics, and equity. These are basic tools for ecoforestry.

Alas, the easiest definition of ecoforestry is by contrast with industrial forestry. For instance, where industrial forestry has a corporate structure, ecoforestry has a community structure. Where industrial forestry tries to dominate the forest, ecoforestry tries to simply participate. Where industrial forestry strives for economic perfectibility, ecoforestry accepts satisfactory conclusions. There are differences in scale, logic and value, also. Yet, it is Aristotelian logic that makes comparisons like this. By the way, although we may be smart enough to go beyond Aristotle's logic in our actions, we should remember that, for Aristotle, the goal of action was always contemplation, that is, knowing and being, rather than getting and having.

An extended, morphogenetic logic is nonpolar and multicentered. That means that ecoforestry, which uses this kind of logic, should not be considered ecocentric, but drymoperipheral. We can define ecoforestry much like a crab approaches a morsel of food—by walking sideways and looking at it out of the corner of our eye. Ecoforestry is the management of the human use of forests for necessary goods at appropriate scales to the forest. It puts the health and continuity of the forest first, asking what to leave before what to take. It uses all science, philosophy, technology, cosmology, economics, and history—and from other cultures as well. Ecoforestry is a framework for forestry—it does not center on anything, it perceives and comprehends the whole forest system, centers and peripheral. The forest is polycentric anyway. Every being in every place is a center. Their cycles and patterns have centers. Our interests have centers.

Ecoforestry has been defined as selection forestry or restoration forestry. It has been defined as a context-based, community forestry that is based on traditional wisdom combined with scientific knowledge. It could be defined as the pragmatic study and use of genealogical actors in ecological roles (in the evolutionary play in the ecological theater)—the activities of forestry certainly make good theater, with protesters trying to save redwoods, crashing trees, ant-lines of full logging trucks.

Ecoforestry is the first forestry to consciously consider limits as well as other scales and other dimensions (history, ethics, and equity). It is scientific as well as intuitive and emotive. It is profit-making as well as preservationist. Most importantly, ecoforestry is adaptive and responsive to change. Ecoforestry is grounded, dynamic, realistic, cyclic, far-sighted, sophisticated, complex, diverse, ethical, and responsible. Its goals are to restore degraded forests, preserve representational old-growth, educate the public, balance exploitation with human need and equity, and provide a framework for human relationships with the forests.

The main objective of ecoforestry is keeping and restoring fully functioning forest ecosystems, continuously addressing forest needs first as a prerequisite to addressing human needs. This means keeping and restoring the elements that help a forest serve its own needs: By maintaining a multi-species, multi aged forest with healthy soils, water quality and flow, stable climate and habitat for wildlife, the forest continues to be a forest, which is self-ordering and self-maintaining.

Ecoforestry has to be practical, so that it works. It has to be generous , for it to be defended against more damaging varieties, and it has to be radical, for its changes to be meaningful and effective.

Although ecoforestry uses applied ecological sciences and technologies, it uses also an applied philosophy; thus, ethics and valuation are as important as the production of commercial goods. To some extent, ecoforestry is a crisis science, much like conservation biology, from which it has taken many ideas and principles. It addresses the whole forests from which billions of trees are taken each year, displacing billions of insects, animals, and other plants. Ecoforestry has a crisis strategy, that is, a way of life, for redesigning whole forests—preserving or restoring natural processes, respecting the opportunities for existence of other beings, and building upon the productivity of forests for our needs.

Ecoforestry rests on four practical pillars: The character and health of forest ecosystems, the health of the human communities living in or dependent on forests, the timber itself and other products, and the practices of acquiring an assortment of products.

I. In addressing the first pillar, it is important to preserve the characteristics of the wild forest, from its historicity to its productivity, complexity, and stability. That requires understanding the interactions in a forest as a mature community, including competition and cooperation, available energy and real limits. For instance, the interactions of individuals in a food chain results in a unique pattern and structure of the community—that can be disrupted by bad practices; proper standards, such as preserving minimum habitat for animals and birds, can preserve the pattern and structure.

Forests have been changing and evolving for thousands of years, with human influence and without. In cases where human impacts have become a routine disturbance in the forest, such as Swidden practices in Amazonia, the forest has adapted. Where changes have been rapid and large-scale, the forests have been simplified or have collapsed.

The cornerstones of forest health are the regulation of climate, water and soil conditions—without these intact, we lose everything else. We can do a lot to insure good climate, and water and soil conditions, by maintaining adequate cover of a mixed species, multi layered and aged canopy; by encouraging the diversity of understory trees, shrubs, herbs and ground cover; by encouraging nitrogen fixers, deciduous leaf fall, standing and down dead logs and woody debris from dead and rotting material—in short maintaining the components and structures of a forest.

We always want to keep subsurface water and other cycle flows structurally intact. Understanding cycles, flows, structures and functions in relation to individual elements and popular species keeps us focused on causes rather than chasing symptoms. To insure healthy forest conditions, we consider primarily ecological components, structures and functions.

We look at the energy movement from the sun through plants, herbivores, carnivores and detritivores; the cycling and recycling of dead and living material by herbivores, carnivores and detritivores; dispersal of pollen, seeds and individuals; climate and microclimate effects; and communication systems and paths such as chemical, energetic, or behavioral.

We focus on minerals, topsoil and topography, hydrological systems and cycles, transport paths of nutrients, and population checks and balances in food webs. Indicator species can demonstrate failure in the food chain, hence forest structure and functioning. Suitable habitat should be left for all naturally associated species. Biodiversity and genetic richness ensure a greater degree of stability in adapting to changing conditions and empowers the ability of the forest to do its own checks and balances. We keep in mind the relationship of each stand to the larger ecosystem surrounding or adjoining, planning at the landscape level first.

We identify functions, keeping in mind that each structural and compositional feature of a forest serves many purposes and is connected to forest health in many obvious and not so obvious ways. A tree in a forest is multifunctional in that it can play a variety of roles that keep a forest healthy and productive

Different forest ecosystems are fragile or robust to different degrees. Ecoforestry recognizes that unique or important forests must be preserved or limited in varying degrees. Other forests will require varying degrees of protection.

II. The human communities in forests, their health and characteristics, are equally important. Human beings participate in the forest and its cycles. Although they exploit forests, human beings have characteristics that must be maintained for those communities to be healthy, for instance, levels of material equity and personal equality. Human beings need to use appropriate images, share natural wealth, and set limits on their needs and desires.

Forests provide stable places for human communities. They provide the processes that clean water and air, regulate air and water flow. They provide productivity, of which humanity can claim a part. Forests provide many other benefits. Nothing can compare to the economic and ecological interest gained from the combination of the annual growth interest from softwoods; the annual growth of hardwoods and special forest products ; the sum total of benefits accrued from priced and priceless valuables in cultural categories, such as recreation or spiritual experience, associated with the continuously growing, fully functional forest ecosystem; the sum total of benefits from the complimentary nature the forest ecosystem has on nontimber based industries like tourism and fishing; the interest gained due to scarcity of quality grained, #1 peeler sawlogs, as well as intact native forest ecosystem with its genetic storehouses; the stability and security of the diversified resource base of a fully functioning forest ecosystem; the stability, security and grand productivity associated with the establishment of industries that give our native forest ecosystems a break, focusing on better land use of already cleared and converted lands close to the urban areas, for example, intensively managed organic agroforestry sites in and around urban areas growing fruit and nut trees, as a restoration plan for old dairy farms, grazing lands and poorly utilized cleared areas. Finally, there is basic aesthetics—people like to work in the woods.—and physical health—touching trees, seeing animals, and walking lowers blood pressure and drives imagination.

Despite having to work within the dominant form of economics, with its rigid metaphors and misapplied myths, ecoforestry recognizes the dangers of that economics and emphasizes a small-scale, alternative economic niche.

III. Timber is a small, but significant part of the diversity of forest products. The forest produces many things. The older the forest the more things it produces, although it may produce less of each.

In addressing human needs, we should selectively harvest forest goods and products and mind all other activities in such a way that we always leave behind us, in the wake of what we take, a forest that can feed, seed and balance on its own. We emphasize what to leave over what to take. The practice of ecoforestry is based on and demands an understanding of ecological processes and patterns, so a forester knows how to work with a forest and not against it. Timber from mature and old-growth trees is more desirable, although not necessarily priced higher, due to the length of wood fibers and narrowness of rings.

The practice of ecoforestry is based on recognizing the full range of intrinsic and instrumental values of the forest. What we humans are demanding from forests, as consumers and citizens, is dramatically shifting according to a few "gigatrends" that will affect how we manage our forests, based on changes in the value system, information and communication system, technology, income, changes with substitutes and complimentary services and goods, changes with competition and changes with environmental, political, social and economic conditions.

Practitioners of ecoforestry can earn more from fewer resources by bringing out the full value of the forest and each of its products. Not just the utility of wood, but the importance of wild herbs and of other kinds of fiber and chemicals. These things have to be protected from overexploitation. We have to use many species, limit the use of each kind, and respect and recycle as many materials as possible. When we understand the connections and the effects of exploitation, we might create a complete product cycle in a form of natural and industrial ecology that will lead to more stability in forests and human communities. Imagine someday growing our own tree houses.

IV. The fourth pillar is the practices and techniques for getting the products while preserving the forest process. This requires a whole education, from play to community. Education means leading out. For ecoforestry this means leading out of the walls and into the forest. By developing ecosystem observation skills, through observing all these patterns and processes and how they change over time, one can begin to generate the kind of observation pertinent to an ecosystem's well being. Education requires information, which in a forest can be gotten from observation and monitoring. Ecoforestry encourages observations obtained from many perspectives, from crawling on the forest floor to climbing into the canopies.

This requires the design of the forest community on at least four levels, from components and products to systems and communities. With respect to the characteristics of good design, from frugality to respect, using proven design principles, from the proper use of scale to an optimum diversity. Design principles have to be identified and applied, from maintaining the completeness of the forest to balancing paths and margins.

The forest is wild, large-scale, long-lived, self-making, and self-designing, but we humans now influence every natural system, taking what we need from some ecosystems, enhancing a few, misusing others, and interfering with the rest. We need designs to restore the balance between human needs and natural processes.

Ecological designs focus on whole communities that work in the same self-sustaining and self-limiting ways as nature. By consciously creating meaningful order, we can develop ways of producing widespread community wealth while positioning the community for a long, sustainable future in a healthy environment.

Not just design management needs to be addressed, but forest management, planning, and operations, also. Operations also must be included, from surveying and monitoring to harvest timing and techniques. Monitoring systems are established prior to any harvesting in order to study changes in the forest due to the impacts of human activities in the forest and to provide data to adapt forest management practices in the future. We can monitor at the landscape level and subregions, gathering site-specific information. Geographic information systems (GIS) serve the need for storing and accessing information that deals in questions like "what is it?" "where is it?" and "what is it next to?"

Data recorded for a field map includes: Changes in land shape, tree sizes, species, changes in density, understory plants, soil and water quality, quantity and integrity of structure and function, animal and insect populations, critical areas such as steep areas, endangered areas, riparian areas and the surrounding areas which influence riparian areas, unstable areas, and cultural sites.

When combined in a map these various areas can be color coded, clarifying where harvesting activities are not permitted or are unusually limited. We initially monitor to guide our practices and then we continue to monitor in order to evaluate the effects of our practices both positive and negative, giving us continual direction. Although effective monitoring is labor intensive, it becomes part of our entries into the forest as we take care of many things at once. Also, monitoring and assessment can be achieved through various education and training programs, using volunteer work that benefits both the manager and the community.

Standards must be comprehensive and high. For instance, the number of snags and fallen trees must be determined. Buffer zones and preservation areas have to be identified. Types of roads, equipment, and damaging activities, such as slash burning, have to be identified and limited. Goals for the forest and its use have to be identified. Other related activities, such as certification and restoration, have to be considered.

Management itself has to be adaptive and flexible; it has to be updated seasonally or annually. A plan is crucial. Things to include in the plan are: Management objectives for activities; inventory analysis of all plant and animal species; ecosystem evaluation at landscape and stand levels; zoning and identification of key land features and protected areas; production level limits; future desired condition; monitoring requirements; work plans for achieving objectives; equipment choice; local community, native culture and employee provisions; and a process for the periodic review and revision of the plan.

For any application there have to be short-run and long-run decisions. The decision to shift from an industrial model towards an ecoforestry model calls for an integrated planning approach and support from a number of areas. There have to be set principles. There are certain general standards that can be identified: Do not clearcut in most cases; only take less than the annual growth rate, from those things in abundance.; use natural regeneration with good diversity and distribution of parent seed trees to protect native gene pools; do not bring in fertilizer, herbicides, pesticides, or any toxic pollutants; use the lowest impact removal methods possible, combining sustainable selection with careful directional falling, and a low impact access system that insures all features of a fully functioning forest ecosystem are left intact; and restore, maintain and or keep all values associated with fully functioning ecosystems, intact. Water, soil and climate conditions, along with indicator species, offer us cues. Respect the forest. "Think like a forest," in other words. Forest practitioners will address the need for local employment and community stability and will respect workers' rights and native cultural sites.

One way to reduce costs is though the downsizing and decentralization of your operation. A small-scale, decentralized operation means the use of small, easily maintained equipment that requires a small initial investment that can be cooperatively purchased and shared and which does less damage to the environment in that the overhead doesn't demand more than what the forest can provide; the roads and trails can be smaller. Small scale also means doing more with less land. It has been theorized that small operations have the lowest costs for lowest output and that big plants have lowest costs for greatest output. This is due to mechanization, which lowers human employment and specialization, and which singularizes skills and activities.

Small scale, decentralized operations allow for full employment, flexibility and innovation that creates an environment for value-adding and discovery. The scale is small enough that individuals can afford to work. Small mills are more flexible and able to create value from unique wood sizes, shapes and species. Small mills also have great recovery from each log. Small mills are portable and can be milled on site, saving transportation and middleman costs. Small mills can be cost shared, owned and used cooperatively, so that overhead doesn't dictate practices. Small mills are relatively easy to maintain and are cheaper to certify.

Certification is such a simple concept: It is the provenance of wood made visible and known. Yet, it is part of a complicated network and has far-reaching implications. Certification is driven by consumers, but consumers are reacting to forest degradation and destruction, which are partly caused by consumer demand for products made from wood. Consumer pressures have often influenced the extent of exploitation in a landscape. Consumers have indicated that they want forests for wood products, but also for habitat for other species, aesthetics, recreation, and other values. Certification itself is a demand. As consumer demands change, however, the uses of landscapes may change. Certification may bring about other far-reaching changes.

To work in wild forests, certification has to be done by independent third-party groups. Certification requires the knowledge of forests gained from observations, inventories, and monitoring. It will encourage the acceptance of the limits and constraints of the forest ecosystem. Certification will address the problems of human equity and the distribution of forest goods, and, finally, it may shift forestry toward ecoforestry practices.

In summary, using an ecoforestry approach means that you have to continue to learn from forests, through observation and participation. It means questioning what you do not understand. People ignore large central questions about meaning because the questions are frightening. Justice or survival can be set aside for happiness and contentment. An ecoforestry approach always asks the basic questions, such as: What is there I n the landscape? What could be there (or what is missing)? What should be left? What should be conserved? What should be used? How should it be used? How do we get it? How do we sell it?

Ecoforestry tailors its harvest rates to below the annual growth rate per year, reducing costs as the forest plants, fertilizes, controls pests, climate, and water flow and quality, builds soil, and produces itself with the enhancement of intelligent human use. Meanwhile, with restoration costs down, profits can be increased through value added products and expansion of the market into special forest products, both of which benefit from the continual maintenance of all the values associated with a fully functioning forest ecosystems.

Ecoforestry is a concept-based, process-oriented, crisis-aware, community forestry based on traditional wisdom and scientific knowledge. It provides an ethical path, through its oath, and always creates goals for forest health. Ecoforestry keeps in mind the Buddha's definition of a forest: "A peculiar organism of unlimited kindness and benevolence that makes no demands for its sustenance and extends generously the products of its life activity; it affords protection to all beings, offering shade even to the axeman who destroys it."

If you are interested in the practice of ecoforestry, we invite you to participate in our courses, workshops, and field experiences. If you have questions, comments or criticisms, please call me at our Oregon office.

Chapter 10

Global Planning Flaws?
Savory Thoughts about Grazing

I am always really impressed by anyone who studies whole systems and then makes grand hypotheses and deductions. Alan Savory has certainly done this; his hypothesis about Texas and Southern Africa is quite likely very workable. His ideas are inspiring and thought-provoking. They should be used to address management problems as an umbrella.

However, there are problems with it as a global hypothesis and in its general nature as well. The hypothesis is too general, even divided into two, moist and arid, categories. Furthermore, the hypothesis assumes adequate knowledge, which Savory demonstrates that he does not have. Finally, the particular tests, while necessary for short-term feedback, are not adequate for the long-term.

His program is too general. Two categories are not enough. High elevation and low elevation would not be enough either. Even the IUCN biome categories are probably not enough for any kind of holistic management. Specific management styles have to be tied to local ecosystems.

Inadequate knowledge. Savory's knowledge of many systems is outdated or incomplete. Some of his errors include:

1. Grazing animal populations are not limited by predators in general. In almost every case the limit is food supply. For example, North Slope Caribou in Alaska are limited by grasses and forbs. Wolves and Inuit only serve to keep the herds healthy and alert. David Klein at the University of Alaska has studied this for 25 years or more.

2. Some grasslands, such as the Palouse (the large dot in the upper left of the erosion slide), absolutely require long rest after any grazing. The Palouse grassland has not been grazed for over 10,000 years (after bison moved east)—only by a few antelope and deer; it rarely had fire—the native peoples, Palouse, Nez Perce, Coeur d'Alene, and others never needed to use it for herding or seed dispersal. The first use of fire started to destroy the vegetation and the soils, a process that still continues. Savory's approach would have destroyed the original grassland.

3. Human fires are natural because humans are natural. Native people's fires shaped and maintained the central prairie. Excessive human fires, especially in the last 20 years, may interfere with ecosystem processes, but that does not mean that all fires are bad or destructive.

4. Prey can outrun predators almost every time. Except for short bursts, ungulate prey is always faster than predators. Herding is primarily a protection for the young, old, or weak animals, not normal adults. A healthy, single adult moose, caribou, or elk rarely has any trouble with wolves, for instance.

5. More than just demonstrating that a community is a whole, the starfish example showed how predation forms and maintains diversity. Steven Stanley has written extensively about this.

6. Cities can be self-sustaining; they are not necessarily sinks only. Paolo

Soleri and John Todd and others have been working to demonstrate that urban agriculture and city design can produce "fully functioning cities."

The kind of tests he suggests are an improvement on our tragic management styles, but fall prey to many of the same limitations. One of those is trying to think, monitor, and manage only for the short-term. The short-term tests he proposes provide only short-term feedback. Many long-term trends are counterintuitive and not obvious at all in the short term. We should be setting up management for hundreds of years.

One thing I really wanted to see in Savory's presentation was those ecosystems that are increasing in biodiversity as a result of a balanced resident human culture or of benign neglect. That might has altered the form of his hypothesis.

Traditional human science has always had difficulties with complex situations, from the three-body problem in physics to explaining linkages in ecology. Our success has been with simple systems, except where human cultures stabilized themselves in a complex system.

Agriculture, by definition, was a reduction of diversity through selection—older or newer styles, e.g. natural farming, encourage more diversity, maybe more than natural ecosystems. What he calls brittleness sounds like nonresilience.

One problem with paradigm shifts that should be emphasized is that the old paradigm has all the economic and political power. That is why the changeover is so slow and painful. People are profiting from the old paradigm and they will fight to keep rich. Lynn Margulis has written an excellent essay on just this thing.

Margulis argues that Neo-Darwinist fundamentals are derived from the mechanist worldview. Such fundamentals include a bodiless, linear concept of evolution and uncritical acceptance of adaptation. These fundamentals, burdened with pre-evolutionary legacies, are at odds with a nonmechanistic systems philosophy. They ignore physiology, metabolism, and diversity. They fail to describe the reciprocity of a living environment (where life does not adapt to a passive prior environment, but it produces and modifies its surroundings). Furthermore, the technical terms of the Neo-Darwinists, such as species, individual, units of selection, fitness, and genotype, she describes as "battle cries." She suggests that the use of these terms continues on the mechanistic "thought-collectives" because they are consistent with the major metaphors of the industrial civilization, which are concerned with technology, power, and wealth. We can expect change only after these living metaphors change (or evolve or shift).

Chapter 11

Is Forestry Not Sustainable?
(Comments on a Letter from Andy Kerr)

Your article was passed on to me by a colleague forester. I have taken the liberty of commenting on it.

The planet has always been forested to some extent, as well as desert, grassland, ocean, rivers, swamps, snow fields—it is an ocean planet first and foremost, but forest are a significant feature. But I agree, we should be reforesting much faster than we are cutting. I would like to see a better balance between humanity and nonhumanity (the human-producing matrix, or what we used to call nature); smaller human populations, larger bear and fish populations.

The public forest land is too valuable to log—we should enforce zero cutting; there are vast areas of private land that could be logged sustainably.

We do need more forests and we do need more wilderness. We have not been smart about setting aside wilderness areas for viable populations of wild species or for primary peoples who would prefer to continue their cultures in place.

Since modern industrial forestry does cut 90 percent of a forest, then it should be reduced by 80 percent or more.

I would beware of transferring private lands to public hands. Private hands have a better record for preserving land than public hands. Maybe we need more of the Nature Conservancy and less federal government. It might make more sense to incorporate forests, as I have suggested elsewhere.

Yes, it is true that private tree farms will not be able to meet the per capita wood demands now (sustainably), much less in a decade or so—even more so when land is reduced as it should be. The solution, however, is to reduce population and demand, not find substitutes like steel (remember during World War Two, if you're old enough, how houses were built of steel plate on i-beams because wood was rationed? Those houses had some odd characteristics in electrical storms) or concrete, which also have tremendous environmental impacts on a large scale.

It is true that a tree plantation is not a forest. Many of us think tree plantations are absurd. A managed forest, however , is not a tree plantation. A properly managed forest, in which only the net ecosystem productivity or less is removed annually, is a complex (not simplified) sustainable forest, as I have also argued elsewhere. The removal of biomass in large amounts does impoverish a forest. The removal in small amounts may stimulate the system. For example, wolves and caribou remove biomass from a forest ecosystem (to grassland ecosystems); migratory birds also carry biomass away. Chris Maser has written about the number of trees that end up in the ocean; this also transfers biomass between ecosystems. No ecosystem is a closed system. The difference, however, is that humans sometimes (too often) take up to 90 percent of the above-ground biomass out of a forest ecosystem. That's simple silvicide.

If by forestry, you mean industrial forestry practiced on a global scale,

I agree that it is not sustainable. Industrial forestry is not the only model of forest use, though. Native peoples use the forest without destroying it. Ecological forestry uses the forest without destroying it.

From the badness of economic reasons it does not follow that all forestry is bad. Human civilizations have almost always, as Gregory Bateson noted, reduced the "flexibility" of ecosystems, sometimes causing their own failure and collapse. Ecological forestry is concerned with keeping flexibility in forest ecosystems.

It pays to mine trees in the modern economic system, as it pays to mine whales or copper—because the system is set up to depreciate the value of equipment and potential of the resources without valuing the resources that highly—and because it is based on misleading myths, e.g., substitution or infinite growth. Neither the agriculture model (forest are domestic) nor the mining model (forests are resources) is proper for forestry. We should provide a third, don't you think? A wild forest model.

On a global scale, reforestation is a losing proposition because it does not pay in the economic system. Because clearcutting is economic, it is done. I would think the solution would be ecological economics with total cost accounting. And this would be a better solution than just putting forest off limits in parks or preserves. Why? Because it would be too expensive to cut down forests and replant a few percent. Why? Because the corporation would have to pay for all the watershed and atmospheric costs, as well as forest functions lost.

I have to agree that people abuse their forests—all my neighbors would rather have sanitation cuts to finance vacations than to invest in replanting or sustainable practices.

It is true that time is not on the side of forests, because we value our short human time much more highly. Although I could envision circumstances where sustainability could work; for instance if the company you mentioned had 1200 forest ecosystems to manage and each one were partially cut in a different quarter. Private individuals have it worse. Although a corporation is theoretically immortal, individuals are not. Furthermore, individuals are less bound to the plans of their parents than are corporations. I suspect that our heirs might just sell the land.

The risk from natural processes, such as insects or fire, is really only a risk in the current economy and if you plan to cut 90 percent of the trees. Otherwise, it is just the tithe of nature: 5-10 percent of the productivity. Furthermore, the market for wood has lasted at least 3000 years; it should last another 3000, if we do.

It is unlikely that many of the traditional uses of wood—fine furniture, sturdy beams, weathered siding, artistic papers, creative gifts—will become obsolete. Merlo is obviously demented, even if millions of other demented people agree with him. Technology is great, though, and fiber chemistry can duplicate some of the structural and functional features of long-fibered wood from old growth. The functions might eventually be duplicated, but it is unlikely that the texture or aesthetics (or even snob appeal) will be.

We do have evidence of careful forest practices in primary peoples (Campa, Desanai, et al.) and industrial societies, by individuals (Merve

Wilkinson et al.), communities, and corporations (Tall Timbers et al.). Real evidence takes 500 years or more for certainty. We have no evidence that any large-scale culture is successful for that long.

Sustainability should be the norm, but it should be rigorously defined and monitored. Most people have more than one objective for anything they do; that is not a valid argument against trying for sustainable practices. Many of us are benefiting from higher prices because we did not liquidate our forests; we have husbanded them and limited our cutting to below the ecosystem productivity (not just the growth rate). To say that we will not profit in the future assumes that the price will not keep rising (as they have for over 100 years) and that the current economic system will still be in place.

Forests are too important to destroy. They can be logged, however, without being destroyed, by respecting ecosystem limits.

Taxes are a good way to direct the economy and save important parts of the environment. I would like to see a Maine Woods Park, and a Cascades Park, and a Palouse Grasslands, and a New Jersey Pines Park. Ecosystems need to be preserved as part of the operation of global chemical cycles, as homelands for nonhuman life—for many reasons, with and without native human populations.

I agree with your three policies. Kenaf and Hemp have proven their usefulness. But, how much land do you suppose would have to be converted (from asphalt or native ecosystems) to grow fibers for just a Sunday NY *Times*? Probably more than we would find acceptable. There seem to be tradeoffs for every solution. Population and consumption should be reduced as a first step. Wood should be valued as if we knew the real costs of acquiring it, and products should be made to last 500 years or more (some antique furniture is already 500 years old and still useful). But, none of these policies require the abolition of all forestry practices, especially archaic or ecological ones, which can be sustainable.

Chapter 12

Is Forestry Education Necessary?

Introduction

It's time for education in forestry to make a leap into an ecological paradigm. This kind of education must be based on a new ecological cosmology, a new ecological philosophy, a new ecological ethics, a new ecological science, a new ecological economics, as well as on historically-beneficial indigenous knowledge This ecological paradigm in forestry education and practices we call Ecoforestry. In this paradigm the forest is considered first, in its entirety, before being regarded as a source for materials to satisfy human needs.

A true education need not be bound by conceptual and economic limits. A minimum education may train students for an economic role in society, but a good education teaches them how to enjoy living among other human beings and to perpetuate a good society. A complete education requires intense effort, discipline, patience, and a tolerance for failure.

Alternative institutions, such as the Ecoforestry Institute, can offer great benefits for participants, with a teacher "leading" a pupil to an education. By encouraging participants who are already working outside academic walls, alternative institutions can foster a kind of synergetic education. The most valuable qualities of an alternative education are personal contact, which allows noncompetitive constructive criticism, and flexibility, which allows the educational process to fit an established and meaningful life-style. Alternative institutions provide for education within the larger community, in the larger context of work and recreation.

To be effective in this age education needs to be rooted in conceptually rich philosophies, such as deep ecology, and in relational, holistic sciences, such as conservation biology. Furthermore, education has to be rooted in a local culture, and it has to consider appropriate ways of behaving in a local ecosystem.

Deep Ecology

Deep ecology takes its inspiration from the science of ecology. As a science, ecology describes the interrelationships of organisms and environments, that is, the experience of living together in the biosphere. Ecology is not a reductive discipline and is not readily amenable to complete quantification. Even scientific ecology is an integrative discipline that extends beyond the boundaries of science. Ecology can be considered an amphibious discipline, with the authority of science and the force of moral knowledge. Studied through its components and relationships, ecology is a way of seeing, a perspective of the human situation in its interconnection.

Deep ecology is a movement, formulated by Arne Naess, that goes beyond a concern with pollution and resource use to consider humanity in a relational, total-field image. The movement promotes human equality, conservation, and local autonomy. In principle, it proposes a biospherical egalitarianism, that is, the equal right of all beings to live in place. It adds a normative dimension to ecology in a framework of ecosophy, literally the wisdom of the house. As a philosophy, deep ecology investigates the normative aspects of living together, that is, ethics, and the maintenance of the affairs of communities, that is, economics and politics. As a noetic discipline, deep ecology provides information on the state of nature, recognizing that human beings are participants in nature, as well as participants in human societies.

Deep ecology emphasizes biological equality. When Charles Elton transformed the "Great Chain of Being" into a chain of eating, ecologists realized that the bottom link of the food chain, plants, was the most important. Humanity is part of the food chain, although it appropriates a large amount of the productivity of most ecosystems. The exploitative competition of humans in ecosystems is an important part of biogeochemical cycles. Humanity cannot unparticipate by choice.

Deep ecology argues for diversity. In nature, variety emerges spontaneously, as the capacities of species are sorted by the environment. Variety provides flexibility in systems. The diminution of variety through human interference may debase the wholeness and stability of systems. Furthermore, aesthetic, ethical, and utilitarian reasons all support the efforts to conserve the diversity of nature.

Deep ecology incorporates a broader scientific method that might be called patient practice. There are ways of dealing with the earth that are not scientific or technical; they are aesthetic or ethical. These alternatives are not incompatible with traditional science.

Conservation Biology
Conservation biology is a synthetic discipline that draws on history, science, and philosophy and applies itself to the widely documented crises in forestry, resource management, and preservation. It is science combined with action. The origins of conservation biology can be traced in traditional religious and cultural beliefs, early naturalism, and various ethics: Muir's preservationist ethic, Pinchot's resource conservation ethic, and Leopold's land ethic.

The ethical principles of conservation biology can be presented as a series of statements:

1. The diversity of organisms is good. (Diversity provides greater variety, buffers against catastrophe, and aesthetic appreciation for humans.)

2. The untimely extinction of populations and species is bad. (Extinction is a natural process, but humans have increased it a thousand-fold.)

3. Ecological complexity is good. (A complex natural environment supports complex interactions, such as coevolution—that probably would be lost in artificial communities.)

4. Evolution is good. (That is, the evolutionary process leads to new species and increased diversity—relatively large natural populations in natural cycles are required for it.)

5. Biological diversity has intrinsic value. (Species have value to themselves, as well as to others and to the whole, all of which are conferred by evolutionary history and supersedes human economic valuation.)

The goals of conservation biology are to investigate human impacts on biological diversity and to develop practical approaches to prevent the extinction of species, by providing scientific conservation principles, by identifying conservation problems, by establishing corrective procedures, by making scientists and managers responsive to problems and issues.

Conservation biology addresses problems in forestry. Current logging practices reduce the complexity by knocking out key players in the ecological web, players such as spotted owls, fungi, and dead trees. Conservation biology can make several recommendations which would preserve diversity and complexity (and avoid numerous extinctions) in forests:

1. Promote cutting practices that respect the productivity and complexity, leaving snags, logs, and many-aged forests.
2. Grant timber leases that are contingent on the maintenance of productivity and diversity of the land.
3. Reduce fragmentation through the design of forested areas, taking into account the genetic diversity of the trees, catastrophic conditions, minimum viable populations, corridors, and edge effects.
4. Recommend that reserves be made large enough for minimum viable populations and minimum viable ecosystem areas.

Because current reserves are usually too small to hold viable populations, corridors must be planned to intersect with the larger areas set aside, and highway routes and underpasses must be modified. According to Harris, highways are a major force in fragmenting habitats. Conservation biology recognizes the need for planning at a landscape level.

Ecoforestry Education
The Ecoforestry Institute in the U.S. and its sister organization in Canada, the Ecoforestry Institute Society, offer a variety of Education and Training Programs and Workshops at their headquarters at Mountain Grove in southern Oregon and outside of Victoria, British Columbia, as well as across the U.S. and Canada. The primary objective of their programs are to educate and train foresters, forest owners, forest workers and others who seek to understand and practice ecologically responsible forestry in ways which sustain and restore fully functioning forest ecosystems, while harvesting a range of forest goods and products.

The Ecoforestry Institute's Education and Training programs are based on a pedagogy which seeks to mirror the characteristics of ecosystems: open rather than closed, dynamic rather than fixed, diverse rather than monocultured, complex rather than complicated; synergistic rather than reductionist; and creative rather than destructive. The discipline of ecoforestry starts by considering a forest first, in its functioning entirety, before being regarded as a source for materials to satisfy human needs.

The Ecoforestry Certification Program consists of 4 major components: a 9-day Intensive Course, Apprenticeships and Internships with master ecoforesters, a Distance Learning Course, and Workshops on ecological forestry theory and practices. 1. A nine-day Ecoforestry Intensive Course is taught by leading ecological scientists, philosophers, and practitioners of indigenous, sustainable forestry.

The Institute's faculty includes leading ecological philosophers, scientists and practitioners, such as:

- Chris Maser, a research scientist with over 20 years studying forest, range, subarctic, desert and coastal ecology;
- Carolyn Merchant, who teaches Environmental History, Philosophy and Ethics at the University of California, Berkeley, and is the author of *The Death of Nature, Ecological Revolutions*, and *Radical Ecology: The Search for a Livable World*;
- Merve Wilkinson, who has spent the last 50 years living and working on the Wildwood Forest in Lady Smith, Vancouver Island, B.C.;
- Alan Drengson, who teaches ecophilosophy at the University of Victoria and is the editor of *The Trumpeter*;
- Dennis Martinez, a Native American ecological restorationist and permaculturalist;
- Richard Hart, the author of the *Forested Landscape Assessment & Monitoring Handbook*, who directs research & monitoring for the Headwaters organization in southern Oregon;
- Reed Noss, the author of over 70 papers on ecology and conservation biology, who is the editor of the *Conservation Biology* journal and serves as the science director of The Wildlands Project.;
- Herb Hammond, a British Columbian forester, forest ecologist and author of *Forest Among the Trees*; and others.

The Ecoforestry Intensives seek to provide a whole-systems understanding of forest ecology as applied on a site-specific basis within a landscape context. Much of the curriculum is presented in-the-forest during the days with evening lectures and discussions. Topics covered will include: Landscape Forest Ecology, Ecoforestry Principles and Practices, Deep Ecology Philosophy, Indigenous Land Ethics, Ecological Restoration, Forested Landscape Analysis & Monitoring, Conservation Biology, Ecological Economics, Fire Ecology, and 'Special Forest Goods and Products': Medicinals, Mushrooms, Decoratives and Edibles.

Ecoforestry Courses are held in the 420-acre forest of the Mountain Grove Center, a two hour drive north of the California/Oregon border, in southern Oregon. Mountain Grove is located in the 2,000 acre Woodford Creek Watershed, which drains off the north slopes of King Mountain and drains into the Umpqua River and then flows to the Pacific Ocean. The Mountain Grove Center and the Ecoforestry Institute have formed a partnership with the Glendale Resource District of the Bureau of Land Management to develop a Woodford Creek Watershed Landscape Design and Ecosystem-based Forest Management Plan which will serve as a public demonstration and teaching forest to demonstrate ecologically sustainable forestry practices on a watershed basis.

Summary

Education takes place in communities; it is the means for communities to continue. As Plotinus and Novalis recognized, education has an outward, social and civil, aspect as well as an inward, personal and self-revealing, aspect. Education has at least four ends:
1. the appreciation of the richness of nature
2. the comprehension of human existence
3. the understanding of the nature of human society
4. the training for a position in human society.

Education has become more universal, but its goal, the well-rounded individual, has been distorted by its fourth aim, training for the economy. Ethics, in the second and third aims, has been neglected, since it might limit or contradict its economic obsession. In fact, the first three aims are restrictive to a growing, industrial economy. With its emphasis on play, liberation, and community, an alternative education provided by the Ecoforestry Institute integrates all four ends.

Education in ecological forestry (ecoforestry) is a movement growing out of a circle of learners. Being a learner, as opposed to a knower, necessitates an attitude of humility towards life. If one enters a natural forest with an attitude of openness, one will begin to learn about the complexity which surrounds you. Forests are more complex than we can understand, much less "manage." While industrial forestry simplifies forest ecosystems by objectifying, quantifying and computerizing, ecoforestry does not seek to "manage" forests but rather to work with the processes and patterns of the forests. At its best, ecoforestry is able to use human creativity to augment the natural abundance of forests and other ecosystem types—and live from the increase. At least, ecoforestry seeks to minimize the destructiveness of human intervention as we harvest forest goods.

The Ecoforestry Institute is only beginning its third cycle of its Certificate Program for Practicing Ecoforesters, but already there are some 40 ecoforestry projects: doing forestry based on ecological criteria, which include all aspects of this new ecological paradigm. the cultural, social, political, environmental, economic and others.

When human beings cooperate spontaneously, because they understand what it means to be human, because they understand how to treat their places on earth, only then can education be considered successful and, perhaps, lead to a more peaceful and humane world.

Chapter 13

Why are Old Growth Forests Cut at All?
(A review of David Wilcove and J. T. Olson,
"The ancient forests of the Pacific Northwest:
A case study in conservation and economic development,"
in C. S. Potter, et al., eds., *Perspectives on Biodiversity*)

Second Thoughts
Although Wilcove and Olson mentioned several costs of preserving old-growth, they, and we, have always assumed that it is right to consider lost opportunities as costs. Yet, we are not losing logging jobs in old-growth areas because we never *had* logging jobs in old-growth areas; what the industry is losing is a chance to extend its unsustainable rush for low-cost fine wood to the few remaining old-growth areas. The jobs lost are lost due to automation and extensive ecosystem destruction, not because of *not* cutting forests that have been set aside for nonhuman purposes. This kind of economic fantasy is the result of market economics logic.

Oddly though, such logic does not extend to humans yet, although according to this logic, our body organs, at least those that have built-in redundancy like the liver, or duplication like the kidneys, could be profitably sold to hospitals for large quantities of money. (Which reminds me, did any of you ever see the movie, *Jekyll and Hyde — Together Again*, in which one of the characters sells one of his testicles to a billionaire—"Hell, I'd give my left nut for capitalism, but not both of 'em." [Yet, for an extra million, he does]) The price for eyes, hearts, kidneys, and other organs is extraordinarily high, and managing them as valuable resources is quite logical, yet almost no one sells their "extra" organs. Why not? Could health be more important than income?

All members of an ecological community contribute to the integrity of the whole, which is vital to maintaining what we humans consider important: visible animals, pharmacological sources, a moderate climate, and clean air and water (in a sense, nature has meta-economic values). Until old-growth is recognized as a vital organ in the functioning of ecosystems, as important as a human heart is in the functioning of bodies, it will continually be assaulted by the logic of economics, and it will not be safe until economics is placed on an ecological basis.

The idea that old-growth needs to be managed is based on the myth that all of nature is a resource to be processed. As Aldo Leopold pointed out, the weakness of relying on economic motives to determine value is that most members of an ecological community, such as fungus or songbirds, have no economic value—and those beings with no economic value can be labeled as weeds or pests and destroyed so that crops and animals with short-term advantages for human ends can be substituted. Until people are educated about the importance of old-growth, about the real costs of industrial forestry, about the immense losses of forests, and about their role in ecosystem dynamics, forests will have to be protected politically, by laws and duties.

Third Thoughts

Having discussed the long-term value of old-growth forests, the question remains as to why corporations liquidate them rather than keep them as long-term capital. The answer might be provided by recent economic history: After the second great depression starting in 1932-3, the US government tried to flatten out the boom and bust cycle of our capitalist economy by manipulating taxes, interest rates, and money supply, by subsidizing jobs, and through corporate welfare. Although the government discouraged monopolies, it did nothing to regulate oligopolies (control of the market by a few large companies); prices are set by the costs of production (including bloated bureaucracies and executive salaries), rather than by demand and supply in the free marketplace. As the automated production of goods became more efficient, goods-producing jobs declined. On the other hand a large, well-educated labor force (mostly women) was available at relatively cheap salaries. Information processing and human services, however, proved to be far less efficient than automation. Marvin Harris suggests that corporate bureaucracies wasted labor and lowered productivity faster than automation could save or raise it. Corporations were no longer efficient enough to expand their production out of sale-generated income and took on additional debt.

As economists point out, debt is inflationary because borrowing puts more money into circulation. Rising interest rates caused cash-flow problems, which many large corporations responded to by lowering quality and by passing the costs of borrowed money on to consumers—which allowed them to borrow more at even higher rates (which the government had raised to cool off borrowing). Corporations have gradually increased their dependence on deficit financing to supply capital. Harris notes that corporate debt has gone up 14 times while federal government debt has gone up only 3 times (as a percentage of gross national product). As more money was owed, more was paid to service the debt and less was available for cash flow. More short-term debts were taken on to keep up cash flow. Corporations borrow faster and pay more to borrow, but can keep raising prices to consumers so they can keep borrowing—or liquidate some of their assets, which is probably why Burlington Northern and other land-owners allow massive clearcuts: Cash flow needs due to long-term inefficiencies and massive short-term debts. The impulse to save a young squander-fast corporation may override the need to save old growth forests.

Chapter 14

The Long-term Implications
of Forest Certification

Certification is such a simple concept: the provenance of wood is made visible and known. Yet, it is part of a complicated network and has far-reaching implications.

Certification is driven by consumers, but consumers are reacting to forest degradation and destruction, which are partly caused by consumer demand for products made from wood. Consumer pressures have often influenced the extent of exploitation in a landscape. Consumers have indicated that they want forests for wood products, but also for habitat for other species, aesthetics, recreation, and other values. Certification itself is a demand. As consumer demands change, however, the uses of landscapes may change. Certification may bring about other far-reaching changes.

1. Certification will develop beyond a simple dichotomy, which recognizes two basic approaches to certification. The systems approach (1) verifies whether a landowner has adopted quality management processes that are consistent and can be improved. It is considered by forest industries to be more goal-oriented and less prescriptive that the product approach (and possibly less restrictive and more profitable, depending as it does on self-assessment). This approach is more in line with the efforts of the International Standards Organization, a business-dominated group.

The product (or practices) approach (2) measures specific characteristics related to the origins of the product with performance indicators (ecological, economic, or social)—the primary goal is the health and continuity of the forest itself. This type of certification involves contracting with a third party for an audit to rate performance in order to receive a seal of approval. It is associated with low-intensity and uneven-aged management. A Society of American Foresters study group called this a prescriptive approach (meaning rule-guided) in their April 1995 issue on certification. Prescriptive forestry is considered unnecessary by some foresters, since there are already federal regulations prescribing the practice of forestry in the US.

Since existing laws do not protect forests from being clearcut, however, something more is needed. The environmental management systems (EMS) offered in place of prescriptions only avoids conflict with traditional industrial forestry by ignoring the health of the forest for the meaningless continuity of the management process. It gives timber addicts a gold star on the forehead and lets them have a cheap fix

Many of the distinctions between these two approaches are based on misunderstandings or faulty logic. Furthermore, the product approach does not use a prescriptive logic but a proscriptive (prohibiting) one. That is, from 'what is not allowed is forbidden' to 'what is not forbidden is allowed.' Thus, in a product approach, it is forbidden to destroy the forest to get products; we must leave the components, structures, and functions that satisfy the constraints of survival and reproduction.

The product approach is associated also with low-intensity and uneven-aged management—the very systems that have forest health as their goal. Of course, a product approach must include good management, even if it is benign neglect, and its goals are far more ambitious than simple management.

The product approach (just certification hereafter) not only uses a proscriptive logic, but it is an inclusive program rather than an exclusive one, that is, it affirms rather than denies—it affirms the integrity of the forest and human needs. The product approach strives for balance within important ecological limits, which cannot be compromised. Ultimately, it is an on-the-ground approach that is sensitive to changes in the forest.

2. Certification will be predominately performed by independent groups, after strenuous opposition. Certification is classified by the US EPA as an environmental label, or "ecolabel," used to communicate information to consumers. Ecolabels can be issued by three kinds of certification organizations: (1) first party, those made by producers about their own products, (2) second party, which are endorsements made by affiliates of producers (often with financial interests), and (3) third party, where approval is issued by independent entities based on objective assessments.

Independent, or third party, labels seem most likely to make public participation part of their process (which has the advantage of increasing the independence of the approach). Third party systems, however, are criticized for making decisions based on arbitrary and unscientific indicators and criteria—this is in fact its strong point, since not all criteria are or can be scientific or nonarbitrary (unnatural). The forest not only contributes to human populations and the human mind, it is a living system. Third party systems are also criticized for removing the motivation for improvement once a "green" label is awarded, but there is no more incentive to improve in first and second party situations, other than lip-service to improvement.

Everyone (certainly all three kinds of certifiers) will be offering certification at first, until the differences are sorted out. What is certified will be different in many cases. The certification may be based on scale (local, regional, or global), or on the kind of forest (wild or plantation), or on managerial or ecological concerns. Eventually, some universally accepted principles will be held in common. More than basic principles, however, uniform definitions, procedures, and a common moral ground are needed.

Certification will be blamed, as the US EPA and ESA are now, for management failures and social problems. Certification will be seen as a restriction on the freedom to harvest at will. GATT and other agreements will protect destructive harvesting under the guise of equal access.

By denying the use of fertilizers or pesticides, the ecological product approach of PCC certification will seem to be antagonistic to production objectives. Those objectives, however, may be revealed to be detrimental to the long-term health of the forests.

Standards will seem to be set too high, and arguments will be made to lower them on country-by-country situations. The Long Beach Model Forest Society suggests that certification could lead to a "leveling [sic] of the playing field internationally." But they may just mean enforcing the lowest common denominator for standards. It will be argued that certification will accelerate deforestation in poor countries. And, under current economic practices it may. This means that the practices need to be changed at the same time for certification to succeed ultimately.

Certification will become more refined as it develops. At first it will seem that every aspect of forestry will need to be modified. Later, certification will identify particular areas where improvements can be made in management.

3. Certification will require more knowledge about the forest, as well as improved inventory and monitoring levels and techniques. Certification implies that a complete inventory must be taken of the forest to provide a baseline for further assessment. It also implies that monitoring must be continuous.

Certification requires much better scientific research to determine minimum, optimum or maximum numbers of features, i.e., the optimum number of overstory trees that can be removed, or the optimum percentage of trees to be left to maintain forest health.

Many unidentified and unmeasured ecological and cultural forest values may be identified and addressed. These values would then contribute to the importance of forests. With wild forests, many other incidental benefits that do not have monetary amounts associated with them can accrue.

4. Certification will encourage the recognition and acceptance of ecological forest limits, and possible reduce wood use. The uses of the forest may then be more compartmentalized, e.g., hardwood, pulp, poles, preservation of ecosystem values (hydrological and atmospheric services), recreation. Waste would be reduced due to the improvement of practices.

Certification acknowledges environmental limits and costs. It implies acceptance of limitations that cannot be expanded or erased or ignored. It implies acceptance and understanding of environmental costs. Yields may be lower to match the productivity of the forests. There will probably be an increase in the dollar costs of production.

The use of wood itself might be reduced on a per capita or total basis. A wood use reduction program could dramatically reduce the global draw on tropical, temperate, and boreal forests. Certification could identify and recommend optimal policies that promote improved forestry practices, reduce the demand for wood and wood products, and encourage cooperation between corporations and local communities. Certification could expose the roots of wasteful practices and propose improvements that could lead to ecologically sustainable practices.

Many demands are used to support unsustainable lifestyles, which increase the demands for more forest products. Certification should be tied to some extent to more frugal, but not necessarily less rich, cultural practices.

5. Certification may increase prices dramatically at first, but the prices, for the first time, may reflect the previously "free goods" of nature and the

"wage slavery" of workers. Independent certification may mean a disruption of business as usual, which has invested many millions of dollars in setting up plantations. Of course, tremendous one-time profits were realized in clearcutting forests to start the plantations.

As certified wood replaces or competes with plantation wood, prices may rise dramatically. However, these prices may also be reflecting the true costs (previously externalized) of growing trees for hundreds of years, as well as harvesting wood with unfairly recompensed human labor.

Catherine Mater's research indicates a majority of people would be willing to pay a premium price for certified products. Over 90 per cent of wood product manufacturers stated a preference for certified lumber for their operations. Over 70 percent of consumers said they were looking for independent certification as opposed to government or industry certification. Consumers also indicated a willingness to pay a premium for certified wood products.

Plantation wood has an advantage in the short-term: it has a positive cash flow. Wild forests often take hundreds of years to develop. Of course, many nonindustrial woodlands are fairly old—as are their owners (average age 67)—and may have started to acquire some old growth characteristics. These forests are the ones the players will be competing for with certification proposals.

6. Certification may cause a shift in forestry. The increasing market for certified wood may reduce the demand for plantation wood if the total demand is static. If per capita demand continues to grow, certified wild wood and plantation wood (certified or not) may both increase. If demand for certified wild wood exceeds that for plantation wood, plantations may be gradually converted to wild forests, either by natural regeneration (unlikely) or by new planting regimes.

Certain perverse incentives, such as massive road-building, may disappear. Other special industry subsidies, such as land grants and low-interest loans, may also fall as barriers to community forestry.

The kind of certification stressed by PCC may result in a shift from artificial plantations to naturally regenerated forests. Perhaps this shift is necessary, if plantations are doomed in a generation or two—some plantations cannot sustain more than three rotations of trees. This shift may protect forests in the very-long-term. This would mean in general, less human influence on and intervention in wild forests.

There may be a long-term shift from industry forestry to community forestry.

7. Certification will address concerns of human equity and social justice. Land tenure is one barrier to meaningful certification in many parts of the world. Poverty is another such barrier; as long as the compensation of workers depends on the style of the ruling regime, there may not be enough stability for certification.

Certification can contribute to a needful awareness among political representatives, who have been hypnotized by the flow of money.

Certification has the potential to increased desperately needed revenue for regeneration, management, and research. It has the potential to increase revenue to local communities, rather than contributing to the flood of cheap raw materials exported elsewhere. In a very real way, it can encourage community forestry.

Certification should also create millions of jobs and generate billions in expenditures. By comparison, almost 4 million environmental protection-related jobs were created in the US in 1992. Environmental protection expenditures totaled $170 billion, more than the largest US industrial corporation (GM) that year ($124 billion).

Finally, the public expects certification to assure that future generations will have equal opportunities to benefit from forests. Michael Toman in *Defining Sustainable Forestry*, suggests defining a "safe minimum standard" to address intergenerational fairness, resource constraints, and human scale in place of some vague, unenforceable intergenerational contract. Damages to a forest, for instance, are characterized by two attributes—expected cost and degree of irreversibility, which can be treated economically. The standard draws a fuzzy dividing line between the "moral imperatives to preserve and enhance natural resource systems and the free play of resource tradeoffs."

Living generations are responsible for limiting their actions within a reasonable framework of cost and irreversible change. The standard requires conversation and some consensus about the limits, which are never exact. Ecological value has to be balanced with socially optimal resource allocations (that consider future generations as well).

Summary
Certification is one way for consumers to discriminate against forest products that do not enhance or are not "friendly" to other values. Ecologists may see certification as a way to protect forest ecosystems and their nonmarket benefits, such as cleansing water and air. Foresters may see certification as a justification for labor-intensive (nonindustrial) forest practices. Business people may see it as a way to sell more wood or furniture. Politicians may see it as a way to avoid the social conflicts of limited resources.

Certification is a way for ecologists, foresters, managers, and retailers to interact with consumers to define and refine demand. But, it cannot be isolated from the large-scale economic and political trends that shape human culture. Problems are part of a matrix of industrial social practices and policies, but there are now pressing economic and ecological reasons to revamp them.

To be successful, certification must address economic shortcomings and industry expectations. Recognizing that some attitudes and problems lead to unsustainable forest practices and demands, certification can contribute to greater efficiencies and reduced demand.

Certification can identify issues that must be examined to bring about sustainable forests and societies. The concept of sustainability is being incorporated into every level of forest planning and management, from forestry cooperatives to global strategies. The concept of sustainability can be used to develop practices that decrease depletion and waste and reduce the threats to future generations. One goal of an independent product approach to certification is to establish a framework within which meaningful revisions and realignments can be formulated and examined.

In summary, forests are critical for human welfare. They provide materials that we consume—of which timber is a small part, as well as provide recreation, historical and mythical values, immense ecological services from water flow to wildlife, educational and scientific opportunities, and land for living and grazing.

The language of certification must address economic and management concerns, but it can never forget that the forest is a living, breathing being, and not a resource or a machine. Certification works to let us see forests as living systems composed of interacting beings and material components in self-maintaining cycles. Furthermore, certification contributes to the image of forests as one ecological system, connected by water, birds, weeds, humans, and numerous other elements. This is the image we need to guide us in our work in forests.

Chapter 15

Reforestation Goals—Are Planning & Gambling Enough?

Deforestation, with forest decline, is a planetary crisis resulting from numerous cosmological, social, economic, and demographic events. The crisis must be addressed through changes in policies and institutions. Saving the remaining wild forests is urgent, but not sufficient; massive reforestation is required—replanting, regeneration, and the restoration of all forest components, from soil and microorganisms to herbs, trout, and wolves. And, that will require the involvement of communities, in addition to universities, governments and individuals.

Although forest cover in industrial countries is more or less stable, it has been declining in the US and Canada since 1963. It is declining dramatically in the southern hemisphere. Even in countries partially reforested in the 1960s, like India and China, forests are being cut again unsustainably to improve temporarily the standards of living (see the Mangrove Action Project news).

A lot of planting is done in fact by industrial forestry—but it is done primarily to increase supplies of pulp rather than to reforest areas that have been cleared. Some industrial forestry advocates ask if planting is even necessary—after all, almost two-thirds of the original forest area of the planet is still forested, although a far smaller (4?) percentage is in primary forest. Ecoforestry advocates ask how much of the planet should be forested. Answering questions like these is the domain of planning. Planning in general means deciding on goals to be achieved in specific situations.

Planning becomes more important as demand increases. According to Worldwatch, the world's commercial timber harvest expanded by more than 50 percent since 1965. In the Pacific Northwest, in the first half of the 1980s, timber harvests on industry-owned lands exceeded the sustained yield by 25 percent—by 61 percent in US national forests (sustained yield, remember, is a subset of sustainable forestry, since it refers only to stocks of wood and not biological diversity and environmental services of the forests). Part of the problem is that few planting goals and almost no permanent goals for coverage, are ever set, by anyone.

For central planning by state or province or federal governments, the goals are usually small and not comprehensive, such as a cutting level or a single species preservation, and usually end up being a compromise in cost-benefit analysis. For ecoforestry the goals are large and comprehensive—to the extent that small goals will be dictated by the large ones.

Goals, in one sense, are horizons that we travel towards, but in another sense they are the tools that shape us as we travel—as we are one of the tools, along with natural disturbances, pathogens, and other agents, that shape the forest. In industrial forestry, global and national goals are often more important than local goals. That is true for ecoforestry only to the extent that global cycles are to be protected; for economic reasons, local goals must be treated as more important than regional or global goals.

The reforestation goals for ecoforestry are divided into three levels of implementation, from the local to the global. Some of these goals are common sense, some will require legislation, and a few may require behavioral or cultural changes.

Local Goals

Forests must be restored at the stand level, with connections to the landscape level. The local watershed must be restored, with connections to larger watersheds. Local human communities have to participate in planning and planting for any restoration to be successful; they inhabit the forest, also.

For example, our Moscow office is in a *Pinus ponderosa/Physocarpus malvaceus* association in the Pinus Zone in the eastern Palouse Hills in the Columbia Basin physiographic province in the Coniferous dry woodland and xeromorphic scrub biome; it is within the Paradise Creek watershed within the South Fork Palouse River watershed, which is part of the Columbia River watershed. Paradise Creek has been stripped of trees (by the first Caucasian homesteaders); other forested areas have been cleared for agriculture, and most recently, for construction.

Suitable goals for this local area include: Replant trees in all riparian areas, with at least 30-meter buffers. Reforest formerly forested agricultural lands in the foothills. Preserve all small old-growth areas—no plantation can be a substitute for keeping the remaining old-growth, and no secondary forest can match old-growth for biological richness or ecological importance. Protect the long-term health, integrity, and ecological balance of forests, that is, the ecological and evolutionary processes that make forests.

Protect the health of human communities. Broaden local economies from timber cutting—timber employment fell 15 percent between 1979-1989, during a time of record cutting levels and record corporate incomes—to alternative products. Set up cooperatives to refer and share work. Combine forest and agricultural crops where appropriate (in tree crops or permaculture). Tax timber and products so that benefits stay local. Expand local capital in forests.

Educate all people to feel their connections to the forest, because, until they feel them, they will not act ethically or ecologically. Educate people to realize that long-term sustainability requires healthy forests, and that protecting forests protects jobs

Regional Goals

Planning at the landscape level is necessary to avoid wobble at that level (wobble is the instability that can lead to dysfunction). Planning is critical to reduce habitat fragmentation from clearcutting patterns. It is also critical to reduce the cumulative effects, such as erratic streamflow and climate change. And, it is critical to design and zone functioning reserves and preserves.

Specific goals would include: End logging in national forests; in the US, for instance, Clinton's 1993 FEMAT reported that the only way to ensure a chance of maintaining viable populations of all species in Pacific Northwest coastal forests was to halt all logging, whereas the approved Option 9 (from outer space?) allows for the extirpation or extinction of 800 species.

Reforest areas large enough for minimum critical habitat or minimum viable populations of the native species necessary for native biodiversity (virtually no existing areas are large enough, even Yellowstone Park in the US). J. M. Thiollay concluded that, to maintain a complete bird community in French Guiana, with raptors, rainforest reserves would have to be between 1 and 10 million ha (1 ha = 2.47 acres). In Canada, for viable populations of wolves and elk, 10 million ha should be set aside.

Restore forest cover in North America to pre-European levels. For the US, replant 142 million hectares of forest lands—up to 438 million hectares total (the approximate level of cover in the year 1600). For Canada, replant over 80 million hectares of forest, up to 530 million hectares total. For the Northwest (Pacific NW coast forest, US and CND), replant up to 47 million ha. For the Northeast (northern hardwoods), replant up to 11 million ha. For the Southeast (Oak-pine forest), replant up to 129 million ha. For other ecoregions, replant to a high percentage of 2000 BC levels, especially around the Mediterranean.

Global Goals

The Worldwatch Institute estimates that to supply current global demands for fuelwood, lumber, and pulpwood, 9 million ha of trees need to be planted each year starting in 1995; for soil and water conservation, another 6 million ha (at an estimated cost of 6 billion dollars); and 110 million ha just to catch up with cutting (these trees would also sequester 700 million tons of carbon, reducing the greenhouse effect).

Global goals include: Reimplement international initiatives to slow deforestation—the UN notes that previous initiatives *accelerated* deforestation, as in Cameroon, where log production is to double in the forest (home of 50,000 Pygmies with a unique and valuable cosmology and life-style). Employ initiatives to reforest areas, such as in Burundi, where only 1% of the forest is still virgin; at least 1,296,000 ha of forest should be restored. Plant and maintain forests sufficient to guarantee indefinite support of known and unknown global biogeochemical cycles. Protect fragile ecosystems with global importance. Reduce threats to forests from acid rain and other nonpoint-source pollutions. For the planet, this means reforesting 1.4 *billion* ha to restore the 30-40% forest cover removed in the past 3000 years.

Reforestation goals only make sense in the context of other ecological or social/cultural goals, such as reducing demand for wood products, increasing efficiency in use, fitting human populations to biological limits, and educating people about the roles of forests. Personal goals, such as ensuring personal security and fulfillment, living frugally, or questioning industrial practices, are also relevant to the goals of the profession.

Conclusion

The importance of forestry goes beyond profits and employment. For humanity, trees have always been a place to hide, a place to go and live outside of society. People in traditional societies or first nations lived in the forest as home. The poor in industrial countries have always used forests to make their homes and heat them, to cook food and make tools.

Most forests are now under stress. Forestry as a profession is under stress. There are separate angry, noncommunicating groups, such as foresters and preservationists. Ecoforestry, however, is not anti-forestry or anti-preservation. Furthermore, ecoforestry is not just a reaction to industrial forestry; rather, industrial forestry is one small perspective in the framework of ecoforestry (although industrial forestry has been distorted by excess profits and economic fears). Ecoforestry is a framework for forestry—it does not center on anything, it perceives and comprehends the whole forest system, centers and peripheral. The forest is polycentric anyway. Ecoforestry is based on a more comprehensive logic than the two-valued, dichotomizing logic of forestry and on a deeper philosophy than savage capitalism. Ecoforestry, as Arne Naess has said, is spacious enough to accommodate the viewpoints of all these groups, because it has emerged from a broad scientific and philosophical foundation. Ecoforestry is the first forestry to consciously consider multiple scales and other dimensions, as well as limits as well as other scales and other dimensions (history, ethics, and equity).

Even today, our ignorance of the forest is incredible. We are not sure if mycorrhizal fungi can survive without squirrels; we are not sure what the productivity of the forests is or whether we are cutting the right trees or too many of them. We do not have facts to base our actions on. Nature is a stochastic process, always changing; forests are always changing. There is a profession, however, that acknowledges the operation of chance and makes conclusions in the absence of facts: it is called gambling—few people are successful at it. Even though it is guided by a better ethics and practices a more benign, sustainable exploitation, ecoforestry is also gambling. We do not know for sure what effects our actions, even individual tree selection or preservation, will have on the forest. The proper virtues for gambling with nature are humility and courage, not arrogance and fear. With these virtues, eventually we might develop wisdom.

A fir tree, like a pine marten or bark beetle, has biological wisdom; it is clearly self-making, self-governing, and self-choosing (within limits of its treeness). Forests are books of biological wisdoms. Through our cutting and decision making and overuse, we are tearing up the sacred books, literally leaf by leaf. Rather than telling the forests what to do, rather than controlling their growth, we need to watch forests to see what they do (this used to be the function of natural history), and we need to let them do it. Abraham Maslow called this attitude as "taoistic," and the way to forest health is letting the forest do most of the choosing and working. A taoistic approach to forestry means being nonabstracting, nonimproving, nonintrusive, and noninterfering.

The current trend in desacralization is a human defense mechanism against the loss of meaning caused by industrial civilization. Ecoforestry is concerned with resacralizing forests, with restoring them to their extents and grandeurs, by regrounding science in ethics (as ways of living together), by changing our attitudes from utilization and flat efficiency towards awe and appreciation, and by participating with our hands and hearts in the lives of forests.

Chapter 16

What is a Forest Worth?

How do we even measure the value of a forest—for us, for other species, for itself? Considering economic values alone has become complex. Many economic kinds of value are difficult to quantify. For instance, until people were asked in surveys how much they would pay to have wild forests, existence value (that is, just knowing that a forest exists without having to visit it) did not have a dollar value. Some economists have estimated also numerical amounts for *option values* (retaining options for the future) and *bequest values* (leaving as-is for future generations). For instance, a typical Colorado household was found to be willing to pay annually $4.04 option value, $4.87 existence value, and $5.01 bequest value for 1.2 million acres of wilderness in Colorado.

The kind of value measured most often for forests, other than stumpage, is recreation. The costs of recreation can be identified and added up; although the forest is a "free good," transportation, equipment, or entrance fees are often required to visit it. A typical visit may cost $15.00 a day in 1992 dollars. But this does not reflect the value of the experience. Some economists have proposed a "contingent valuation" by subtracting the actual costs from what the consumer is willing to pay—this surplus value makes the recreational value more objective, because there is a dollar amount.

Another category of economic value to consider is *conservation value*. Natural capital such as forests function to regulate climate, produce topsoil, and cycle elements through the ocean and atmosphere. These environmental services (the wild infrastructure) support the economy without providing direct economic benefits.

Consider, a "what-would-it-cost-to-replace-it" game that analyzes the replacement costs of natural services in terms of human labor and technology. Buckminster Fuller once calculated that it would cost just over a million dollars per gallon to manufacture gasoline using chemical processes and electrical power (California Con-Edison 1972 rates). The following list from a 1994 Ecoforestry course is a thought experiment with very approximate numbers.

Table 5. Costs of Replacing a Forest

Function	Human cost
Pure water	$0.70/liter
Pure air	$0.04/cubic liter
Climate Moderation	$22,000/day
Wind protection	$6000/hectare
Wild genes	$11,000,000/gene
Recreation	$2,000,000/park
Flood control	$24,000/hectare

For a small forested watershed (and airshed), the costs of replacing basic forest services would be billions per year. One thing business can do is put a price on nature, but it should be a real price, reflecting the real cost of replacement; for example, the costs of raising and planting trees, and then monitoring them.

Just last year, at the annual meeting of the American Association for the Advancement of Science, scientists were told that the goods and services provided annually by natural ecosystems are worth many trillions of dollars in conventional economic terms, according to Stanford ecologist Gretchen Daily. These services are the life support functions performed by ecosystems, such as purification of air and water; detoxification and recycling of wastes; generation and maintenance of soil fertility; pollination of crops and other plants; regulation of climate; and mitigation of weather extremes like flood or drought. Ecosystems also provide goods, like timber, whose harvest and trade represent an important part of the human economy.

Ecosystem services operate on such a grand scale and in such intricate and little-explored ways that most could not be replaced by technology. They are priceless, but because they do not have a price, they are not traded in economic markets or considered in calculations. Government and industry urgently need to incorporate these life-support values into their policies and planning. Estimates of the value of nature's goods and services is critical to informing decision-makers.

Yet, it is as impossible to capture the value of a forest as it is to define a human life in strictly economic terms (either as $0.98 or $78,000). Even assigning monetary values to nature, homes and lives is problematic. Ask anyone who has lost a family member, home, or favorite woods if the money gained was reflective of the real value. Many people do not allow their ancestral homes to be assigned even the highest monetary value, since they have the "right" to occupy them. For example, Thai *muang faai* communities refuse to let sanctions against cutting trees in community forests be interpreted as prices on the trees, which are necessary to supply water to rice fields. Many people in industrial countries refuse to participate in questionnaires that ask them how much money would be acceptable to compensate for the loss of visibility by air pollution or the loss of forests by clearcutting. Refusal to discuss the price of a forest is legitimate communication; it should serve to channel forest policy into larger noneconomic realms.

Even if we cannot put a discrete amount on aesthetic values, common mathematical sense tells us that as the dividend (number of forests) approaches zero, its worth goes to infinity—no one can experience it if it is gone. The infinite value of forests is only a constraint on economic exploitation and reasoning.

All the true costs of any technological process must be internalized even if it means assigning arbitrary dollar values to resources. This does not necessarily mean that an aesthetic resource is worth $1 billion, but that there is a potential economic and spiritual loss to the system if the resource disappears. Value systems are the driving variables in all economic systems, not just peripheral attachments. And patriarchal nations are devaluing these values and cultural wisdom with their dominant, abstract, quantitative economics. Economic objectives have to include new concepts of value.

We get our values from knowing what is valuable in nature. Values usually encode information having survival or prestige importance. Perhaps the most valuable thing is living time, the experience of life—aesthetics (from the Greek meaning perception). This may be why humans value walking in the woods and acquiring rich sensory experiences from direct contact. But economists rarely mention these values. Light, wind, dirt, plants, birds, all act during a walk, but not with the meaning of crops or sheep, which is for their utility—they just are. People do not live without these things. All values are based ultimately on a healthy ecology. It must be kept healthy.

Chapter 17

Can Forestry Stand Alone?
Can Forestry Afford to Wait?

Throughout this journal [*Ecoforestry*], we have talked about saving old growth forests, but not about plans to restore them ecosystematically, regionally or globally. We have talked about sustainability, but not about the endurance of all beings in a forest. We have talked about sustainable consumption, but not about reducing both demands and populations. We have talked about the use and abuse of corporate power to decimate forests, but not about strategies for revoking the charters of corporations and reforming them as the public service groups they were originally licensed to be. We have talked about rural community empowerment, but not about meaningful community anarchy. We have talked about national processes, but not about a second Jeffersonian revolution (in the US at least). We have addressed international trade and various political problems, but not the revitalization of the United Nations with the power to conduct immediate disarmaments, to guarantee equity in education, and to support for nations based on traditional cultures and values.

There is no doubt that we are tending towards global markets in a global society. The crucial thing is to have a global society composed of vital local societies. Without local energies, the global will become dull and rigid; without the global, the local will become dull and separate.

One way to integrate our concerns and the two levels would be in a Eutopian (the good interpretation from Thomas More's pun on Greek words meaning good place or no place) structure. A Eutopian structure is a complete environmental protection and human survival package. Where activists play for stands or forests, or industry plays for regions, Eutopias plays for the planet and the solar system. Ecoforestry, and other separate efforts, cannot work well without major shifts in economics and politics, and international relations. Why not try it all at once?

We do not know what will happen in the future. We are not sure whether we have passed some critical threshold with forest health or amphibian populations, with chemical waste or carbon dioxide. We cannot make long projects, nor can we seem to understand the complexity of a myriad of unknowable and unseen circumstances or to peer very far into any future.

There are things we can do daily, besides living well and enjoying life, the first of which is to promote inner and outer change in the present. We need to challenge unsupportable greed and lifestyles now. To learn how our life-support systems work, then become politically involved and push our leaders in the right direction, and always protest racism, sexism, religious prejudice and the gross economic inequity that make it so difficult to preserve and restore the natural services upon which we depend. To keep good cultures. And, to make good places. We have to do it without fear or guilt, and with love and peace. We have to do it now.

Chapter 18

Lies, Dirty Lies, and Dirty Photographs

The book, *Clearcut: The Tragedy of Industrial Forestry*, published by The Sierra Club, and sponsored by the Ecoforestry Institute and the Foundation for Deep Ecology, has made many people "angered and saddened" according to a booklet, *Closer Look: An On-the-Ground Investigation of the Sierra Club's Book, Clearcut*, sponsored by the American Forest & Paper Association. Although the people quoted were affected by its "misrepresentation," probably everyone else who read it was angered and saddened for other reasons having more to do with destruction, ignorance, and greed. *Clearcut* has been reviewed favorably in *American Forests*, *World Watch*, and the *Whole Earth Review*. Like the dueling reviews in the *Journal of Forestry*, however, the AFPA booklet is primarily negative. This orientation can be understood when we examine their assumptions, goals, and simple errors.

Exposing Themselves
The anonymous authors of *Closer Look* state that, "After brief examination, foresters ... felt that many of the photos used in Clearcut were a distortion of reality." They requested foresters around the United States to review the book and identify areas with which they were familiar. A book like *Clearcut* should be examined critically, and detailed analysis is a good idea. *Closer Look* tries to do this.

After an exhaustive search, they were able to locate 10 sites, which they present as "healthy, diverse forests." In these 10 photographs, they claim that the Sierra Club "falsified" the record or presented "misleading" information on forest conditions. For example, they state that areas described as industrial deforestation "were in fact salvage cuts following devastating wildfires and insect or disease infestation." Perhaps an even closer look at the history of those sites is needed; some of the wildfires and infestations were caused by previous cuts. Victor Menotti, of the Foundation for Deep Ecology, a sponsor of *Clearcut*, is preparing a detailed response for each site mentioned.

Although their primary purpose was to expose the Sierra Club's misrepresentations, they included a short piece by Thomas Bonnicksen defending modern forestry and the proper use of clearcutting. The 10 photographs are examined on 20 pages. There is not very much documentation, beyond anecdotes. Even if all 10 photographs were errors, the remaining 140+ photographs in *Clearcut* offer an undeniable indictment of clearcutting.

Most of those 10 photos are not errors, however. The AFPA admits that they contain clearcuts, either old clearcuts or recent relatively small clearcuts of 75 acres. Their real complaint seems to focus on the context, on the fact that not every square centimeter shown in every photograph was an intentional clearcut.

This book erroneously states that people opposed to forest management "further their agenda" by focusing on clearcutting. It never occurs to the authors that people are opposed to clearcutting *and* in favor of better forest management. They also state that the tragedy resides in the "many distortions, omissions, and misinformation" in *Clearcut*, and that the Sierra Club has chosen "fiction over fact" and "hyperbole over reason" by discouraging rational discussion (but, fiction books rarely have photographs when you think about it).

Closer Look admits that some photographs do depict bad practices, but claims that such practices are illegal and no longer used. Then, they defend clearcutting by using the familiar and fallacious argument that, when properly applied, clearcutting "mimics nature" but without the "lasting and wholesale destruction" of hurricanes, volcanoes, and fires.

Specific Sites

The first photograph they analyze is of Mt. Shasta. What is the native vegetation of the west of Mt. Shasta? Is it pine forest or mature manzanita chaparral? Was it cut before 1900? The areas shown seem oddly geometric for a topographic or edaphic maturity.

The West Maine Mountains were clearcut in the 1930s, so inclusion of that photograph in *Clearcut* seems justified. *Closer Look*, however, after admitting the clearcut, blithely points out that half is still being clearcut (as recently as 1991), but in 75-acre patches with irregular shapes to "mimic nature." The 60-year old firs are described as "past their prime" and if they are taken for important paper products, a "new forest will appear." Of course, a new forest would appear if they were not taken. Tiny patches and narrow corridors have been left, but one wonders about their effectiveness. Too bad the money that went into this book was not spent on research on minimum effective corridor size.

The Blackfeet Indian reservation forest in Montana is another site that was cleared for salvage in the 1960s, but *Closer Look* claims it is fundamentally different from a clearcut because the *reason* was different. The text assumes that lodgepole pines live in even-aged, single-species forests, and that only by human intervention can a new forest be regenerated. Bad assumptions. George Wuerthner has written a much more detailed response to these claims regarding this site and Houghton Creek.

On the northern California coast, "nature visited" this area and burned parts of the forest. One wonders where nature is now—visiting some other forest perhaps. After nature's visit, "rehabilitation "was completed within two years." Yet, it is unlikely that the forest is the same as it was before the fire, with old-growth redwoods. And, it is equally unlikely that nature required the roads in the photograph for her "visit."

In the Targhee National Forest, bordering Yellowstone, the Forest Service clearcut to reduce the fire hazard of beetle-killed pine. *Closer Look* contends that the clearcutting here is the proper management technique because it "imitates the natural life cycle of the lodgepole pine." Where in the natural life cycle does every stem get carried into an urban ecosystem? If this is a healthy forest, as the anonymous ones claim, why can we not see any trees? *Closer Look* compares this clearcut with Yellowstone, where the Park Service did "nothing." Yellowstone is called a "catastrophe" where the scars are terribly evident today" and recovery is "many years away."

They make the mistake of regarding the 1988 Yellowstone fire as a terrible disaster caused by the absence of forest management—even though fire suppression and fragmentation are the results of forest management. Furthermore, they seems unaware that these fires are natural events on 200-300-year cycles in Lodgepole pine forests, and that management made very little difference to the extent of burning. It was not a disaster ecologically; 45% was burned, and its remarkable "recovery" has been documented in the *Journal of Forestry* (11/94), as well as in *National Geographic* and *Current Health 2*. They do not seem to understand the important roles of fire: regulating oxygen on a planetary level (according to James Lovelock), reducing the acidity of and increasing the organic matter in forest soils, and increasing species diversity (especially nitrogen-fixing legumes under the canopy).

They ascribe fragmentation of Oregon's Blue Mountains to "unproductive soil conditions" and southern exposures, and not as a result of harvesting. Soil conditions do not cause fragmentation, which is a human artifact, but they do shape natural mosaics. Too often, harvesting on fragile sites ensures that the forest cannot come back. They condemn the Clearcut editors for implying that practices, such as salvage harvests following insect infestations or restocking poor sites, represent bad forestry. In fact, many salvage harvests of insect infestations and bad plantings are unnecessary and do represent bad forestry practices.

On Weyerhaeuser land south of Mt. St. Helens, the harvest rate in one drainage was increased to compensate for timber destroyed in the eruption. (Was clearcutting the harvest technique used?) Thus, "nature" placed "extra burden" on lands not destroyed directly. What is wrong with this sentence? Weyerhaeuser replanting is said to contrast sharply with part of the blast zone not rehabilitated. Although some areas still seem devastated, the recovery of much of the land around Mt. St. Helens has been documented in scientific journals, such as the *Journal of Ecology*.

The criticisms in *Closer Look* are based on the many unconscious myths, logical fallacies, kinds of ignorance, and professional agendas that underlie their activities.

Unconscious Myths

Closer Look purports to address the myths that surround clearcutting and forest management, but it never even acknowledges the myths on which its own outlook is based. Myths affect every element of our individual, social, economic, ecological, and spiritual lives, usually all at once. For instance, in the nineteenth century, a central myth (founded at the 1851 exposition) was that industrialization would bring universal peace. Did it? Violent confrontations are escalating. Later, in America, the myth of the frontier (as the physical frontier was being closed in the 1890 census) promised that all Americans would prosper in the frontier way. This myth of limitlessness has become stronger than the logic of limits. Corporations and banks have used the myth to offset the dismal flavor of their economics. The victims, especially loggers and truckers, have been tricked into passionately defending their own exploitation and that of the environment.

Two other myths, progress and nationalism, arose with the industrial revolution. Aldous Huxley described progress as the theory that one can get something for nothing; progress assumes that all consequences could be foreseen, and that the ends justify the means, even theft, destruction and murder. Nationalism is the theory that the state was the only true god; all others, especially other states, were false. Other old myths, such as " man is lord over the earth," have proven to be dangerous. They should all be retired to the scrolls of dead myths.

The myths of the mutant modern economics have tremendous impacts on how forests are treated. The old analogy of the economy as a machine leads to unhealthy and dangerous myths about forests:

- everything is a resource (and its corollary, everything has a price [and its corollary, everything that does not have a price is worthless]): Every forest can be cut to provide wood for human needs. The essence of a resource is that its existence acquires meaning only as it is necessary for human needs and wants. Furthermore, if everything is a resource then everything can be used—the forest, not just the trees, can be used.
- resources are unlimited: If forests are unlimited then we can keep cutting them (and even if they are not unlimited, by the tenets of modern economics, being a scarce resource makes them even more valuable.. Although many economists admit that forests may not be exactly unlimited, the assumption surfaces again in advertisements about the forest industry, where the industry credits itself with planting billions of trees.
- the future is less valuable than the present (discounting): A forest that may have some value now in a few useful species is worthless beyond a certain time (two years, maybe 20). Modern economics has enshrined this one form of selfish behavior and pretends that it is rational and optimal. Unfortunately, as Costanza and Daly have pointed out, short-term self-interest is usually inconsistent with the long-term best interests of the individual or society. Economic discounting is not in the best interests of keeping forests complete and healthy.
- the myth of innocence (after Patricia Limerick), or it's not our fault— since it is not us, we can blame a natural catastrophe, sudden market forces, government trends, or the stars. Give us subsidies, or let someone else pay for the damage.

By contrast, Bonnicksen erroneously claims that ordered, self-regulated, stable forests are a myth, that "nature's clearcuts" constantly created new forests. Wild forests are ordered, self-regulated, and stable; there is evidence that many forests have been stable for tens of thousands of years, especially in the tropics. Forests do change dynamically, but the changes are usually too long for human observation. For instance, palynological evidence shows that 5000 years ago alders went into a rapid decline, becoming rare for 1000 years, before recovering over another 1000 to their original levels.

Curse of the Fallacies
The arguments of the anonymous ones have many errors and logical fallacies. They use *ad hominem* arguments to try to undermine *Clearcut*. Name

calling by any group is a traditional invalid argumentative device; it occurs in *Clearcut* and in *Closer Look*.

The fallacy of complexity occurs throughout—where arguments have multiple assumptions, and attacking one has the appearance of attacking others, which may be adequate. When *Clearcut* attacks clearcuts, Popovich claims that forest management is being attacked. When *Clearcut* points out problems with the inbreeding of education and industry, the anonymous ones claim that forestry schools and the profession are being "denounced."

There is a fallacy of the false dilemma, where only two alternatives are claimed and one is considered unacceptable, as in: Either A or B; not A; therefore B. Obviously, there are more than two possibilities. Preservation is not the only alternative to clearcutting; the Ecoforestry Institute has been promoting other alternatives, such as individual tree selection.

Begging the question (*petitio principii*) is a popular fallacy. In this fallacy the premises are insufficient to establish the conclusion. For instance: 'When a forest has an insect infestation, it is unhealthy; Spruce bud worm is killing Umatilla National Forest, therefore the forest must be clearcut.'

The anonymous ones all use a logical fallacy of the appeal to authority (*argumentum ad vericundiam*) by constantly referring to foresters as having all the facts. For example, one forester exclaimed that he would be "proud" to hang a picture of the excellent management of the Rangeley Lakes harvest on his wall.

Flight of the Euphemisms
Much of the discussion of clearcutting and industrial forestry takes place on a euphemistic level. A euphemism is just a polite way of expressing bad news (from the Greek words meaning 'to speak the good'), thus a dying patient becomes "vitally challenged." Many euphemisms can be spotted in reviews of *Clearcut*:
- "birthing scene" instead of total destruction of an ecosystem (not yet replanted) In his review of Clearcut in the Journal of Forestry, Beaufait states that the editors have chosen to focus only on the "pain and bloody residue of birthing scenes—on the primal scream rather than on the future forest at their feet." A clearcut is not a birth scene, especially without parents or inheritance or loving attention—the death of a forest is not necessary to ensure the life of a new forest.
- "misinterpretation" meaning that the good aspects of clearcuts have not been considered
- "misinformation" for a disagreement with prevalent views on forest management. In his review in the *Journal of Forestry*, Luke Popovich uses this phrase.
- "overmature" meaning a valuable, large tree
- "crop" instead of wild long-lived trees in a wild ecosystem
- "rotations" instead of successive clearcuts
- "thin" for logging
- "salvage" for clearcutting

Bonnicksen, like many foresters, is stuck on the idea that industrial forestry is restricted to the agricultural model, where natural vegetation is replaced with a crop. Clearcuts, which are based on this model, are constantly referred to in *Closer Look* as "not pretty" or "ugly." They recognize that clearcuts are ugly, not because they are disorderly, as the reviewers think, but because the structure (and therefore the function) of the forest has been destroyed to remove the boles of trees.

Joseph Meeker judges that a burned forest is ugly because it is truncated; but it is as beautiful as a baby, and for the same reasons: Potential, development, and being in the process of renewal. Our sense of beauty, of what is intrinsically meaningful, dominates our grasp of what is real. Until we understand that all phases of the forest have beauty, we will not appreciate the meaning of the process. Understanding beauty could even allow a deeper understanding of the utilitarian. For example, what is the use—or the beauty—of a burned forest, as related to the function of lightning or the planetary carbon cycle? Lightning is essential to life by breaking down amino acids into ammonia, methane, hydrogen, and water. Burned forests are necessary to maintain diversity, among other things.

Other euphemisms clog our arteries of communication. Industrial forestry itself is sometimes a euphemism for killing forests (silvicide) for profit.

Flavors of Ignorance

Clearcutting is presented as mimicking natural processes. It does no such thing. Mimic means to 'look the same as' without having the same function; for instance, sweet-tasting butterflies that look like sour-tasting butterflies (to the eyes of predators) live longer and breed, a superficial biological resemblance gives possible survival advantage. The intent of silvicultural practices is to "mimic" the development of a wild forest—but speed it up for economic convenience. The function of a forest is to "be" and to provide places for participation of the beings that make it up, whereas the function of an artificial forest is to be cut. Artificial forests that look like wild forests (to the eyes of the public) are pronounced the same, even if the former only live for 25 years. In a sense we are unconsciously mimicking because we put in enhanced trees without a supporting web of relations. We should not try to mimic in this sense, but to preserve the functioning of the original forest.

What the reviewers are trying to say is that clearcutting is the same as, that is, has the same function as, natural processes. This is not true either. Clearcutting is the large-scale removal of a significant part of the biomass, destroying the structure of the forest. Fire does not "sweep" away old trees. We cannot duplicate the structure of the forest in a tree plantation because we are uncertain about all of the parts of the structure. Far more than natural events such as fires and hurricanes, we are cutting, clearing, and wasting at a scale that threatens the stability and existence of forests. At the same time we degrade habitats, rapidly add novel elements to the forest cycles, compact soils, cause transient perturbations in energy relations (from burning), manipulate species, and interfere with wild species. None of these things by itself is exclusive to humans as a species, but they are excessive, rapid, compounded, and very large-scale.

The reviewers state that the *Clearcut* editors attribute natural catastrophes—wildfire, storms, insect infestations—to the perfidy of forest management. Many catastrophes only seem natural because of our ignorance of our effects on forests; 90 percent of fires are human-caused, according to Charles Stoddard, and various percentages of blowdown and insect infestations result directly from poor management techniques.

Catastrophic disturbances are not nature's clearcuts, as Bonnicksen and the others emphasize. Wildfires, which are presented as "nature's clearcuts" in the discussion of Houghton Creek, are haphazard and partial—rarely burning more than 10 percent of a forest, unlike a clearcut, which is a complete with removal of stems on a short cycle. Even catastrophic disturbances like hurricanes rarely damage more than 5 percent of a forest. Furthermore, damages from hurricanes and fires are not "wholesale" and "lasting" as the anonymous ones claim.

More than being agents of mortality, insects, diseases, and animals are native components of complex food webs in ecosystems that contribute to the selection of certain kinds and ages of trees (determining the composition of the forest, which changes over time). Insects pollinate some trees and overwhelm others—rarely more than 1 percent of a forest. Diseases remove stressed trees, also probably a low percentage on the order of 1 percent. The effect of these disturbances on the long-term health of a forest can only be regarded as positive, unlike a clearcut, which destroys the forest.

In general, industrial forestry denies forest processes in an attempt to control and regulate production for economic reasons. By trying to prevent one kind of mortality, industrial forestry merely sets up another kind. Ecoforestry, by contrast, accepts a typical percentage of death as the normal condition, necessary for the renewal of the forest. The rate of death per year in an old forest is remarkably consistent at about 1-2 percent, even with wind storms, fires, disease outbreaks, and animal damage. In spite of Boise Cascade's recent television advertisement about our public forests ("Let our public forests rot or burn again? What a waste!"), rotting and burning do not produce waste (certainly nothing like pressed-wood furniture in landfills) and are an integral part of the cycle of life and death in the forest.

Conclusion
This book is too short and general to be a detailed rebuttal of possible errors in *Clearcut*. Many statements are tossed out without any kind of support, e.g., "Quick assumptions don't work very well for assessing forest health or forestry practices which are often complex and very site-specific." So, what did knock down those trees in the photograph?

There is little historical understanding of a site. Previous forestry practices are never connected to drought and infestation. In many cases, it is those new forests that are under attack from the (very real) ghosts of previous practices. Rocky slopes, meadows, and agricultural land are presented as if they were all that had ever occupied a site. In fact, many agricultural lands, as well as grazing lands, meadows, and rocky outcroppings were once forested, have been clearcut, and cannot get reestablished. The reviewers are deceiving themselves, not by the objectivity of their facts, but by their triviality and unrelatedness. Parts of problems are identified and analyzed, but the conclusions are trivial, weak, and irrelevant.

Everyone calls for a balanced presentation of forestry, although forestry has been polarized since Pinchot and Muir. The Ecoforestry Institute is working to integrate ideas and perspectives into a comprehensive approach. Part of the problem is that we have not found any convincing scientific arguments for clearcutting, so it is hard to give balanced arguments. The anonymous ones mention the overwhelming record of successful regeneration, but US and Canadian government statistics do not show any such thing.

Without meaning to, *Closer Look* ends up supporting *Clearcut* by reproducing pictures of clearcuts and admitting that they have been clearcut and are ugly. While explanations for some of the photographs were helpful in understanding the contexts and reasons, none of them seemed to justify the clearcutting.

The peevishness of the reviewers seems to be a general response to their perceptions that they are being attacked and misrepresented by "anti-forestry" forces and not a response to valid criticism. The debate on clearcutting needs to be framed as a dialogue so that the people who make the decisions will think of other alternatives. *Closer Look* started in the right direction, but derailed itself with name-calling and its own misinformation. All it can offer is weak entertainment value for knowledgeable readers.

Chapter 19

Dying Forests are Silent

In his book, *The Dying of the Trees*, Charles Little presents details of a pandemic of sick and dying trees all over North America. This book focuses on a topic often ignored in forestry: Death. Death is usually not addressed in literature or research, other than as a limit or a temporary medical shortcoming. In fact, a forest is as much dead as it is alive; there is a rhythm of death and replacement from the cellular to the ecosystem level.

Trees in a forest are always dying, either individually or in groups, waves, cohorts, or systems. Forests also may die, if enough of the trees die, as a result of catastrophic change or of too rapid or too much change. Mortality is a normal part of the life cycle. Mortality in forests usually occurs from a combination of factors. Lightning and wind cause tree death and injury. Injury and disease cause many tree deaths. The causes, rates, and patterns of death in tree species are poorly known, according to he Franklin, despite a hundred years of forest research. That presents a problem: If we do not know the normal rates of mortality in a forest, how will we recognize abnormal ones?

Part of the life cycle of a tree is death. The dead trees keep contributing to the life of the forest, standing for a while (1 to 150 years), then falling and decaying (over 20-200 years). Ecological forestry accepts a typical percentage of death as the normal condition, necessary for the renewal of the forest. The rate of death per year in a mature forest is remarkably constant at about 1-2 percent, even with wind storms, fires, disease outbreaks, and animal damage. Rotting and burning are an integral part of the cycle of life and death in the forest. Tree mortality from pathogens occurs on various scales: gap phases (small scale), forest development (large scale), and landscape patterns (immense scale). Yet, even catastrophic disturbances like hurricanes rarely damage more than 5 percent of a forest. More than being agents of mortality, insects, diseases, and animals are native components of complex food webs in ecosystems that contribute to the selection of certain kinds (including healthy) and ages of trees (that determine the composition of the forest, which changes over time). Mammals and birds disseminate seeds. Insects pollinate some trees and overwhelm others (rarely more than 1 percent of a forest). Diseases remove stressed trees (also probably a low percentage on the order of 1 percent). Pathogens are one of the determinants of growth and development. Their effect on the long-term health of a forest can only be regarded as positive.

In compiling anecdotal evidence, Little shows that trees are dying throughout every continent, in greater than normal percentages and for a variety of reasons: Acid rain in New England, New York, North Carolina, Tennessee, Georgia, Ohio, Indiana and Kentucky; smog in California; excessive ultraviolet light (through a damaged ozone layer) in Arizona and New Mexico; rising temperatures and sea levels in Alaska and Florida; destructive forestry practices, such as clearcutting, in Colorado, Oregon and Washington; pesticides or toxic heavy metals (released by burning coal and oil) in many other places, or combinations of all these factors.

Little cites studies by Hubert Vogelmann, a botanist at the University of Vermont, who wanted to study an undisturbed forest; in 1965 he made a thorough survey of Camel's Hump in Vermont's Green Mountains. He thought he was describing a healthy ecosystem; he measured the types and sizes of the trees, as well as various other aspects of the ecosystem. Periodically, he resurveyed Camel's Hump, and a pattern began to emerge: The trees were dying. His survey in 1979, compared to the baseline study of 1965, showed a 48% loss of red spruce; a 73% loss of mountain maple; a 49% loss of striped maple, and a 35% loss of sugar maple. Vogelmann was able to show that the health of Camel's Hump had begun to decline in the period 1950-1960. Similar studies in the Black Forest of Germany, and in southern Canada, revealed that the most likely cause was acid rain. (Acid rain was not "discovered" until 1972, by Eugene Likens and F. Herbert Bormann, although it had been falling on New York, New England, and southern Canada for about 20 years, as a result of the massive rise in use of fossil fuels, coal and oil.)

Vogelmann was able to show how acid rain affects the soil and thus the entire ecosystem, including trees. Soil contains large amounts of aluminum, in the form of aluminum silicates, which is not available to the roots of plants in that form. But acid rain dissolves the silicates, releasing the aluminum and making it available. When trees get aluminum into their roots and their vascular system, the roots clog, preventing them from taking up adequate nutrients and water. The trees are weakened, and may then be susceptible to extreme cold, insects or pathogens. Acid rain not only releases aluminum, it also releases other minerals —calcium, magnesium, phosphorus—required by trees; the minerals are washed out of the soil, leaving it depleted of nutrients.

Furthermore, acid rain kills mycorrhizal fungi, thus further reducing the ability of trees to absorb water and nutrients. The tree roots provide sustenance to the mycorrhizal, and the mycorrhizal helps the tree roots gather water and nutrients from the soil. Acid rain also kills off portions of the detritus food chain. The detritus food chain is composed of microscopic creatures that compost leaves, twigs, and pine needles, turning them back into soil. Because the detritus food chain is damaged by acid rain, forest litter builds up on the floor of the forest. In deep litter, seedlings cannot take root in the soil. Furthermore, the litter promotes the growth of ferns, which give off substances that inhibit the growth of red spruce saplings.

Throughout the book, Little describes studies and statements by the US Forest Service that downplay the importance of tree disease and death. For example, in 1991 the Procter Maple Research Center at the University of Vermont pinpointed acid rain and other air pollution as an important cause of decline of sugar maples in Vermont. The following year the US Forest Service issued a report saying that 90% of the sugar maples surveyed were healthy and the overall numbers and volume of sugar maples were increasing (but they had counted only standing dead trees, not those lying on the ground).

Although Little cites many instances of damage to trees by pollution, other scientists, such as J. Innes in his book *Forest Health*, states that pollution is not proven to be a major cause of tree death (while admitting many of the declines from ozone pollution that are also in Little's book); Innes suggests that evidence points to climatic stress and poor site matching. David Perry, in his book *Forest Ecosystems*, argues that pollution is involved in the decline of many forests, as a contributing stress, especially in Germany, where 52% of the forests are classified as diseased (in 1985). Beginning in the 1940s, it became evident that the plantation system, with single species even-aged trees, was susceptible to catastrophic change—wind, pollution, insects—in a way that natural forests were not. The forests began to die as forests—the Germans called this disturbing phenomenon *Waldsterben* (forest death). Even into the 1980s, scientists were trying to find the causes of forest death to preserve the plantation system.

One central question of forest decline or death is whether or not human influences have accelerated natural mortality rates or caused new mortality. Several scientific studies have claimed that the evidence for decline due to human causes is insufficient.

The occurrences of forest decline are well-documented from prehistoric to present times. Over 5000 years ago, elms declined in NW Europe, perhaps from disease or forest-clearing. From the 1930s to the 1950s, birches in eastern Canada and northeastern U.S. declined; no single cause has been identified, but stresses from severe weather are suspected. Starting in the late 1960s, there was a decline in Ohia trees in Hawaii, possibly from cohort senescence (especially in a short-lived species colonizing recently disturbed areas, where even-aged stands get stressed by conditions in a mature forest). In the 1970s and 1980s in Germany, forest death in firs and spruce—the primary factors in this "forest death" seem to be air pollution, acidification, and toxic metals; it also occurs in Austria, Czechoslovakia, France, Italy, Switzerland, Scandinavia, Poland, and England. In the 1970s and 1980s, declines of balsam fir in "fir waves" in Newfoundland and New England in the US were probably caused by ice damage on the leading edge of the dieback. Symptoms vary between species, as they should. The causes seem to be combinations of biotic and abiotic factors. A major factor might be rates of forest clearance, destruction, and fragmentation, which have been accelerating in the last 100 years.

While scientists admit seeing declines in the vigor of mature forests, leading to stand-level mortality, they say they cannot make conclusive statements about "causal factors." They mention that forest declines occur in contaminated airsheds, but also where pollution is not important. There is overwhelming evidence of pollution-caused death in Europe and eastern United States. The fact that there are other causes that work independently or with pollution does not invalidate the other evidence. To argue so, as some scientists do, is based on a logical fallacy, the semantic fallacy of complexity.

The problem with scientific reasoning is that proof may take many decades and cost billions of dollars. Although more long-term studies are needed, it might be better to shift the burden of proof on the safety of industrial processes and products instead.

The concluding paradox that Little identifies: Trees are abundant everywhere, but dying everywhere. People see trees everywhere and perceive that there is not a problem with trees dying. The grass is green everywhere too, but there are fewer species, fewer natives, younger plants, and lower quality material. Perceptions need to be changed.

Little offers details, although some reviewers have criticized him for not offering enough scientific evidence. Even if Little has not supported his argument adequately, even if he may be wrong is some instances, the preponderance of evidence is not easily dismissed. Regardless of the exact causes or interactions of causes, trees and forests are dying.

His overall suggestions, such as controlling human population growth or stopping cutting forests, are good common sense suggestions. The greatest threats facing forests are not just disease organisms or pollution, but the synergistic effects of fragmentation, pollution, and climate change. And these are best addressed all at once.

Little has described conditions that we need to know about, that we need to correct or to address. Maybe forest death has not been conclusively demonstrated by his arguments, but the evidence is disturbing, and there is sufficient reason to take corrective action now.

Chapter 20

Trillium Marches on Tierra del Fuego

Dear *Seattle Times* Editor:

I am responding to the article in last week's Sunday *Seattle Times* by Tom Dietrich, entitled, "Harvest with care," with the following subtitles spread over the pages: "lumber baron backing technique that could turn logging on its ear," "New way to log a virgin forest," and "No spotted owl." Each of which implies that this operation will be radically ecological and careful instead of the same old destructive exploitation.

The Indians are gone from Tierra del Fuego, hunted down by ranchers as Dietrich point out—and now the forest may follow. Dietrich says that the "exploitation of this largely virginal landscape ... may be one of the best things to happen to the environment in the history of the world: the largest sustainable logging operation in Latin America, and possibly the world." With that kind of advertising hyperbole, it may be facing impossible expectations by foresters and conservationists.

Dietrich tells the story of how the Trillium corporation went to the ends of the earth to find wood, but the story he tells is the simple one of bargain-hunting, trying to get the most for least; otherwise, Trillium would have bought more rights in the US, Canada, or Russia, in spite of the uncertainty and higher prices.

I am surprised that Jerry Franklin, inventor of sloppy clearcuts, thinks that this would be a credible or "incredible" model for resource development. We have used it here in the Northwest and it has not been especially successful. It is certainly not a "radical recasting of logging." Shelterwood cutting as a method is over a hundred years old. It was developed in Germany in the 1800s and applied extensively in the early 1900s to the Black Forest. A shelterwood cut is sometimes referred to as a sequential clearcut. It is an end cut that effectively ends the life of the forest. Many aspects of a shelterwood cut are arbitrary: the size of trees to be removed, which trees are removed, what percentage to be removed, and how long between stages. The timing of the cut after a good seed year, which may only be once in 17 to 70 years, is imperative—the success of the regeneration depends on it. There are some advantages: protection for seedlings leading to better survival, conservation of a few select species, and the potential for modification to mixed species on long rotations. But there are also obvious disadvantages:

- simplification (depauperization) of the forest ecosystem—the end result is a kind of clearcut over a even-age tree farm.
- minimal diversity, minimal stability
- need for artificial inputs to maintain the farm form
- damage from machinery
- advanced roadosis (too many roads), with compaction and impermeability.
- potential (and history) of abuse—clearcutting if regeneration is poor

Of course, habitat diversity after shelterwood cuts could be increased by extending the rotation and applying it to a greater variety of trees, including many noncommercial ones—also by leaving some old-growth for wildlife, which Trillium has planned to do. Many shelterwood systems, especially in the tropics, try to control regrowth of valuable trees only. And, of course, shelterwood cuts can be improved in a number of ways: By cutting in small blocks and by minimizing edges exposed to wind (very important in Tierra del Fuego). Perhaps a closer look at the fate of German forests, and what is being done there now, is in order.

Trillium has a bad record in the Northwest, as Dietrich points out. One wonders properly if any of its lands here are sustainable. And one wonders if they cannot be sustainable here, why go to Tierra del Fuego—unless of course, none of their forestry techniques are truly sustainable. Dietrich mentions that Syre can "cut slowly and carefully and still make a good profit" but will he? Or will he succumb to the need for fast cash?

The debate is presented as a simple split in the environmental community, but it is not simple or limited to the environmental community. Although deep ecology does advocate preservation of wild places, especially old-growth forests, it also advocates using selection systems in natural forests instead of clearcuts or shelterwood. Sustainable development also applies to many groups, from environmentalists to archaic peoples and industrialists. Satellite surveys show that over 30% of the land area of the planet is still relatively untouched. Economic development is a single and inappropriate approach to every ecosystem, especially those we humans depend on for basic services.

The fundamental question Dietrich proposes is easily answered: Sustainable development (as opposed to single-minded economic growth) is quite possible—but not every forest or grassland or wetland should be developed, only enough to support a reasonable human population. Although developing some lands can improve them—there are a few examples, development does not improve every landscape—there are many examples.

Perhaps the root of Trillium's problems are found in Syre's love for farming. Forests are not farms, however. They are wild ecosystems. They cannot be treated like annual crops without destroying the soil and forest itself. I am pleased that Syre has had good economic luck. I will be very pleased if in fact he is a fast learner and an ecological doer. Forests, especially old growth forests, are far more complex, however, than complicated financial situations. Development and land are good, that is true. Perhaps another of Trillium's problems can be found in Syre's influence by Orville Vogel, the sparkplug of the green revolution— the green revolution has done more to destroy agriculture than any other change; native germ plasms have been replaced by vulnerable super wheats, and the equipment and fertilizers required for the green revolution have bankrupted many farmers in developing countries.

Dietrich mentions that Syre has a vision, but not what it is. What is it? To get rich? To cut more trees than anyone else? To build malls and resorts? What are his values, if not rate of return? We are told Syre had a dream, but in context it seems to be the old dream of getting rich. The motto of his company, "leadership, artistry, and hope" is no help. Leadership in what way: Making money? Artistry in what sense: The minimalist landscapes of clearcuts? Hope?—Ben Franklin thought that those who lived on hope would die fasting; common sense would be better.

I am sure that environmentalists would encourage Syre to use more benign forms of harvesting, such as selection, and would support Syre if he were to try it. Certainly his openness with his plans and stewardship review are steps in the right direction, but why not go further and try to meet standards established by the Ecoforestry movement, for instance (through the Pacific Certification Council).

Syre's understanding of wood demand is incomplete. Other things, such as style, values, and culture, also determine demand; culture, in fact, can lower the demand, as it will have to do for true sustainability in the future.

Chile was a good choice for several reasons, such as low cost and economic problems. But Chile has other problems with its forestry. In fact, we might say that Chile is repeating the mistakes that the US repeated that Germany made, without learning enough. Plantations have been encouraged by decree since the 1930s; most are privately owned (possibly a legacy of the Spanish conquest, which imposed a new language, Catholicism, and a two-class social system, as well as the belief that large land holdings impart great status and prestige to the owner). Most roundwood production comes from second-growth forests, especially plantations. Twenty-five years after the Chilean timber boom began, virgin forests suitable for industrial exploitation are becoming scarce—Tierra del Fuego is one of the last. The Los Angeles Times reports that Chilean timber firms themselves are looking to the lush soil and virgin forests of northern Argentina to maintain their rapid growth. Fully intact forest ecosystems are finite, and rapidly diminishing in terms of quantity and quality.

Foresters do know a lot more about ecology, now. The trick is to get management to learn how to manage ecologically. Given Syre's message to Manne, "Here's a whole forest, untouched. Harvest and regrow it right," knowledge of ecology has not penetrated very far. I was curious, how much of the 15 million dollars is marked for research and how much for roads?

I seem to get a mixed message about this "environmental scrutiny" thing: Syre invited it, but does not like it. Where does the real responsibility lie? In being visible or in being sustainable? If Syre wants to be sustainable, he should listen closer to ecologists and environmentalists. Harvesting on a sustainable basis can only mute criticism if it is planned and then shown to be sustainable. Most shelterwoods are not sustainable.

Although the Beech forest in Chile is a simpler ecosystem than temperate Northwest rainforests, it still has indicator species (as the Spotted owl is here), such as the mountain lion and fox. The understory was rich with 6-7 shrub species and an incredible diversity of herbs. Lama browse grass and leaves. Other predators include weasels and skunks. There are vultures, condors, owls, falcons, hawks, and the giant hummingbird. There are many ants and hoverflies, but few bees (possibly due to climate and full canopy). There are many lizard species (twice as many as in California) and frogs.

The profile of Douglas Tompkins thrown in as an aside in contrast to Syre, is unfortunately incomplete. Tompkins does not threaten Chile's economy any more than Syre does. Tompkins is working to save large extents of old growth forests. As we know in the Northwest, more jobs are provided by preservation than through forest mining. If Syre the farmer runs the forests as a farm, many more Chileans will be out of work after 10-15 years than if the forest is treated wisely. Furthermore, Tompkins has supported other logging and forestry alternatives, such as those espoused by the Ecoforestry Institute. You should give equal time to Tompkins and his ideas.

The plans for Tierra del Fuego include "drastic thinning," many new roads, a mill, and a nursery. These impacts could all effect regeneration and the health of the forests. Certainly, other possibilities should be explored, such as managing the old-growth or selection cutting.

There are many things touched upon in this article or just barely mentioned. We are definitely a long way from a sustainable balance—the way to balance, however, is more in the direction of reducing our populations and their demands than in striving to satisfy every addiction for wood, oil, and entertainment. These Chilean lands are not considered beautiful enough for parks and tourism, according to Dietrich, but they have many other functions, not the least of which are carbon deposition, environmental services, and being home for many ultrahuman beings. Perhaps, by incorporating many advanced standards, Trillium may do more good than harm, but the report on sustainability of forests may take well over a hundred years to answer. Are we prepared to gamble with another old growth area? Do we have enough knowledge and humility to gamble wisely? Do we? Mr. Syre does not know. Neither does anyone else.

Chapter 21

Fun with Numbers, Part 1
A Response to the Financial Analysis
of Initiative Petition 20 for Oregon

Clearcutting is one form of harvest that can be beneficial under limited circumstances in a few forest ecosystems, although it is not based on any complete scientific studies. It has become the favored form because it is economical and easy. However, there are reasons to believe that it is harmful to many forest ecosystems in Oregon and cannot be sustained for very long.

Therefore, revenues from clearcutting will decline eventually as good forests are converted to young monocultures with dubious success. Any economic analysis, that is good for more than three years, has to consider this. The financial impact statement authored by Jacky et al. offers a number of assumptions that are questionable.

- For instance, that "timber harvest values remains (sic) constant." Such values are never constant, but can be increased with good practices. Timber volume alone does not determine revenues—the price of timber is an important component.
- Or, "there will be no shifting in the level of harvest activity" In fact, the harvest activity should shift to private lands that are managed well.
- Or, the data forecasts total harvest reductions "based on the elimination of 30" and greater diameter trees from harvest and maintenance of basal minimum areas. Since clearcutting would eventually eliminate large diameter trees from harvest, keeping some now can only increase future harvests. Furthermore, keeping minimal basal areas should also encourage larger future harvests.
- Or, the "effect of the measure on forest management and thus future yields is ignored." Rather than being ignored, the act would encourage better management and higher future yields. Selective forestry is not only more conservative of forest systems and values, but is far more labor-intensive.

Many other assumptions are never mentioned, such as:
- that forestry is based on an agricultural model, which does not really apply to forests which are wild, long-lived entities, rather than annual grassland crops.
- that regenerating after clearcuts will produce healthy forests, with large trees with strong wood, after short rotation times. There are too many instances of failed replantings and weak growth.
- the economic value (and revenues) from outdoor recreation, which is directly related to the amenities of healthy forests, is not considered (as it is by Niemi and Whitelaw in their Forest Service report).
- Revenues from nontimber forest products, most of which are destroyed by clearcutting, are strangely absent from the analysis.
- Spiritual values, which have a direct effect on human well-being and health, are not considered at all.

Numerous other considerations are never taken into account. For example, clearcutting, according to Professor David Perry et al., results in increased erosion and flooding, and many of these events result in the loss of human lives and physical infrastructures, which are never factored into the costs of clearcutting.

Alternative forest practices also generate more wages and thus more income taxes. Many of the practices, such as using herbicides and pesticides, that we use to control competing species, can be substituted with labor-intensive practices, from prescribed burning to weeding and pruning, that not only have fewer negative effects, but reengage our large, under-used labor pool.

The forest ecosystems themselves, at least in healthy conditions, offer ecosystem services worth many billions of dollars, according to Gretchen Daily and others at Stanford. Cutting forests imperils or reduces these services, which then have to be provided by human effort, such as water purification and air conditioning. These costs are not factored into the analysis, yet they are a direct result of harvest forms. Wild species and natural amenity values in healthy forest ecosystems increase their value, and potential for revenue, greatly.

A few last things to consider:

- None of the changes so far, such as the Endangered Species Act, have eliminated the profits of large companies.
- Furthermore (at least in parts of California, according to Norgaard), total timber payments and timber payments as a percentage of county revenues, have shrunk substantially since the early 1980s, even though volumes have increased—thus, more cutting does not automatically lead to increased revenues.
- Finally, soon the extreme costs from clearcutting, in terms of loss of services and amenities, as well as those costs now distributed across society, such as loss of lives and infrastructures, will outweigh the economic benefits from logging. The initiative petition 20 should preserve other values, reduce costs, and permit sustainable harvesting by selective systems.

Fun with Limited, Imaginary, and Irrational Numbers, Part 2: A Response to the Financial Analysis of Initiative Petition 20 for Oregon

The estimated fiscal impact from Initiative Petition 20 looks bleak indeed, but these numbers are very limited. Elsewhere and elsewhen one can imagine how other fiscal impacts from other declining industries or one-crop efforts looked equally bleak—for instance, slavery, cotton, beaver pelts, buffalo tongues, or tobacco. Nevertheless, society learned to cope with change and even to prosper.

I can find no fault with the numbers as they are, but they do seem so isolated on the pages. The total impact is estimated at a negative 74.59

million dollars.

What would the numbers look like if we continued clearcutting for 10 years? The income looks good, but the real costs seem extremely high and make the whole sum negative.

Overall timber *income*	+4,000,000,000
Tax revenues	+400,000,000
Legal fees	-95,000,000
Earth First! employment	+1,000,000
Restoration employment	+75,000,000
Costs of roads	-25,000,000
Loss of infrastructure	-25,000,000
Loss of life	-90,000,000
Loss of jobs to machinery	-13,000,000
Decrease in prices	-15,000,000
Loss of nontimber products	-38,000,000
Loss of recreation revenue	-3,000,000
Loss of amenities	-90,000,000
Loss of community wealth	-1,000,000,000,000
Loss of species/diversity	-1,000,000,000,000
Loss of ecosystem services	-2,200,000,000,000

Perhaps if we used a different system of harvesting, such as selection, or a different kind of forestry, such as restoration or ecological forestry, the numbers would look very different and the sum would be positive:

Overall timber *income*	+2,900,000,000
Nontimber income	+3,000,000,000
Recreation income	+4,000,000,000
Tax revenues	+900,000,000
Legal fees	-5,000,000
Earth First! employment	+500,000
Restoration employment	+75,000,000
Forestry employment	+200,000,000
Costs of roads	-5,000,000
Loss of infrastructure	-2,000,000
Loss of life	-9,000,000
Loss of species/diversity	-203,000,000
Loss of ecosystem services	-600,000,000

There might still be some losses, but they would not overwhelm the benefits. These numbers, with a positive sum and not a negative one, would be consistent with the clearcutting ban and indicate the possibility of the indefinitely sustainable use of healthy forests.

Chapter 22

Why Forests Cannot be Saved?

Why Forests Cannot Be Saved
Regardless of our design sophistication, forests cannot be saved if humans do not stop spreading out (our expansive style of life). Building upward in vertical cities (or arcologies) could save vast areas for wilderness and agriculture and recreation. We have the technology for arcologies, but not the will to abandon our spreading edge cities and private lawns.

Forests cannot be saved if there are too many people making demands on them (our unplanned population size). Human nations must be self-sufficient, as much as possible and as soon as possible, which implies fewer people. Smaller populations of people will always be able to harmonize more successfully with wild ecosystems, such as forests. For long-term survival, optimum-sized human communities are dependent on the health of ecosystems.

Forests cannot be saved if humans approach them with inappropriate images (our simplified industrial cosmology). The image of the forest as a machine with interchangeable parts is ridiculous and incomplete. Large industry is inappropriate to deal with local ecosystems, such as forests. We must achieve a mature view of technology, mixing handicrafts, intermediate technology, and even heavy industry (for transportation systems or large building, for instance).

Forests cannot be saved if we cling to our remoteness, watching violence and misery behind a glass wall of television (our insular indulgent psychology). Understanding the ecological relationships in a forest and with other ecosystems requires deep attention and participation.

Forests cannot be saved if humans cannot acknowledge and enforce limits (our four-year feel-good politics). We know what the limits are, but we are addicted to luxury and growth. We may have to induce people to adhere to limits with economic incentives, or simply legislate limits to growth and use.

One goal of ecoforestry is to save forests. Many things distract us from our goals. (In *The Meaning of Life*, one of the characters in a Monty Python skit was discoursing on the care of the soul and how to avoid distraction from it, when he was interrupted by a question on sales figures for hats, to which everyone turned their attention. Discussion of forest health is like that.) Our path to self-sufficiency and awareness, to forest preservation and right livelihood, is interrupted all the time by trivial concerns. Ecoforestry requires concentration and dedication. Working together, perhaps we can restore forests to their place in the body of the planet and in the heart of our lives.

Why Forests Can be Saved
The negative reasons seem so convincing, but there are positive reasons. Forests can be saved because human beings are willing to sacrifice their own good for other beings, because human beings are willing to change to ensure a future for their own offspring. Human beings are capable of learning about, and observing, limits; some people can acquire wisdom. New images can be forged and perhaps influence a healthier perspective. With forests there is still so much to learn. For instance, many good forestry books spend countless pages outlining the extent of our ignorance.

Here are just a few of the things we need to get to manage wisely:
- Accurate measurements of trophic level productivities
- Greater understanding of the complex relationships among plants and soil organisms and how they influence successional dynamics (Perry)
- The ability of long-lived organisms, such as trees, to genetically track short (less than a generation) environmental intervals (Perry)
- How trees pick up minerals, e.g., how trees get iron from siderophores (Perry)
- How plants tolerate very low water potentials without leaf death (Perry)
- The role of bark beetles and other pathogen-carrying insects in forest health and succession (Stocek); the insects have coevolved with trees— interactions between them may help maintain genetic variation and cross-fertilization
- Understanding of factors affecting forest growth (Morris and Miller, in W. Dyck et al.)
- The environmental effects of harvesting, long and short-term; the ability to predict consequences of harvesting
- The study of below-ground processes and productivity (above and below processes are linked, but we usually only consider above)
- The assimilation of carbon by trees from the atmosphere
- The role of forests in human thought processes (Fowles, Shepard et al.)
- How much is enough, in number of trees of a species, area, etc. How much should be exempt from exploitation. What are limits and thresholds
- Key species in a forest ecosystem—they all seem to be key players, although you could say that larger, more complex individuals, e.g., bears, may be less critical than smaller, e.g., fungi (but see below).
- What is the loss in biodiversity; how many critical pieces may be missing. In the absence of demographic data for species, perhaps ecological criteria, e.g., frugivores missing in a system, is the only way to assess ecosystems and species
- A measure of the extent of damage to forests from harvesting or clearing to determine which are vulnerable, endangered, critically endangered, or gone
- Effects of parasitoid insects on populations of their hosts under natural conditions. As well as competition among them and between them and other secondary consumers such as birds. avoid cage effects by excluding birds?

- Does competition among decomposers minimize the quantitative effect that their predators have on them? How abundant are detritivore populations? How much competition do they have among them?
- Native eucalyptus forests have a great percentage of GPP consumed by herbivores. Does this reduce competition between trees? How much competition between herbivores? These forests also have fires that are eliminated from plantations. How does this change the fitness of the trees?
- Branches do not fall off immediately after they die. Dead branches can act as conduits for fungi and insects. Merve Wilkinson, for instance, suspects that the powderworm (Western redcedar borer, *Trachykele blondeli*) enters the tree through dead branches; Merve is actively engaged in an experiment to match a control group (cedars with the dead branches left on) with a large number of cedars with the dead branches removed. Allowing dead branches to be removed by weakness, wind, rubbing, or precipitation is referred to as self-pruning; removing them with a saw is called artificial pruning. The life history of branches is a complex and fascinating history by itself.

Unfortunately, we do not seem to have the will or resources to protect all species in all forests. We can reduce the complexity with models and limit risk to a meager 100 years. Even though we have ways of reducing complexity to a keystone species or to a pyramid of value to understand, we have not reduced the complexity itself.

When you think about it, there are many keystone species in a forest: Trees, squirrels, millipedes, mycorrhizal fungi. In a way every species is a key to the whole. Some species, such as mountain lions, can be removed from the forest with little gross change, but there are changes. *Any species* is an expression of variety, a niche maker—that enriches the ecosystem and expands the habitat for others— and locus of feeling. Any heterotroph, according to Eugene Odum, that consumes autotrophs and excretes matter (with inorganic ions) contributes to the circulation of nutrients and minerals. Furthermore, any being contributes to the energy flow of the system and is a link in the web. Although an ecosystem can survive without a species, it is reduced accordingly.

We also use the pyramid to show importance in an ecosystem. Complex individuals are considered more valuable. There will always be conflicts between forms of life. So we will have to value lives differently. Complexity can be related to value; this is why we can kill millions of mosquitoes to protect hundreds of humans. A healthy biotic pyramid could be used as a basis for calculating the value of living organisms. For example: 1 human = 7 wolves = 48 deer = 2500 willow shoots = 7,000,000 root bacteria. Of course, many people would say it is wrong to equate nonhuman lives with human lives, but they seem to have no reluctance to destroy ten million mosquito lives for one human life. Living organisms cannot all have absolute values— even humans. Although this scale may be marginally accurate for the value of beings in a living system, it would cause problems at the human end.

However, the forest habitat is more complex than a human habitat.

When we invert the pyramid, the least valuable species becomes humanity, followed by mountain lions and bears. The most valuable are fungi and bacteria. That does not mean that we should not care about bears and humans (or about fungi and bacteria, depending on how the pyramid turns). Ecoforestry is concerned with saving the bottom of the pyramid as well as the top, regardless of how it stands.

How Forests can be Saved
Throughout this course there have been many suggested practices and principles that you can follow to remove some materials from forests and protect the functioning of the forest. Design, for instance, combined with conservation biology and other disciplines, can suggest patterns of cutting that reduce fragmentation, windthrow, and still appeal to people's aesthetic sense. The history of forests records vast changes, migrations, and shifts of forest species; knowing that, we can effect the direction of change by cutting certain species, such as grand fir in a Ponderosa pine zone. The personal experience of Merve Wilkinson, Orville Camp, Herb Hammond, David Perry, and many others, can guide the intensity and scale of cutting on Northwest sites. We can distill a few general principles that you can keep in mind as you work; please feel free to add to these, since they are not complete:

- Learn as much as you can about the history and ecology of the forest you are working in, especially about archaic practices and the previous use
- Be humble and cautious about cutting trees; when you take a tree use everything you take and make sure the rest gets returned to the forest floor
- Minimize your impacts on the structure of the forest; if you cut over 10% of the gross primary productivity of a forest you compete with the trees; over 10% of the net primary productivity and you are competing with the animals and decomposers; over 50% of the net community productivity and you are risking losing sustainability (this is what the conference participants in Moscow last week could not understand— they thought all productivity was ours to take)
- Consider the watershed context of the forest
- Protect the soil and protect the water
- Emphasize labor-intensive, community-building practices
- Monitor, monitor, monitor
- Keep your forest and human community healthy
- Keep yourself healthy—not just the core self but the larger, extended self.

Forests and Wisdom

The importance of forestry goes beyond profits and employment. For humanity, trees have always been a place to hide, and a place to go and live outside of society. The poor have always used forests to make their homes and heat them, to cook food and make tools. Although forest cover in industrial countries is more or less stable, it has been declining in the US and Canada since 1963 and it is declining dramatically in the southern hemisphere. Even in countries partially reforested in the 1960s, like India and China, forests are being cut again unsustainably to improve temporarily the standards of living. For the planet, forests have been critical organs in the balance of carbon and water. Many scientists argue that forests are the lungs of the earth. Most forests are now under stress.

Forestry as a profession is under stress. There are separate angry, noncommunicating groups, such as foresters and preservationists. Ecoforestry, however, is not anti-forestry or anti-preservation. Ecoforestry cannot be derived from industrial forestry—the paradigms are different. Industrial forestry is a failing system, a small-scope, once-through, temporary process for transforming forests into commodities, plantations, or deserts; ecoforestry is a maturing stage that bases forestry in a community context and limits the use of the forest to that which the forest, through its abundance and growth rate, can afford to provide and remain healthy. Ecoforestry is based on a more comprehensive logic than the two-valued, dichotomizing logic of forestry and on a deeper philosophy than savage capitalism. Ecoforestry, as Naess said of deep ecology, is spacious enough to accommodate the viewpoints of all these groups.

Where forestry is repeating the damaging quasi-isolated history of agriculture, ecoforestry recognizes the interdependence of all beings in an ecological community. Where forestry is imitating agriculture by trying to replace complex natural systems with artificial simple ones, ecoforestry works within the system to exploit a small percentage of it. Where forestry and agriculture claim adequate knowledge and power, ecoforestry admits its limits and strives for greater understanding.

Even today, our ignorance of the forest is incredible. We are not sure if mycorrhizal fungi can survive without squirrels; we are not sure what the productivity of the forests is or whether we are cutting the right trees. We do not have facts to base our actions on. Nature is a stochastic process, always changing; forests are always changing, seemingly as the result of chance actions. There is a profession that acknowledges the operation of chance and makes conclusions in the absence of facts: it is called gambling—few people are successful at it. Even though it is guided by a better ethics and practices a more benign, sustainable exploitation, ecoforestry is also gambling. We do not know for sure what effects our actions, even individual tree selection or preservation, will have on the forest, which used to live so long. The proper virtues for gambling with nature are humility and courage, not arrogance and fear. Eventually we might develop wisdom.

A tree, like a pine marten or bark beetle, has biological wisdom; it is clearly self-making, self-governing, and self-choosing (within its limits of treeness, e.g., it cannot walk over the ridge to get better sun). Forests are the books of biological wisdoms. Through our cutting and decision making and overuse, we are tearing up the books, literally leaf by leaf. Rather than telling the forests what to do, rather than controlling their growth, we need to watch forests to see what they do (this used to be the function of natural history), and we need to let them do it (this requires patience and temperance). Abraham Maslow regards this attitude as "taoistic," and the way to forest health is letting the forest do most of the choosing and working. Taoistic means being nonabstracting, nonimproving, nonintrusive, and noninterfering.

Maslow also pointed out that most of the action is at the growing tip of a plant. Ecoforestry is the growing tip of forest care. To be a good forester, you almost have to be part of a good culture, and that culture must almost certainly have a good image of the world. Otherwise, the culture may destroy whatever good work you do. What I am saying is that you, as an ecoforester, cannot just care for the forest. You must also participate in society, and if it is blinded by greed and bad ideas, you must try to influence it to the good. Your being good may contribute to a social good. If it is hard to read all this stuff about goodness, it is also hard to write it, but I cannot think of any other way to communicate the ideas.

Goodness & Practices in Forests

Chapter 23

Gigatrends in Forestry

Some trends are readily evident in the human present; other processes or cycles take many human generations to complete. Many trends in forestry can be fit into the megatrends identified by John Naisbitt over a decade ago. Naisbitt identified ten larger patterns in society, including the move from industrial society to information society, the economic interdependence of the human world, and the restructuring of society from short-term considerations to long-term time frames, that is, he says, from two-year horizons to a "very long-term" time frame of "six to ten years"—long by economic standards, but not long enough for trees and ecosystems. Some other counter-megatrends (resist calling them negamegatrends), such as the denigration of reason and science in the popular press and media, or the incredible explosion in information — without a corresponding increase or spread of wisdom—are ignored by Naisbitt. Negative trends in general are left out of the picture. The things that are most popular to society, such as money, real estate, insurance, and politics, are the things that are treated as the most important. Italian Foreign Minister Gianni De Michelis identifies the most important megatrend of the century as the availability of free time; he claims the US economy will remain the most important economy in the world because its GNP is increasingly geared to entertainment, communications, education and health care, all of which are about individuals 'feeling well.' The things that will ultimately be most important, such as directing the course of civilization, limiting human activities, or preserving wild ecosystems, are relegated to a sideshow.

There are megatrends in forestry. Forest companies exemplify economic interdependence as they move to countries with weaker regulations to mine their forests. Forest companies have learned to restock some sites, but not to plant well or to nurture the trees. Furthermore, rather than moving from industry to information, forestry is moving backwards, to providing raw resources. And rather than moving towards multiple options, forestry is backsliding into one option: logs.

The real long term trends in forestry—and in forests—occupy the entire human calendar. They are sometimes invisible in the present. We might call them gigatrends, since they are larger and more involving than megatrends. Gigatrends (*"giga"* from the Greek for very large or giant) are long or very-long-term trends, usually ignored by science and economics, such as atmospheric temperature increases or global deforestation. Some gigatrends may only cover a span of fifteen years, but some of them have continued for thousands of years. In general, the larger the unit of study, the longer the trend; for example, changes in the physiology of a tree can occur in days, while changes in the forest can take decades or centuries. Some gigatrends are beneficial, although most of them seem to be detrimental to human well-being as well as to the health and stability of forests.

Because of their length and scale, most of these trends can be represented graphically with simple lines, which show gradual or rapid (exponential) changes (see Figures 3-6). Several trends seem contradictory or inconclusive in the short-term, but are evident with long-term study. For instance, atmospheric carbon dioxide increases and decreases with seasonal change in vegetation, but has been rising slowly and steadily for at least 35 years, according to Keeling and Whorg; forest succession is also misleading in the early stages, as pioneer species take advantage of a disturbance—here the short-term abundance of intolerant nitrogen-fixers prepares the site for a mature forest. Other trends are actually complex. For example, wood use as fuel decreased from the 1800s until 1970, when it started to increase again, due not only to the fashion for stoves in industrial countries but to the rising prices of petroleum and coal.

Many of the negative trends have been noticeable for thousands of years, but nothing has been done to halt them. Environmental factors have shaped the course of human history to a greater extent than has been realized. The decline of Rome demonstrates that ignorance of forest ecology can have important consequences. There have been environmental catastrophes in the Tigris and Euphrates valley, Greece, Khmer, Maya, Cahokia, and others. These civilizations were very successful before they failed. Failure from success is tragic. For the Greeks, the operation of tragedy resulted from success taken to great lengths, that is, where successful behavior in one context is applied to all contexts, with the result that the opposite action occurs from the one desired. For example, humans in moderate numbers were able to take what they needed, such as wood, from natural ecosystems without interfering with the processes. Our dominance, once so successful because of our big brains and tool-using hands, has now become self-destructive. When human cultures adapted to ecosystems over long periods of time, the ecosystems also adapted to human cultures; when the human impact has been rapid and intense, as it has been in North America recently, the ecosystems collapsed or stabilized at a simpler state.

It has been argued that humanity is not adapted to live everywhere. Since the human species emerged in a subtropical climate, where it acquired certain biological characteristics, it may lack a degree of fitness to survive in the tropics, the arctic regions, or even temperate forests. The species may remain genetically best adapted to a certain type of subtropical savanna. Rene Dubos presents this development as explaining some of our present behavior patterns: subconscious fear of forested wilderness (where good vision was little use for avoiding danger); the commonness of design features in landscape architecture; the preference for a narrow range of temperature; and the biochemical similarity of nutritional requirements. Perhaps this may explain why people modify their surroundings the way that they do, as well as why forests are less valued than lawns and gardens.

Although many gigatrends are interrelated, they can be discussed is several categories: human populations, ecosystems and forest ecosystems, human economics and technology, and forest technology and management. Positive trends, often smaller and more recent, are discussed after negative ones. This list is not meant to be complete or detailed; please add to it.

General Human Populations and Needs

With the success of the human species has come human domination of ecosystems. Human beings have modified animal and plant associations, simplifying patterns of energy and chemical exchange, and solidifying ourselves at the end of many food chains as a dominant species. Our domination is related to our large biomass, our large annual increase (over 2 percent annually), our high energy use, and our high structural organization (information and matter). These very large-scale effects relate to basic gigatrends having to do with population size and dominance:

- Human populations increase exponentially—at 2 percent per year the doubling time of the entire population is 35 years. The growth of human populations for 500,000 years was minuscule; the agricultural revolution (10,000 years ago), which increased food supplies, and the industrial revolution (in the 1800s), which decreased the death rate, led to dramatic increases in population numbers and the rates of growth.
- Humanity takes over the habitats of other animals and plants, regardless of loss of function of ecosystems, limits to carrying capacity, or deficits. Kenneth Boulding suggests that eventually humans may perform the functions of other species.
- The material goods of human societies have been increasing. Few of us in Canada or the US can carry everything we own on our backs, a bicycle, or a horse. The impacts of a small percentage of people (the wealthy) increase exponentially as the result of heroic consumption.
- With the social avoidance of hard decisions, no one wants to take responsibility for meeting ecological or social limits; no one can say no to their constituents, representatives, or business partners.
- Societies seem to be working toward a minimum for human existence: a box for everyone, with sufficient heat, light, air, and plumbing. Mere existence has become an acceptable option. We have gone from the green forest to the gray box. There is no place to hide, no place on earth that the air cannot stink and the rain not burn, that ugliness cannot reach and misery not touch.
- Human interactions become more violent as a result of competition, inequity, and limits in the distribution of resources.
- There is an increase in the scope of ethics, from family, tribe, nation, humanity, to include reverence for all living beings, identified by Albert Schweitzer. An increase in the scope of ethics to include land and forests, identified by Aldo Leopold. An increase in the scope of law to include legal rights for forests, identified by Christopher Stone.

Human population pressure pushes a lot of trends. The pressure from exponential population growth means that remaining forests will be depleted to meet basic human needs. Existing financial and political resources may not be enough to stop it.

Eric Eckholm concludes that the United Nations must identify, analyze and marshal world resources against negative trends. A scientific method would take a long time, however, and poor countries cannot wait. They must attempt a rural regeneration of some kind, to stop urban drift and ecosystem destruction. Eckholm interprets negative trends as indicating the sinking of marginal peoples on marginal lands into a quiet helpless poverty, later leading to urban deterioration—which may perhaps be less quiet.

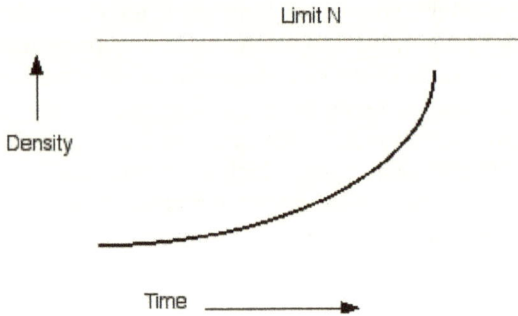

Figure 3. The J-curve (one of four basic curves) is a graphic representation of long-term trends that show exponential growth, such as human population growth, pioneer species after a clearcut, bank interest for a tree farm compounded daily, consumption of paper and plywood in the United States, annual visits to national parks in the United States and Canada, annual sales of recreational vehicles in the United States, daily Canadian Water consumption (since 1910), or membership in the Audubon Society. Such unrestricted growth is usually halted suddenly when a critical resource, such as food or water, is exhausted or an ecological limit is reached.

Forest Ecosystems
Ecosystems build up information. There are at least three different channels of information in an ecosystem: the genetic (in replicable individuals); an ecological based on interaction between cohabiting species (expressed in changes in their numbers); and the cultural, transmitted through individual learning based on experience. Feedback within the interaction of species is expensive memory with little storage capacity. Whenever succession starts again, after a volcanic eruption for instance, old information in the form of interactions has not been saved. Genetic memory has much larger capacity and is long-term. In higher vertebrates, such as wolves or humans, cultural memory is enlarged. The unconsidered use of information results in still more long-term trends:
        Forest destruction was first associated with hunting, when fire was used as a technique to drive animals into the open. Grasslands maintained by burning allowed safer hunting. The technique is still used today by a few small groups of people. Centuries of burning in Africa reduced forest cover to less than 40 percent of its original cover by 1948, according to Eckholm.

Agriculture and forestry have been related in many civilizations. The expansion of agriculture is directly related to the shrinkage of forests; other trends in agriculture, such as opening southern lands, have resulted in forests reclaiming some of their northern territory. Nevertheless, the overall trends have not been affected:

- Forest cover has been reduced since 1100 B.C., constantly reduced since the 1500s, and drastically reduced since the 1950s. No one knows the exact rates of reduction. Overall the world forest cover has been reduced over 35 percent. The actual amount also is very uncertain, since inventories are rare or crude—half the land reported as forest land in many countries is also labeled "unstocked," according to Eckholm and the World Forest Inventory, with the result that grasslands, scrublands, and wastelands are labeled as potential regeneration areas.
- The long-term health of forest ecosystems declines as people fight over access to specific resources for short-term economic gain. Forest health is rarely measured or addressed.
- Natural regeneration is declining, due to interference in the operation of forest ecosystems and the destruction of some of the necessary structure, for example, clearcutting kills mycorrhizal fungi necessary for nutrient uptake.
- Tree planting has been decreasing. According to Robert Mangold et al., 2,419,691 acres were planted in 1993, a five percent decrease from 1992 and almost a 30 percent decrease since 1988. The decrease is in all geographic areas and by all ownership groups.
- The forests that remain are simpler (tending to monocultures), of lower quality, and much younger. Forest ecosystems are kept at early seral stages to benefit from increased production. There is no guarantee that they will ever become mature.
- Ecological limits are ignored, even where they are known. Forests are cut at rates greater than the net primary productivity.
- Ecosystems are simplified and degraded; deforestation, desertification, and exotic take-overs occur on a large scale.
- Vegetation becomes a social artifact. In Scotland, for instance, forest cover was reduced from 55% of the total area to 5% by primitive stock-keeping and agriculture; the moors decreased by half, but meads increased eight-fold.
- Global biogeochemical cycles are disrupted; for instance, atmospheric carbon dioxide has increased since the 1700s.
- Human cultures mutually adapt with ecosystems over time in Asia, Europe, and parts of Africa, resulting in domesticated landscapes.
- Larger areas, such as preservation of ecosystem processes, reservation of archaic cultures, and conservation lands, are set aside from industrial interference.
- Some forests are restored from abandoned fields, anthropogenic deserts, and ruined ecosystems.

Wild forests exist in some kind of balance. Insects damage trees, by attacking different parts of the tree (often during different stages in their life cycle), but not usually at a level more than 10 percent—far smaller than the waste from logging, which approaches 50 percent. After being used and "improved," forests are not being regenerated. They are not being replanted nearly as fast as they are being cut, either in terms of biomass or quality of wood. These trends contribute to a shifting planetary system that has far less flexibility to respond to environmental changes.

Forest Economics
Economic pressures, which are derived ultimately from population pressures, force farmers to intensify their efforts to increase crop production—and foresters presumably to increase their production. This instigates an "utterly dismal cycle" of population expansion, environmental deterioration, and poverty, according to Eric Eckholm: as the population expands, forests are cleared for land, and arable lands are used to capacity— and sometimes beyond. As the soil deteriorates, it requires more fertilizers that cause more hazardous conditions that decrease agricultural capacity; people starve, but the population increases, and marginal lands are used to meet increased demands—or food is imported from other lands—which are also experiencing stress.

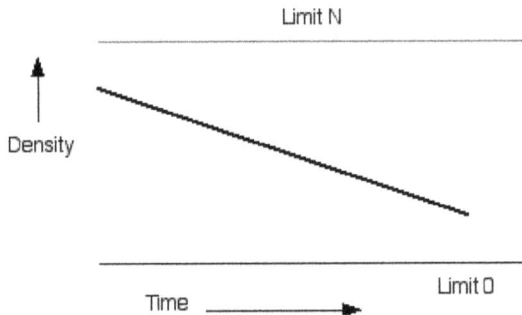

Figure 4. Graphic representation of a linear decrease over time. On a
large scale, the curve may appear smooth, while detail may show
daily, annual, or other fluctuations. This curve could represent
the size of logs used in mills since 1860, employment in the forest
sector of society, gross profit to the forest industry, age of trees
harvested, number of trees planted, area of wild forests remaining
worldwide, or potable water sources worldwide.

Eckholm describes how the usage of such marginal lands could result in dust-bowls, when climactic conditions change. After the US dust-bowl in the 1930s, national conservation programs were able to restore some of the mythical fecundity, through pasteurizing, strip cropping, terracing, and contour plowing. However, current production efforts are causing greater losses of topsoil, and farmers are abandoning some of the conservation methods for economic reasons. Eckholm concludes that free market conditions encourage dangerous trends. The lesson of the latest dust bowl seems to have been forgotten—until the next one. As forestry copies the agricultural model, it contributes to soil losses and destabilizes ecosystems.

There are specific economic trends related directly to forestry. Economics has been defined as the "way people make their living," although as a discipline it attempts to balance human wants with scarce resources. Over many cultural lifetimes, this balancing has resulted in definite long-term trends:

- Shortages of timber. Most Old-World civilizations faced shortages of high-quality wood, from the Mesopotamians, Egyptians, Greeks, and Romans to the modern countries of Europe. Pressures on British woodlands in the 1300s forced people to turn to coal as a fuel source (a source regarded as inferior to wood). The timber famine reached Europe in the 1700s. It had existed in China and India over a thousand years earlier. Countries that exhausted their wood supplies had to invade other countries or find substitutes such as coal or water power. Each substitute required more energy to produce.

- Eckholm warns that there is a serious firewood shortage over most of the earth due to population pressures on the remaining woodlands. In developing countries, hundreds of millions of rural people strip the trees and shrubs around their homes for fuel. In Gambia, for instance, it takes 360 woman-days per year per household to gather wood. Even when firewood is available for sale, it may be beyond the budgets of the poor. When wood is scarce, the poor are forced to use millions of tons of crop waste and dung, which should be used to regenerate soils. The poor are destroying their means of survival in order to survive now.

- There is an overall growth in wood consumption related directly to population growth.

- Demand for wood for some uses, such as fuelwood, railroad ties, or masts for ships, has been steadily decreasing, although fuelwood has been increasing again since 1970. Wood products are replaced by petroleum-based products, such as plastic bags and synthetic fibers, although this trend depends on inexpensive petroleum, which is unlikely to continue (recently, GM listed petroleum reserves for only 8 years).

- Demand for wood for other uses, such as paper or fine coffins, has been increasing. Paper use has been increasing despite the advances in electronic information technology. So far information technology has proved to enhance output rather than substitute electrons for paper—although the long-term trend may reduce paper usage.

- The per capita demand for wood is increasing as new items are made out of wood.

- The price for wood is emphasized over planting or preservation; this is logical in the current economic system, which is a based on a narrow limited logic.

- A centralization of power in large corporations leads to the elimination of small operators; the large corporations have larger holdings.

- Attempting to control the resources, forest corporations integrate the process from planting to product—partly in response to tax legislation.
- Manufacturing of wood products is released from a land base. Corporations depend more on trees from tree farms (and temporarily from national or provincial forests) than from company managed land.
- There is an increase in the economic value of forests, partly as a result of the decrease of wild forests, and partly due to the growing recognition of the value of the few remaining old-growth forests.
- There is growing recognition of the scientific and aesthetic importance of forests.
- The markets for wood products are expanding.
- Real price of industrial wood has increased 65 percent in the past 50 years, due to increasing demand.
- Economics emphasizes global relationships, which reduces the importance and protection of local forests.
- There is an overemphasis on the microeconomics of situation, such as how much fertilizer to use in a tree plantation, instead of on macroeconomics, which might be more labor-intensive, but costs less to society in terms of welfare, unemployment, food stamps, and depressive addictions. Our infatuation with an economy of scale, which dictates increasing size as a solution to costs, but which ignores real optimal sizes for management units, contributes to our problems.
- There is overemphasis on the control of natural processes. Economists fear that "letting things alone" will lead to stagnation, poverty, and chaos. The technocratic vision is "life under control." The technological imperative is to strive for better efficiency, that is, economic efficiency.
- With the trend to smaller logs, management has shifted its emphasis from growing trees to harvesting them economically. Forests are biological processes that are subject to considerable variation in output; management, however, deals badly with uncertainty and diminished yields. One solution to many of these problems is a reduction in scale for everything from forest use to management units, with local controls and local use primary. Management costs increase with the size of management units; more levels of human hierarchy are required to deal with problems; decisions are slower, and the people who make them are more remote from the site.
- The desire to reach and maintain a maximum harvest is a common goal. But, even a modern, balanced exploitation may destroy forests and fisheries. Currently, many resource managers espouse the ideas of equilibrium maintenance and maximum sustainable yield. These ideas are poor guides to management according to C. S. Holling. By trying to maintain habitats in equilibrium, we often set them up for catastrophic decline, for instance, in fire-climax pine forests, or destroy resident species such as the pine marten.
- Humanity depends on the drawdown of renewable resources to support numbers and styles, which is a temporary fix. William Catton defines drawdown as using a renewable resource faster than it is being replaced.

Control has long been the goal of science and technology. Scientists and foresters are obsessed with the treatment of weeds and vermin. The industrial culture has put humanity at war with forests and ecosystems, which always misbehave. Stability has been raised to sacred scientific state. Furthermore, our western culture has been distorted by the modern emphasis on the scientific method, with its devaluation of philosophical concepts. Rational values are exaggerated and spiritual events are ignored or suppressed. Most philosophies and technologies, however, are not adequate to deal with nature and ecological relationships.

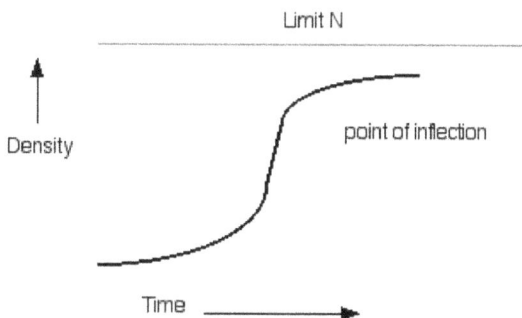

Figure 5. The S-curve (sigmoid) is characteristic of a population reacting to changes or limits, such as wolves or bumblebees in a forest—the growth rate starts slowly, accelerates, and then slows below the carrying capacity or upper limit. Other things might follow a sigmoidal pattern, such as capacity of the forestry industry, the net annual flux of carbon from global deforestation, or prices for forest products. The first part of this curve resembles a J-curve; the final difference has to do with the reaction at the point of inflection.

Every advanced country is now over-technologized. Past a certain point the quality of life diminishes, not improves, with each advance. Big science serves big technology, which supports and is supported by big government. And there is no science like big science, and no administration like big administration. But this cannot last. Scientific advances and technological changes result in unforeseen consequences, good or bad. They cannot be controlled or legislated against before the fact. But the investment seems too big to abandon. Perhaps big science has too much momentum.

Forest Technology , Management and Planning
Realizing that high levels of cutting may be putting stress on forests, human beings respond with economic planning and technological solutions, which lead to further trends:
  • Artificial forests (plantations), requiring high-tech intervention, are planted to replace wild forests, following the agricultural model. Complete site transformation with technology replaces traditional

179

silvicultural care for forests.
- Following the agricultural model even further, forests are treated as monocultures; a single species, such as Douglas-fir, is grown intensively and exclusively.
- There is a shift from addressing individual pests to treating entire ecosystems with an increasing volume of pesticides (more than 50 times the 1940 levels)—this greater reliance is proportional to the resistance of the pests, and it is an escalating cycle that causes greater environmental wobble.
- Technological innovations have been increasing. Computer techniques, such as BOF (best opening face) and EGAR (edge glue and rip), are developed to increase the recovery factor. Construction techniques, such as wood trusses, save wood. New pulping technologies, such as CTMP (chemithermomechanical pulp), also save wood.
- The number of roads in forests increases. Every technological innovation in vehicles either requires or makes roads. In the US national forests, there are 342,000 miles of roads—the US Interstate system only has 50,000 miles. According to Chris Maser forest roads occupy over 18 million acres in the US.
- Machines are used more often for heavy work, ostensibly because of a shortage of labor for forest operations and the unwillingness of the labor that remains to work hard under dangerous conditions, but actually because large complex machines are cheaper than small immensely complex human beings. The use of machines also limits what can be taken from, and causes more damage to, the forest. Machinery for harvesting forests is not flexible enough to deal with nontimber values. Machinery is geared towards artificial forests that must be totally transformed with scarification, fertilizers, and genetically-improved seedlings.
- Mechanical efficiency is increasing. Logging operations have become so efficient that many times fewer people cut many times the number of trees in less time. Jack Hagen relates how logging was in northern Idaho in the 1920s, where the camps resembled migrant cities. The main bunkhouse held 150 men—fellers, swampers, buckers, horsemen, mechanics (for steam donkeys), carpenters (for buildings and flumes), chokers, and laborers. The bunkhouse was built of rough wood—even the sinks were rough wood. There was a separate sawmill bunkhouse with semi-private rooms for 15 men, including the sawyer, pondman, and cant hooker. The cookhouse held 6-10 men, cooks and assistants. Some of the men who were married lived with their wives in small cabins. There was a commissary (for tobacco and a telephone), where the timekeeper, accountant, and record-keeper worked. Nowadays, trees are cut and sent to the mill by machine without much human intervention at all.
- Biotechnology is used to increase profits. Biotechnology is still economic in a primitive sense. Although the myths have changed to include greater manipulation, the technology is still concerned with utility, growth, and efficiency as short-term goals. The problem with

efficiency is that it is defined within such narrow limits. True efficiency means continuity over long periods of time, as with natural processes. Long-term exchanges in nature are not efficient in the industrial sense.

- National and international goals (and trading regulations) have become more important than local goals and considerations. According to industrial economic logic it is better to stockpile local forests and import cheaper wood from poorer nations. This puts stress on forests in poorer—or less greedy—nations.
- Forest management is becoming more intense. Its goal is always to raise productivity per hectare instead of questioning the demand or identifying other options for getting wood.
- Ownership has increased until every forest is owned (one would think that this gigatrend of ownership has ended with the finite territory of earth, but it shows signs of continuing through the solar system). With ownership comes management and control.
- There is a slow, constant reappraisal of forest programs. The new forestry was one of the first formal changes to the dominant program, although many foresters reject even its mild suggestions for leaving a few trees and woody debris.
- More people practice comprehensive approaches, such as ecoforestry, to maintain forests while taking their needs from them.
- Greater numbers of people invest in small businesses in forest products— businesses that are labor-intensive and practice-centered rather than profit-oriented.
- Global and regional planning of forests is almost nonexistent. Local planning ignores historical evidence as well as native species and ecological limits.

Some trends have stopped because planetary limits have been reached. Like ownership, territorial expansion is at an end. There are no "new worlds" to discover and no virgin forests to be logged.

Some of these negative gigatrends are hard to see, much less stop or reverse, because they are based on misunderstandings, fallacies, myths, and psychological blinders. For example, forest planners often treat exponential growth rates the same as linear rates. Thus, if our forests were to last for 400 years at our current rate of use (before extinction), they would only last for 75 years with a demand growing at 3 percent per year, or 50 years at 6 percent (demand is growing now at over 3 percent per year). Also, Dennis Meadows points out the speed with which surpluses disappear with increasing population and increasing per capita demand.

Many predictions about resources are based on fallacies. There is the fallacy of substitution, that states that a substitute can be found for any resource in short supply. This is not always true, especially when cultural preferences are considered. Furthermore, there is a gross underestimation of the length of time that it takes for a substitute resource to attain traditional markets. For instance, the transition from wood to coal as an energy source took about 50 years, despite the fact that coal technology was established and attractive. Andy Kerr has suggested that steel, plastics, and other materials should replace wood in building and other applications. Even if people thought such a shift was desirable, it would probably take many decades.

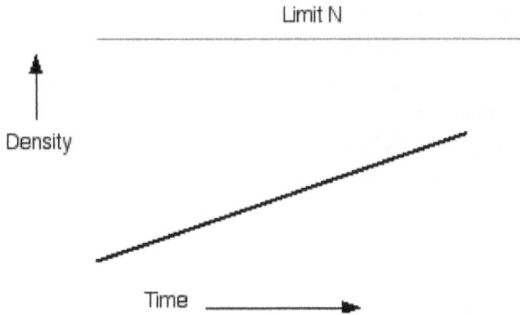

Figure 6. A linear increase over time. Many elements have been increasing steadily, such as the concentration of carbon dioxide in the atmosphere (since 1958, contributing to the greenhouse effect), the annual consumption of lumber in the United States (since 1948), roundwood production in Chile (since 1960), or the growth of gross domestic product in Indonesia (since 1971).

Meadows identifies another fallacy: The expectation that people and institutions perceive problems and react to them rationally, e.g., with the threat of wood shortages, prices would rise and consumers would value the resource more. Yet, the price of wood is still nowhere near the real costs of production, and wood is used for cheap, impermanent goods. Meadows suggests the model of addiction might be more appropriate than adaptation for dealing with consumer demand.

Many of these gigatrends are based on myths, such as the economic myth of forestry, which as it is related by Chris Maser, is based on the rationale of "soil rent" theory, a classic economic theory that assumes, fallaciously, that all ecological variables are constant so that capital investment is easily calculated from the rate of growth of the crop species. Due to the uncertainty of natural processes, we should limit our take to far less than a maximum rate.

These trends are partly the result of our unconsciousness of large-scale, long-term events, partly the result of out cultural amnesia about things that make us unhappy, and partly the result of our cultivated indifference— doubtless due to our remoteness from wild nature, remoteness from the forest as a result of our tools, and the general romantic abstraction of civilization.

Summary

What these trends show together is that humanity is using more forests, and more of each forest, to produce more goods that cost more for more of us. These trends are not fated to continue. Some of them can be slowed, redirected, changed, or reversed. Rather than be converted to plantation forests, remaining old-growth could be preserved or harvested at very low sustainable rates.

None of these negative gigatrends can be modified until our remoteness is re-educated into participation and attentiveness. By making long-term trends visible and immediate, we can understand how they shape our use of forests or our problrmd replanting and renewing forests. Combined with other trends in housing, such as arcologies (Paolo Soleri) and ocean arks and bioshelters (the Todds), and agriculture, such as agroforestry, permaculture (Bill Mollison), and tree crops (Russell Smith), further reforestation of some areas appears likely.

The intent of describing large-scale patterns is to fit human patterns with observed patterns in nature; patterns have a form, sometimes repetition, and sometimes regularity, but each of these is caused by some limiting factor. Fitting the pattern can lead to both continuity and predictability, and both of these things are needed to adapt human activities to natural limits.

Thinking we have conquered nature and that we are omnipotent, we have quit thinking. Satisfied with our comforts, we do not ask enough of ourselves. We seem to be confused over the difference between luxuries and necessities. There may be enough forests for necessities, but not enough for luxuries. We also act confused by the distinction between temporary good and durable goods (nothing is permanent); temporary good are things like cars, entertainment, guns, and drugs (any kind), while durable goods are things like reforested areas, organic farms, well-designed roads, and healthy buildings.

Garrett Hardin used to say that the essence of ecology was found in the question, "and then what?" meaning that everything you did had a primary effect (there were no side-effects) on the system, every action a reaction, or as it has been rephrased by Barry Commoner and others, "you cannot do just one thing." We have to consider the consequences of our actions as much as we can. Even good actions, taken in isolation, can have tragic consequences. For instance, what if, in setting aside forests in North America and reducing the load on them, we put more pressure on forests in Malaysia, causing them to be cut faster and more disastrously? What is the solution then? Is ir social equity with other cultures or peoples? Is it voluntary simplicity? Or, is it global laws?

With these gigatrends possibly ending in tragedy for humanity, we must ask many questions. What kind of forests do we want? Even or uneven-aged? Wild or domestic or both? Managed or unmanaged or preserved? How shall we use those forests? For wood products? To protect watersheds and maintain global biogeochemical processes? As a home for other beings? Recreation? As some kind of balance for domestic landscapes? How many forests do we (or other cultures or the earth) need? What kinds, in what forms? How many should be wild? These questions lead to new strategies for living with the forests.

We have spent most of our infancy fighting nature. Up until the seventeenth century Europeans regarded untamed nature as a vast, hostile desert. Wild nature still remains unwelcome in our cities and gardens. We might understand the historical failure of cities and walls to lock out the forest or nature. After the Sumerian king Gilgamesh killed the great spirit of the forest, Humbaba, he became possessed with the fear of death and tried to lock out nature with the great wall of Uruk. It did not work. The forest must be wooed. The forest will haunt us until we give it a new life in the heart of modern culture.

Chapter 24

Ecological Forestry Research

The mission of the Ecoforestry Institute (EI) is to foster ecologically responsible forest use, through education and related programs and services, to perform and integrate research, as reflected in leading edge work in conservation biology, landscape ecology, forestry, and related disciplines, and to communicate to a wide audience the deepening knowledge of the multitudes of values and functions of natural forest ecosystems. How EI is undertaking its research can be contrasted with main-stream forest research.

Industrial forestry was quick to take advantage of scientific understanding of forests and trees, starting in the mid-1800s. Science, as you know, operates by a method that simplifies and controls a situation over a short term, analyzes it, and deduces information, based on an initial hypothesis, that is, a guess or a premise from which a conclusion can be drawn. By the 1980s, scientific research had resulted in technological applications for growing, genetically "improving," harvesting, and processing woody material on an unprecedented, vast, and unconsidered scale. Industrial companies established tree breeding and fertilization programs, whose goal was usually fast growth. Pine plantations in the southern United States were rated as a success and served as a model for plantations elsewhere; Australia, New Zealand, Chile, and South Africa replaced much of their native forests with exotic pine plantations, using species like Monterey pine. Other countries, such as Brazil, India, Spain, and Madagascar, established plantations with other exotic species, such as eucalyptus.

Much of the research conducted by industry laboratories, faculty, or agencies, such as the Western Forests Products Laboratory or the US Forest Products Laboratory, has concentrated on the efficient processing of more kinds and smaller sizes of timber. Technology here has been used to improve the processing of wood products, using computer-directed sawing patterns, beam lamination, better adhesives, drying schedules, and preservatives. What this means is that less valuable species and smaller size trees can be used. So, in this sense research often follows the armies of workers harvesting the planet's resource—in advance of scientific understanding.

Other research, about the functions of the forest, the structure of soil, the physiology of trees, has greatly advanced our knowledge. The fruits of this research are gradually making their way into forest practices.

Forest managers have adapted newer technologies, including biotechnology, computer modeling, remote sensing, and geographical information systems for monitoring, managing, and cutting. Advances in communication technology permit the rapid exchange of data, information, and opinions (by fax, CD ROM, Internet, computer files, and other media). Other new mathematical tools, such as fuzzy set theory and catastrophe theory, are appearing in research projects. Of course, if the data does not fit the industrial paradigm or belief system, then it is ignored.

Biotechnology

There is an incredibly close genetic fit between trees and their immediate environment—this is why natural regeneration is necessary to retain existing forest genetic diversity (tree seed zones are way too large). No single tree has more than a fraction of the genes in a population of trees (the gene pool). The forest holds the genetic diversity, which is important for the forest to adapt to a changing environment and remain stable enough to continue reproducing. Thus, as individual trees are cut, the gene pool is reduced, and a species gradually becomes more susceptible to extinction.

Foresters need an understanding of the genetic dynamics of tree populations in order to manage forests wisely. It is difficult to develop strategies to prevent the depletion of genetic information, especially with intense public and industry demands on forests, and without knowledge of the diversity and distribution of genes in the tree population. Although study of gene flow and genetic recombination is needed, a lot of this knowledge can come from careful observation of the forest, the mating systems of trees—e.g., pine species require pollination as a mechanism for outcrossing—and the shape and size of a forest area. Observation is not enough, however; genetic work has to be part of the program. Management of genetic "resources" is also critical to reforestation efforts.

The goal of biotechnology is to select genes that allow trees to furnish our immediate needs for fast-growing, straight trees. Although forestry is close to approximating the agricultural model, the shift from the exploitation of wild forests to domesticated plantations is not complete. Whereas agricultural genetic manipulation emphasizes disease and insect resistance, the primary selection criteria for forest "crops" are vigor, form, and wood quality (notice that longevity is missing—there is already a case where a genetically manipulated *forest is dying after 30 years because the trees were not selected for longevity*). Many scientists consider biotechnological applications on agricultural crops to be the best solution for conserving gene pools of wild relatives, multiplying elite cultivars, and resisting pathogens and predators. Russell Haines notes that although there are significant genetic gains for long-rotation species, there has been only a minor impact from biotechnology on the genetic quality of tree plantations. That is to say, trees have important limitations compared to agricultural crops, including long generation intervals (a lot of early work on genes was with annual plants or fruit flies), a long juvenile phase before flowering, and difficulty in assessing the inheritance of characteristics.

Nevertheless, biotechnology is being used to "improve" trees. Cryopreservation (very low temperature) is expected to maintain juvenility and capture gains made by "clonal forestry with industrial species." Molecular markers[1] can be used to fingerprint mating patterns in trees or genetic variation for a few model species, such as *Pinus taeda* or hybrids such as *P. elliottii* X *P. caribaea*. Responses of a culture in a lab (in vitro) to stimulation can be correlated to the expression of desirable traits, such as disease resistance; however, there is little application to forest tree species.

Genetic engineering, the direct insertion of genetic material from other species into the target species, could be applied to forest tree species: To modify lignin biosynthesis through antisense technology (strands of mRNA complementary to a sense DNA strand), making the tree more valuable for pulp; to inserting insect resistance genes into a very short rotation tree species of pine or tropical hardwood; or to insert cold-tolerance genes in exotic tree species such as eucalyptus (thus enabling it to take over the planet). Genetic engineering seems especially risky, since trees have evolved other means of adapting rapidly to pathogens, e.g., mutualistic associations with bacteria that protect the trees from insects. Insect-resistant genes may have the effect of disrupting the mutualism, which would have other chained effects in the forest system (possibly the way fertilizer reduces root mass).

Micropropagation has been used to reproduce plants in cell cultures in artificial media; over 100 forest tree species have been micropropagated despite obstacles such as the high cost of planting the stock and insufficient data about field performance. Haines states that: "Micropropagation has an immediate application in integrated clonal propagation systems featuring the commercial planting cuttings harvested from rapidly multiplied, micropropagated stool plants of the selected clones." One high priority is in *vitro* control of the maturation state through molecular studies; the goal here is the maintenance of juvenility in clonal forestry, that is, postpone maturity. Another possibility is to induce early flowering and reduce generation intervals.

Biotechnology could be used to test some other hypotheses. For instance, tree plantations are greatly simplified forest ecosystems based on an agricultural model; they are often a monoculture of Douglas-fir. If they follow others aspects of the agricultural model, the heterozygosity[2] of the species should be reduced. If there is a reduction, how significant is it? If significant, could it be reversed by mimicking the structure of wild forests?

Most of the genetic engineering is expected to be long-term (5-10 years, which is long for funding), especially in areas such as lignin reduction. Biotechnological research is expensive and companies are trying for massive increases in funding for this research. Haines concludes that biotechnology offers "little to meet the most urgent priorities in forest tree improvement," even though there are many active research programs. The reasons include the variety, range, and size of the forest species, and our poor knowledge about them.

Long-term research
Long-term research is especially important in forest ecosystems, since many of the components live hundreds and thousands of years, and the forest itself can live far longer. The National Science Foundation administers a long-term program with a network of representative ecosystems, including forest ecosystems. Two of the research sites are the Harvard Forest (temperate deciduous forest) and the H. J. Andrews Experimental Forest at Oregon State University, where topics include: Successional processes, forest-stream interactions, population dynamics of stands, effects of nitrogen fixers on soil, disturbance in landscapes, and patterns of log decomposition. Started in 1989, this program collects data on long-term trends in ecosystems, that is, patterns of: Primary productivity; organic matter accumulation;

inorganic inputs and movements through soils and water; disturbances; and populations in trophic structure.

Long-term research requires different levels of monitoring, including environmental, biological, and ecological. Environmental monitoring is an umbrella for many activities, including climatic variables and geological processes; for example, the systematic recording of soil and air temperatures, humidity, air pressure are measured by meteorological organizations to predict long-term climatic change. Long-term research also depends on a stable cultural base and shared values between generations. Future issues of *IJE* will address long-term research in more detail.

Modeling
When we try to represent things that are not directly observable, because they are too small (quarks), too large (the universe), or too complex (forests), we use models (from the Latin *modulus*, meaning small measure). Models can be conceptual or physical; they can be plans, analogies, maps, diagrams, graphics, mathematical equations, or even sentences. Models are indispensable and unavoidable in forest research. A model is a pattern or object representing another object; it is a tool that permits thought to be extended. Models can be mathematical constructs, as well as mental states. Mental models are sometimes based on deeply-held, incorrect beliefs, inadequate information, or conflicting desires (and often ignore new or contradictory information). Decisions based on mental models may be erroneous. The way to improve mental models is through questioning (this is what philosophy does, by combining experience with research) and by setting operational definitions.

A computer model is one that relies on mathematical relations that can be quantified. Computers are powerful tools that can process immense quantities of data and solve incredibly complex equations. Computer simulations are both cost-effective and life-saving, literally. In forest ecology, computers can be used to demonstrate differences in the productivities of ground covers or estimate the effects on hydrological cycles of clearcutting forests. The only cost of a simulation is the time or dollars for computing. This is not true in real systems, where clearcutting or acid rain can destroy forests.

The Finnish Forest Research Institute has addressed artificial intelligence applications for forestry, using object-oriented models. They predict tree-growth models, where object-oriented programs are combined with fractal geometry.[3]

Daniel Botkin notes that the computer now serves as a metaphor for bacteria and ecosystems, but he is careful to distinguish computer metaphors from machine-age metaphors, e.g., "trees are machines." Although the computer is a machine, Botkin is correct in noting that machines have become more complex—new computers with parallel processing can imitate simultaneity, for instance.

With others, Botkin has written computer programs to mimic forest growth, starting with conceptual models to a set of mathematical equations. Botkin claims that this program JABOWA simulates ecological succession accurately and realistically. He also claims that the model predicts species compositions over decades. He suggests that his model works, when others do not, because his program incorporates the results of laboratory studies on photosynthesis and tree growth into the code, that is to say, that nature is not expected to work like a machine in the software model. He claims his model also avoids two traps: (1) continued belief in myths about nature in the face of contradictory facts and (2) believing contradictory facts about nature, e.g., nature is constant and not constant.

I am not sure that either can be avoided due to the nature of nature: Facts are always changing, as are myths, as is nature; and, myths and facts are based on nature and are interpretations. Furthermore, contradiction, like complexity, is a characteristic of nature—nature is actually constant and not constant; certain things about nature, such as light and gravity, seem to be constant; other things, such as the intercepted solar radiation are kept constant by the feedback of life into the atmosphere; and other things, such as hurricanes or volcanic eruptions are not constant, although the chaos seems to be a constant. In fact, Botkin relies on the stochasticity of nature to make his model mimic nature more closely.

Botkin claims that mathematical computer models help us synthesize what is too complex for the "naked" mind to combine. The computer models change our perception of nature by restoring the intricacies of natural history to tools for projections. Computers have gotten faster and more powerful, and mathematics has become more sophisticated in its treatment of chaos. But this just seems to have caught up a bit with intuition. We already knew, for instance, that innumerable events in nature happened simultaneously; the computer models have not really changed our perception of that so much as started to catch up to the perception already existing. Furthermore, the models are not nearly so complex that they can keep track of individual trees interacting with other beings in numerous interconnected processes. We do not have that much computing power, machines that fast, or code that complex. Computer models are still incredibly simple, even with addition of algorithmic chance. That reminds me of a short story by the great South American writer Luis Borges, who described the perfect map (a map is a model in fact)—it was so big (scale 1:1) that it covered the entire land. The perfect computer model would duplicate every link in a forest, and it would be so large that it would be a forest.

Finally, Botkin states that computers have changed our perception of nature by blurring the distinction between life and nonlife: Computers make machines more lifelike, and biotechnology makes life seem more machinelike. I do not believe that this is true either. The qualities of life are quite distinguishable in the standard scientific paradigm. Although many cultures consider everything alive, and a few modern ones consider everything dead, the distinctions are made scientifically here. In forests, the importance of living and nonliving parts is equal.

Needed Scientific Research

A lot of research in forestry is good and timely, and a lot is trivial and trendy. Most research is conducted by scientists funded by industry, governments, or universities (which are supported by governments and industries). This research reflects the interests of the sponsors, not surprisingly. For instance many research projects concentrate on regeneration or erosion after clearcutting. While this research is needed, other kinds of research are needed as well.

Historical Research. We need to collect knowledge about the forest, not just of components but of landscape patterns, from local people, who know some of the recent history, or from tribal elders, who may have cultural knowledge, or from other sources: Written reports (such as Lewis-Clark or early surveyors), maps, and aerial photographs. Many archaic peoples have an exact quantitative knowledge that exceeds modern scientific estimates; for example, the Inuit counts of whales and seals in their territory have been used to correct scientific conclusions. Can we ever know enough history?

Fire research. Slash burning and cheap broadcast burning in the Northwest is considered to be an acceptable method of slash disposal and fuel reduction (and the most compatible with clearcutting). Prescribed burning is used more frequently to introduce fire back into forests where it has been excluded for decades or a century. As a management tool, burning is expected to reduce the hazards of wild fires, although Charles Stoddard says that a forest, no matter how well-managed, can be destroyed by crown fires. In several cases, Yellowstone and Georgia, crown fires have damaged well-managed forests as much as badly-managed forests—but, there has not been enough research on management styles or fire-damage to say that it is always the case.

Chris Maser notes that no real research has been done to find out the effects of natural forest fires on forest functioning. For instance, what role does burnt wood play in the detritus food chain? Does prescribed burning provide the same good to the ecosystem as wild fires, despite the fact that is both low-intensity and regular (no large differences in temperature)?

What is the effect of fire on large boles that become snags, then downed dead wood? Does the wood last longer? Do different species use it? Does it absorb water slower?

Productivity. We have not been able to get accurate measurements of trophic level productivities. We must have these to harvest wisely. Why are there not more than 10 trophic levels in a food web, when most are only 3 to 5, according to Perry?

Understanding below ground productivity and processes, which is critical to the health of the ecosystem. Ecologists have concentrated on above ground productivity, although even much of that, such as tissues consumed by animals, is difficult to measure. For example, are roots an important sink for atmospheric carbon?

Mortality Patterns. Branches do not fall off immediately after they die. Dead branches can act as conduits for fungi and insects. Merve Wilkinson, for instance, suspects that the powderworm (Western redcedar borer, *Trachykele blondeli*) enters the tree through dead branches; Merv is actively engaged in an experiment to match a control group (cedars with the dead branches left on) with a large number of cedars with the dead branches removed by artificial pruning. Allowing dead branches to be removed by weakness, wind, rubbing, or precipitation is referred to as self-pruning; removing them with a saw is called artificial pruning. The life history of branches is a complex and fascinating history by itself.

Old Growth. To learn more about soil topography and water retention, Chris Maser suggests identifying early pit and mound topographies from fresh blow downs and tracking them in long-term plots to keep track of species changes, as well as changes in elements (N, O), moisture, and other variables. What difference does aspect make? How does it age? What is the association between it and microtopography and water retention? How do late successional forests vary over the landscape?

At one time, old growth was the matrix in which we lived and worked, but now old growth is measured in patches. Are there enough patches, in good enough patterns, to support sufficient old growth in the future? Do we need to restore many areas? Where do we start? With old growth structures or species? Dead wood is needed to maintain ecosystem functions in old growth; the question is how much. According to David Perry, we have reasonably good guidelines for cavity users, but not much for soil wood.

Key Species. Key species in a forest ecosystem—they all seem to be key players, although you could say that larger, more complex individuals, e.g., bears, may be less critical than smaller, e.g., fungi. Key species can be identified in direct cycles, such as red vole, spotted owl, mycorrhizal fungi, and fir.

Are there key taxa? Key linkages? Key substrates? Key patterns? Key chronosequences? What are key linkages between ecosystem structure and function? Especially in large ecosystems or landscapes?

Artifacts. Roads shape the landscape in major ways. They act as barriers to the migration of many species, and as corridors for others. They change the shape of the land, through cutting and erosion, as well as hydrological flows. Is there an optimum number of roads and skid trails?

We impose patterns on the landscape, from complete removal of part of an ecosystem to the checkerboards of sequential clearcuts. What effects do they have?

We introduce many materials into forests, from chains and cables to fences, slabwood piles, and bullets. What effects do they have?

Biodiversity. What is the loss in biodiversity from harvesting forests? How many critical pieces may be missing? In the absence of demographic data for species, perhaps ecological criteria, e.g., frugivores missing in a system, is the only way to assess ecosystems and species.

Do we have a measure of the extent of damage to forests from harvesting or clearing to determine which species are vulnerable, endangered, critically endangered, or gone?

Genetics. We need to understand the ability of long-lived organisms, such as trees, to genetically track short (less than a generation) environmental intervals (Perry). We also need to understand the linkage of genes to local conditions (a source of genetic diversity).

Genetic memory is the large capacity and long-term memory of the forest. Symbiosis results in a greater store of genetic information. How prevalent or important are symbioses in forest development?

What is the role of microbial symbionts in enhancing the ability of long-lived organisms, such as trees, to genetically track rapid environmental changes? For instance, the mycorrhizal fungi and rhizobacteria may enable Douglas fir to combat hundreds of generations of insect pests. According to George Carroll, with their rapid evolutionary capacity, the microbes add a fluidity of response to the relatively genetically fixed trees.

Other Questions. Almost every walk in the forest, as well as every discussion and every book, raises more questions about what we still need to know.

- What are the effects of parasitoid insects on populations of their hosts under natural conditions—as well as on competition among them and between them and other secondary consumers such as birds?
- Lertzman et al. suggest research on the interactions among processes that cause change in ecosystems (disturbances, invasions, habitat loss, disease, and physiological stresses) and processes that buffer change (community inertia and competition, the self-stabilizing microclimatic feedbacks in massive forests, active management efforts).
- Does competition among decomposers minimize the quantitative effect that their predators have on them? How abundant are detritivore populations? How much competition do they have among them?
- Native eucalyptus forests have a great percentage of GPP consumed by herbivores that do plantations. Does this reduce competition between trees? How much competition is there between herbivores? These forests also have fires that are eliminated from plantations. How does this change the fitness of the trees?
- The structure of the forest canopy, according to Perry, influences air movements inside and outside the forest and thus the gaseous exchanges between plants and atmosphere, which may have complex effects on productivity.
- We need a greater understanding of the complex relationships among plants and soil organisms and how they influence successional dynamics (Perry)
- Understanding nutrient limitations in forests. Although nitrogen as a limit is fairly well understood, sodium and cobalt limits are not well-defined. Nutrients are difficult to measure; furthermore, according

to Perry, there is no simple indicator of deficiency (as with stomatal closure for water limits).

- How do trees pick up minerals, e.g., how trees get iron from siderophores (Perry)
- How plants tolerate very low water potentials without leaf death (Perry). An understanding of how some tree species are able to tolerate low water potentials without leaf death. Most trees close their stomates (the holes for water and CO_2) when leaf water potentials are between -10 and -30 bars—some trees can transpire (diffuse water vapor into the atmosphere for cooling) at potentials below -40 bars.
- The role of bark beetles and other pathogen-carrying insects in forest health and succession (Stocek); insects have coevolved with trees— interactions between them may help maintain genetic variation and cross-fertilization.
- Understanding of factors affecting forest growth (Morris and Miller, in W. Dyck et al.) To what extent is the all-age structure of forests a function of regeneration by birds or mammals? Ronald Lanner suggests that Siberian stone pine forests could be corvid initiated.
- The environmental effects of harvesting, long and short-term; the ability to predict consequences of harvesting.
- The study of below-ground processes and productivity (above and below processes are linked, but we usually only consider the above). Perry notes that as a general rule, the biomass of aboveground animals correlates positively with aboveground primary productivity; is the same relation true for belowground?
- The assimilation of carbon by trees from the atmosphere.
- The role of forests in human thought processes (Fowles, Shepard et al.). This role is neglected, but there is evidence that it is significant.
- How much is enough, in number of trees of a species, area, etc. How much should be exempt from exploitation. What are limits and thresholds?

Ecoforestry Research
The reductionist path taken by science has yielded tremendous results about how the world is built up out of particles and pieces. Now that we have uncovered the complexity, we need to address relationships. This is where the synthetic path can help, by identifying emergent principles and operations. Science is an open, self-referential, self-correcting system capable of using analytic and synthetic methods. Ecoforestry is based on a set of principles that suggest approaches to research and practical applications. These apply to concepts of diversity, complexity, extinction rates, evolutionary change, and ecological value.

Even if traditional science addresses new areas of research, its approach has significant limitations. The predicate logic it uses, for instance, tends to support either/or interpretations. The invariance of regularities of nature in time and place is required by science. Classes are considered eternal and immutable. Furthermore, science is concerned with knowledge of kinds, not necessarily knowledge of individuals. The limits at both ends, prokaryotes and planet, thus fall out of range of the process. The metaphor of mechanism by Descartes (the world is a machine), which science still assumes, implies that matter is inert and that machines must have a creator, that is they are not self-making. Although the metaphor is no longer used in physics, ecology and forestry are still struggling with machine as well as economic metaphors.

To be truly comprehensive, ecoforestry modeling must start with deductive, synthetic, conceptual, historical models based on data generated from research, the rates of resource use, cultural valuation, minimum wilderness preservation, air and water quality, genetic minima, nonrenewable resources, appropriate technological innovation, the importance of cultural frameworks, adventure, research, beauty, uniqueness, and other intangible experiences. A deductive approach is necessary because accurate measurements of productivities in most ecosystems are lacking and exactness in values is misleading. A synthetic approach is necessary to integrate quantitative and qualitative data. The approach must be conceptual because of the inherent fuzziness of systems (this does not mean that fuzzy systems could not be used to arrive at fuzzy quantities and fuzzy solutions, which are useful and valid). A historical approach is necessary because forests are very long-lived entities, and we do have anecdotal and scientific information about the historical range of forests.

To explore its approach the Ecoforestry Institute has set up several small research projects at the Mountain Grove Center in southern Oregon and is considering several more there and on other sites. Mountain Grove is also hosting students and faculty from Oregon State University, who are conducting research on regeneration, lichens, and decomposition.

Oak-Pine Savanna Restoration. Starting in 1995, a 28-acre plot is being restored to an ecosystem that has almost vanished in Oregon: the oak-pine savanna. Under the direction of Dennis Martinez (see *IJE* V12N3, pps. 279-288), small remnants have been identified and marked; much of the overstory of firs has been removed, and native seeds are being planted.

The goals of this project are to restore a self-maintaining ecosystem and to try to understand the interactions of species between this system and the surrounding matrix.

Pacific Madrone Management. The Ecoforestry Institute has proposed to study the sustained use of Pacific Madrone for veneer and other products within an ecosystem management perspective. Pacific madrone been considered a "weed" tree of little value except for chips or fire wood. Even today madrone continues to be cut and left in the bush, poisoned and burned. Timber companies have attempted to eliminate it with aerial spraying to allow more sunlight for conifers, particularly Douglas fir. This practice needs to be changed; madrone is a natural component of west-coast forests. A greater understanding of the ecological role of madrone and its contributions within the ecosystem to its maintenance as a healthy ecosystem is needed.

The attempts to eliminate madrone from the ecosystems of the region, combined with the historic over-cutting of conifers, have led to an apparent over abundance of madrone compared to what we know and observe to be its 'historical range of variability.' EI is working with scientists from Oregon State University and the Forest Service to study the range of variability of madrone across the region.

EI intends to develop a whole-system, operational design to study, manage and produce veneer from madrone at an appropriate scale which will sustain the resource and its ecosystems, while ensuring economic prosperity and community well-being for present and future generations.

The project will investigate the ecological roles of madrone within an ecosystem perspective: How madrone contributes to the maintenance of healthy ecosystems; its historical range of variability; current inventories of volumes, age-classes and grades; rates of growth on different sites and under different conditions; and ownerships across its range.

The goals of this project are to improve forest health, as measured by species diversity, presence of keystone species, and dynamic stability; Train interested forestland owners/managers and loggers who are committed to the practice of long-term, sustainable forestry to select and cut veneer quality madrone; and to cooperate in the development of markets and community enterprises to employ local people in thinning and pruning of small diameter madrone.

Fire History Study. Working with the Bureau of Land Management in Oregon and the Tree Ring Research Laboratory in Arizona, EI is attempting to reconstruct a fire history of the forest at Mountain Grove. Samples will be taken from a large number of trees using a power corer; the samples will be sent to Tucson for analysis. Dennis Martinez will be trained there in techniques.

EI will host a workshop on the fire history. The goals of this project are to establish levels and frequencies of fires; to estimate contributions of aboriginal burning; and to understand how burning shaped the forest. The information will be applied to the current treatment of the forest, especially prescribed burns.

Old Growth Recruitment. EI is proposing to set up a program to study various aspects of old growth. Old growth has been virtually eliminated from privately owned lands and lower-elevation sites. Although old growth Douglas-fir still occurs on about 25% of national forest lands west of the Cascades, it has been reduced to only 3.3% of the Siuslaw National Forest. Remaining patches of old growth are small and isolated. Old growth from public lands is expected to provide the major timber supply in the immediate future but for western Oregon as a whole there is a projected 22% reduction in cut by the year 2000.

EI is setting up permanent plots at Mountain Grove to measure variables contributing to old growth recruitment. Special attention will be given to pit and mound topography, downed woody debris, and snags. The goals of this project are to try to determine the health of old growth on various properties; to create strategies for restoring old growth structure and function to secondary forests with some old growth features still in place; and to create old growth structure with a minimum of old trees.

Biomass Removal. EI is considering up a broad synthetic approach to the question of the function of biomass in forest, with special attention to the amount of removal over long periods of time. Temperate rainforests are distinguished by tremendous accumulation of biomass. Coniferous forests, with large, long-lived trees, accumulate high biomasses—500 to 2000 metric tons per hectare, according to Franklin and Waring. The increased height of vegetation has importance for allowing bird species to be established. The taller the forest, the higher the biomass, in general.

Biomass is the amount of matter present in living organisms at one particular time. Biomass is a static measure; the productivity that contributes to it, however, has an annual rate. Like entropy and other measures, biomass does not necessarily have a constant rate. With energy flow through the system and production, the biomass changes with time. Values for some compartments, such as leaves, roots, or reproductive structures, vary throughout the year. As the forest ages, it undergoes changes in productivity, structure and population selection. Almost one third of the production and the biomass is underground in roots. Furthermore, as a stand matures, the total biomass tends to increase steadily to a high homeorhetic value (not actually a maximum) if the environment stays relatively constant; otherwise it can decrease somewhat. Likens, Holling and others show that it gradually declines after reaching a maximum.

The biomass does not keep accumulating, it is being recycled by living organisms; the biomass is finite. In other words, there is only so much flesh (or units of potential energy in kilocalories) in the system and it is shared by trees, owls, fungi, and squirrels. Most of the flesh is in bacteria, fungi, and beetles, who use most of the energy up in living. The predators that consume them only get to consume the extra energy—the rest has been spent, which is why the predators are fewer and bigger (more efficient). As more and more energy is used for maintenance, the net community production (NCP) approaches zero. The mature system becomes more efficient, as it supports a larger biomass with the same amount of energy. The food chains become more web-like (dominated by detritus chains as opposed to linear grazing). More mature systems have a richer structure and a lower productivity per unit biomass. There are more steps in the trophic pyramid. There is higher efficiency in every relation. The loss of energy is less, so less energy is needed to maintain the system. This is why old growth forests are efficient.

Biomass is always being gained and lost by the system. Biomass is lost from a system in numerous ways: Grazing by herbivores, especially insects, which are difficult to measure and can fly out of the system. Large mammals, such as elk, can also transport biomass out of the system. Downed trees through erosion or flood. Chris Maser has studied the fate of large trees that end up in the ocean. Fires can vaporize amounts of biomass, which then leaves the system through winds. Biomass is gained in the system mostly by photosynthetic fixation by chlorophyll in leaves, but also by animals and insects entering the system and defecating or dying, or simply by being downstream.

This research project will address the importance of biomass in maintaining forest processes and complexity and in stabilizing the forest. Biomass represents a food source for predators, including bacteria and viruses. This is one of the main reasons that forests become more complex over time, as new species carve out niches in the biomass. This may also be one reason why humans are experiencing new diseases: the human biomass increases dramatically at the expense of forest biomass and the predators adapt.

The system has the possibility of achieving metastable states that are resistant to disturbances. Disturbances, such as windstorms or pollution, may decrease biomass, although disturbances can increase biomass, as when a grazer population is reduced abruptly. Forests, with large long-lived trees and large animals can be metastable, that is resistant to disturbance. Other systems, such as rock outcroppings, are physically stable but have a low biomass.

Old-growth forests, with a high biomass, are being replaced by young stands that are harvested before they reach maturity. Forest biomass is decreasing, especially considering conversion and degradation. Humans may be transforming the biomass of forests into industrial goods or fuel such that biomass consumption exceeds replacement. "The trends and patterns of cutting the old-growth forest have caused widespread concern," according to Larry Harris. Annual loss and removal of timber has greatly exceeded annual growth. This research project will address several questions: What does it mean to a system to lose more than it accumulates? What changes go through the system? In addition to the loss of structures and species, biomass is being removed. Biomass removal is an external disturbance to the forest, whether it is done by fire, animals, or humans. Eventually, the biomass will be redistributed. Chris Maser points out that in geological time, the land that is now forested may become grassland, shrubland, or even desert; the forest is only an ecological snapshot at one time.

The goals of this project are to contribute to an understanding of how biomass removal from a forest affects the long-term regeneration and stability of that forest, to emphasize the importance of biomass (as a shock absorber against disturbance and a source of flexibility), and to relate it to the control of succession by exploitation, conservation, release, and reorganization (after Holling). Relevant questions include: How much wood can be taken from a forest? How much biomass can be taken before the system collapses? 10% 50% How much is taken away in natural processes?

Summary

Advances in science have been quite remarkable. The intent of this article is only to ask in what directions research could be taken to continue being remarkable, but applicable to whole forests. While pure research continues to reveal unimaginable details of the forest, and while applied research continues to support sophisticated forest use, more ecological and landscape research is needed. Certification requires much better scientific research to determine minimum, optimum, maximum, or satisfactory numbers of features, i.e., a satisfactory number of overstory trees that can be removed, or the optimum percentage of trees to be left to maintain forest health. Research is expensive, time-consuming, labor-intensive, and uncertain, however.

The increase in forest research, and in public interest as well, has not been incorporated into management. Especially in tropical forests, the use of management systems has been declining for over 30 years. What has increased are the advertising budgets for forest corporations and the inflated claims by forestry departments about their effectiveness.

A lot of forestry deals with not just the physical resources of the forest, but with the aesthetic and spiritual aspects. Many people are concerned that research will eventually make the forest into a factory, without secrets or magic. William Carr addresses that concern: "If magic be defined as something produced by secret forces in nature, and secret in turn defined as something revealed to none or to few ... then magic is not likely to be diminished by all the science we can muster. Research may provide us with answers, but these answers forever lead to new and more profound questions."

Chapter 25

The Health of Forests

Forest health has become an immediate concern. Health is one of the new buzzwords for ecosystem management. Some timber companies support the Forest Service idea of logging burned forests back to health. Some activists urge the government to set aside more wilderness so that forest health can recover. At several recent meetings, I encountered several different concepts of health. Perhaps they are all correct; but perhaps there is something more to consider.

I attended the Inland Empire Society of American Foresters (SAF) meeting a while ago. The meeting was called Foresters on the Issues. Health was one of the three main issues (along with Certification and Property Rights). The first speaker on Forest Health was John Beuter, Chair on the SAF National Commission on Forest Health. After identifying himself as a forest economist, he admitted he was "troubled by the idea of health." Beuter then made the following statements:

"forest health means different things to different people,"
"forest health is part of management,"
"forest health will change,"
"we cannot say what is healthy,"
"health and productivity are not related,"
"productivity of a forest is standard, enhancing the value of timber,"
"value is from authority, natural law ...," and
"we must focus on productivity ... and abandon forest health."

The next speaker was Winston Wiggins, who described himself as a "bureaucrat" with the Idaho Department of Lands. Wiggins said there was a "battle of values," while repeating many of the gems first polished by Beuter. He stated that the guiding principle of state land management was to create "maximum financial return to the state schools." He did take a stab at defining forest health as the ability "to continue to produce products in perpetuity." That is, productivity.

The final speaker, John Marshall, compared the first SAF Forest Health report (by Logan Norris at Oregon State University) with the new report by John Beuter. Then he criticized forest health as a concept as being "unscientific." Marshall said the concept of forest health was "subjective."

After the three talks, the moderator wondered: "If this is not forestry leadership, then what is?" (—my first thought was "the Girl Scouts—they are less arrogant, more willing to spend time in the forest and learn, and willing to sell cookies to raise money, rather than collect government salaries," but I held my tongue). He then asked for questions from the listeners. Since no one else was at the microphone, I asked the first question: "To all three panelists, why can't you define forest health scientifically?"

Marshall answered first: "Because it is not a scientific concept."
Me: "Yes, it is. It may not be completely defined, but it is a legitimate
  scientific concept."
Marshall: "No, it is not scientific; it is emotional and social. It means

different things to different people, as John said."

Me: "Yes, it is scientific. Let's use an analogy. You agree that gravity is a scientific concept [Marshall nods]; it can be measured precisely, etc. But to someone who fell down the stairs, it has emotional meaning that does not make it any less scientific. Forest health is a similar concept."

Marshall: "Precision is the difference."

Me: "But precision is not required for the concept to be valid; another analogy, with less precision: Medicine and mental health, just 30 years ago, were disease-oriented, not health-oriented; then in the 1970s, both became more health-oriented, defining health in positive terms, not just as the absence of disease. Forest health is where mental health was 30 years ago."

Beuter: "Can you define a healthy forest?"

Me: "Yes, I can, maybe not completely at once, but as continuity, functioning, uh—"

Beuter: "I just bought 200 acres of clearcut land. Is that a healthy forest?"

Me: "No, it does not meet the tests of structural/functional completeness or continuity."

Beuter: "Are you saying I am stupid?"

Me: "No, that isn't necessary. But the forest is not healthy—in fact it isn't a forest anymore."

Beuter: "My clearcut is a healthy forest. We planted 155,000 trees, with help from a Mexican cooperative. Why isn't it a forest?"

Me: "If I eat the pear I got from the break table, it isn't a pear anymore; it has contributed its materials to me. Maybe another pear will appear on the table, but the first one is gone. Cutting, clearcutting, the forest kills it. It doesn't exist anymore. The land may have the potential for a new forest, if the soil fungi isn't dead and the squirrels, owls, bats, driven away. You just said you planted thousands of trees. Would you have had to do that if there was a healthy forest there?"

Beuter: "We planted to advance the succession; there are healthy trees surrounding the cut, the soil is okay."

Marshall: "I wanted to know how you can tell if people are healthy. By their looks right?"

Me: "By their looks as a first impression. Then by whether they are here next year; I would know more with a medical exam."

Marshall: "Can you define human health?"

Me: "Not being a doctor, not very well, but I could give you references to look up [audience laughs]. Well, I suppose I could, in general, um, vitality, uh ..."

Marshall: "My mother is healthy, but not as healthy as me, and I am not as healthy as Dan O'Brien. Are we all healthy?"

Me: "Health is a continuum, like gravity. We could define human health, like we can with forest health, as functioning with a minimum of disease organisms always present."

Beuter: "Would you say Steven Hawking is healthy? [ed. shakes head and says no]. Let me make an analogy. Let's say I am God and Hawking is the land [audience laughs]. Then Hawking is doing what I want him to do—producing, ideas, despite his situation."

Wiggins: "We could say productivity is health."

Me: "Hawking has admitted being rather unhealthy. Ill health is not death, however. He can do what he does only with immense support from others. Another analogy: Badly stressed trees produce heavy seed crops in case they don't make it. Productivity and health have a complex relationship."

Beuter: "Productivity is doable. That's all we can define."

Me: "If I died, became a ghost, and was able to haunt you productively, would I still be healthy by your definition?" [audience laughs]

Moderator: "We have other people who want to ask questions. We should move on now."

This whole exchange was tape-recorded by several people—yes, Beuter did say "my clearcut is a healthy forest." What he was talking about was forestry as deforestation, which was, in fact, its original definition: the removal of all trees. When does clearcutting increase the health of a forest, if at all?

Definition: Health as Productivity and Growth

The important thing that I forgot to say, of course, was that forest productivity has an exact scientific definition (in terms of Gross Primary Productivity, Net Primary Productivity, and Net Ecosystem Productivity—although these are rarely all measured) and therefore, by their own arguments, forest health could be defined scientifically. But forest health is not related only to productivity. The forests they were talking about were plantations, with high GPP and NEP, that is, young growing trees. Old growth, with good GPP and NPP, and an NEP of almost zero, would not be considered healthy by these SAF scions—yet, it would be very healthy by ecoforestry definitions.

In the organic world, growth is healthy only when the rate of change is decelerative in the long run; cancer and population are constant or accelerative. Industrial forestry tries to maintain then cut plantations after the rapid juvenile growth.

To relate health to growth and productivity, we could say that the capital of an ecosystem would be its physical environment and its gross primary productivity; interest would be the net ecosystem productivity. The production percentage would be the amount necessary to keep the ecosystem healthy.

Our measurements of productivity, however, are not adequate. We are
measuring over a year or two only to establish a growth rate or productivity.
We should be measuring over centuries. A forest is a long-term, dynamically-
changing being. We cannot use a short-term industrial approach to measure
a few parameters and then pretend we know enough about a forest to cut
a large percentage of it. Forests are created by slow processes that take
hundreds or thousands of years. In their 36-year study of a 450-year old
conifer forest in Washington, Jerry Franklin projected that it would take the
shade-intolerant Douglas fir 750 years to drop out of the forest. Chris Maser
points out how long it takes for coarse woody debris to decay (200-460
years). Soil formation takes millennia; rates can range from 50-100 years per
centimeter.

The problem with health is that it is such a general concept, as
O'Laughlin says, a judgment "without an operational definition." Since
ecosystems exist at all scales, it is hard to determine their health. Stress
and pathology are normal parts of living systems; an outbreak in a small
ecosystem may be part of a larger healthier system. Furthermore, not enough
data exists about what levels of stress or pathology threaten the health of
ecosystems—there is not enough data on ecosystems, not enough reference
points or control groups.

Often foresters assess stand health rather than forest landscape health.
Both are important, since stand health is a way of evaluating forest health
over time. We do need baseline data, but it has to be from a comprehensive
set of indicators, some of which may be qualitative and other quantitative.
The data has to be collected religiously.

Assuming that forests are limited by the real biological constraints of
ecosystems and biogeochemical cycles, a deductive, synthetic, conceptual
model of forest health could be created, based on data generated from
research on net primary (NPP) and net community (NCP) productivity. A
deductive approach is necessary because accurate measurements of trophic
level productivities (especially underground) in most ecosystems are lacking.
A synthetic approach is necessary to integrate quantitative and qualitative
data. The model must be conceptual because of the inherent fuzziness of the
systems.

Redefinition: Health as Absence of Disease
A useful metaphor for forest health is human health, which has been studied
for thousands of years. Our first clue to unhealthy people is abnormal
behavior—not moving or breathing, for instance. The first thing we do is to
classify the symptoms. Then we measure vital signs: heart, blood pressure,
temperature, and maybe white blood cell count. After a diagnosis is made,
it is verified usually by more measurements. Doctors, who often rely on
their long experience and learning, then discuss a prognosis and prescribe a
treatment. So, health is considered the continuity of normal behavior.

Health was also defined generally as the absence of disease. By this
traditional way, via negativa, health is not having cancer, infections, high
blood pressure, diabetes, or other ailments. The opposite of health is disease.

But, what is a disease? Many diseases, at some stage or concentration, obviously compromise health. But some human diseases, such as sickle cell anemia, make one healthier in the sense of being able to resist another worse disease, G-6-PD deficiency (glucose-6-phosphate dehydrogenase), a genetic disease that causes red blood cells to dissolve.

A definition of health is the condition of being sound in body or being well. The World Health Organization (WHO) defines health as a total physical, psychological, spiritual well-being of an individual. In forests this may be related to diversity. Diversity means species richness, different age and size classes in a population, and genetic differences in a species, as well as kinds of habitats present in an ecosystem and the kinds of communities occupying the habitats; and the kinds of ecological processes that maintain habitats; and the variety and richness of the planet's genetic heritage in general. Forest ecosystems that have diversity are usually healthy.

Aldo Leopold started to describe a science of land health; "Health is the capacity of the land for self-renewal. ... A science of land health needs, first of all, a base datum of normality, a picture of how healthy land maintains itself as an organism." Notice that Leopold anchored this concept in a theory of organism rather than an ecological theory of community. Interestingly, the Buddha also gave an organismic definition of forest: "a peculiar organism of unlimited kindness and benevolence that makes no demands for its sustenance and extends generously the products of its life activity; it affords protection to all beings, offering shade even to the axeman who destroys it."

Problems with Redefinitions
Forests are not organisms in the strict sense, so the analogy is not perfect. The problem with the health of forests is that we do not have the same huge compendium of diseases and symptoms. In fact, we are just starting to compile the stresses and causes of forest illness. Furthermore, we tend to use our human values to judge health in forests; for example, we tend to think that forests should exhibit regularity, but many forests, such as boreal forests, are arrhythmic, that is they are punctuated by surprise events (as Holling suggests).

Human medicine itself is changing from being disease-driven to wellness oriented, from focusing on symptoms to describing the properties of health, such as interconnectedness and self-realization. Some ecologists or foresters, Rapport or Kimmins for example, define health as the potential for recovery after perturbations (such as logging); this definition still reflects human values. Forest health has to be defined without undue emphasis on human needs.

Forests are not just simple organisms; there are other differences. Forests live longer than humans. The difference in life-times makes discussions of health more problematic. Health can only be evaluated over time. The conditions of forests change over decades or millennia. When do we know if the change is succession or symptoms? Have we ever studied a forest for even one forest lifetime?

However, concepts of health applied to organisms can also be extended to communities and ecosystems. Both are complex whole systems with parts and functions. Of course forests are more complex than humans, which is also why we have trouble measuring social or cultural health (there is no standard society or standard forest as there is a "standard human"). That means we should be measuring a large number of variables, starting with soil depth (richness, compaction), then annual nitrogen uptake (often related to leaf litter), trophic flows, species counts, and patterns of activity. In forests, foresters usually measure the number and ages of trees, as well as canopy. Some foresters relate health directly to crown density and foliage loss as a percentage. But, assessing the loss has an element of subjectivity and there is disagreement about how much foliage loss is acceptable. Possibly, the presence of too many pests or diseases is the best indicator of the health of the system.

Redefinition: Health as Vigor and Resilience
To be considered healthy, an ecosystem has to maintain its structure and metabolism (rate of energy use) despite occasional stresses. Robert Costanza proposes an operating index of system health that relates these components: $HI = V*O*R$, where HI is health, V is vigor, O is organization, and R is resilience (a form of stability).

Health is a dynamic measure of ecosystem organization, vigor, and resilience. Organization is described by diversity and connectivity; vigor is related to the amount and speed of productivity; and resilience is a measure of reaction to stress. Too much stress, for example, leads to unsustainable patterns of behavior; continuous stress leads to a breakdown of processes that becomes irreversible—the system dies. Quantifying this to be meaningful might be somewhat subjective, although measures already exist for Vigor (NPP, NCP) and Organization (diversity indices). Since resilience is only one form of stability, that component should be an additive function.

The basic medical definition of health used to be freedom from disease. Part of a new definition is resilience to stress—of course, there is good stress (eu-stress) as well as bad stress. Ecosystems respond to stresses in different ways, but usually through a decrease in productivity and material uptakes. Stress may be related to the rate of change for the system, in addition to loss or gain of components or changes in structure.

Jay O'Laughlin et al. defined forest health as the capacity for self-renewal, the ability to recover from stress and disturbance, natural and human-caused. He also defined it a year later (in 1994) as a "condition of forest ecosystems that sustains their complexity while providing for human needs." Or, in an expanded definition: "the vigor or vitality of interacting biotic and abiotic elements of a system characterized by extensive tree cover that function together to sustain life and are isolated mentally for human purposes."

Vitality can include pests and thinning. As a forest ages, it thins itself naturally. The number of trees decrease as the stand ages; the remaining trees are typically bigger. As the frequency of disturbance increases, the forest becomes adapted to the disturbance—even pine plantations in the southeastern US that are managed with controlled burns are less damaged by cool wildfires; after long periods without disturbance, a catastrophic disturbance is more likely—where wind and fire are absent, the probability of insect and disease outbreaks increases. Yet, even catastrophic disturbances like hurricanes rarely damage more than 5 percent of a forest. More than being agents of mortality, insects, diseases, and animals are native components of complex food webs in ecosystems that contribute to the selection of certain kinds (including healthy) and ages of trees (that determines the composition of the forest, which changes over time). Mammals and birds disseminate seeds. Insects pollinate some trees and overwhelm others (rarely more than 1 percent of a forest). Diseases remove stressed trees (also probably a low percentage on the order of 1 percent). Their effect on the long-term health of a forest can only be regarded as positive.

A healthy forest is flexible in its response to diseases and pests. Most healthy ecosystems have high degrees of flexibility. As Gregory Bateson interprets Ross Ashby, any biological system can be describable in terms of interlinked variables, each of which has an upper and lower threshold of tolerance, beyond which the system acts pathologically. Within the limits the variables can be moved for the system to be adapted to the environment. Under stress, some variables move to maximum values near the upper or lower limits—the system loses flexibility, that is the "uncommitted potential for change" (Bateson's definition), and can be destroyed by further stress. The danger in each case is working near the maximum of the system. In the case of forestry, we and our civilizations are part of the system. If an overpopulated society wants more forest products for houses, furniture, and fuel to be more comfortable in an overpopulated state, then more trees must be cut, and this, because the variables are interlinked, means that the stress spreads to more forests. To keep ecological flexibility in the forests, for forest health, wood resources would have to be budgeted in appropriate ways, or demand would have to be reduced—or forests would have to be expanded.

Signs of ecosystem health include the homeorhesis of the system (a similarity of flow, rather than the stable state of homeostasis, after Waddington), the stability of the system (that is, its resilience after stress, such as floods), the diversity of its components, the continuous recycling of elements, and flourishing. Health is the overall ability of a system to maintain itself under a normal range of environmental conditions (which may include hurricanes, volcanic eruptions, or fires). Obviously, a pioneer community may change the conditions to favor a new level of the system with new components.

Using a concept of forest health as persisting within certain boundaries, but varying periodically or aperiodically within those boundaries, management is concerned with actions that might kick the system out of bounds. Furthermore, as a historical process the forest can be considered as having a trajectory towards a stable, mature state.

In a living system, nothing keeps growing forever. Although growth stops, patterns keep developing. Things die and their parts are taken up in cycles. The continuation of the system depends on these cycles. The cycles are bound by limits. Both individuals and communities are usually bound by one or two specific limits. Ecosystem health occurs within those limits.

Figure 7. Beetle on a Pine (Vets Forest, Bulgaria)

It is quite likely that harvesting a limited number of trees in a predominantly natural forest could reduce the probability of disturbance as well as stimulate new growth, although it may equally possible that the regular removal of trees for human use may result in the gradual and long-term decline of a forest. One task of ecoforestry is to allow natural agents to average out and to minimize human disturbance that could destroy the health of a forest.

Reconsiderations of Redefinitions
With a focus on the size and growth rate of stems as log units, many other forest measurements are ignored: shrubs, forbs, ratios of shrubs and forbs to trees, biodiversity, soil depth and dynamics, direction of succession, very-long-term biological processes, carbon loss or gain, biomass (R. Margalef suggests that an objective of an ecosystem is to maximize biomass), and the role of forests in the biosphere cycles.

Monitoring is crucial to understanding forests. Until we understand how forests change and move around the landscapes, we will not know which changes are important and inevitable and which are the unhealthy result of human interference. Until we understand the changes, we will not be able to adjust our needs to the limits of forests.

Richard Hart's approach to monitoring is based on his observations: "The actual substance of which the forest environment is made consists of patterns rather than things or individual species. The forest environment is generated by a patterning of ecological ebb and flow of energy, substances, individuals and species across a suitable landscape. The distinction between growing and declining patterns is not arbitrary, and can be arrived at objectively."

Hart suggests comparing current patterns with historic patterns. Monitoring for elements of the patterns will reflect their changes in distribution and abundance. A list of the most important variables or indicators is made, including compaction, stream chemistry, landscape morphology, and atmospheric.

Descriptions of each indicator outline the specific attributes which can be measured, as well as an explanation of why they contribute to our understanding of forest health. It is anticipated that as understanding of ecological processes increases the number and type of indicators may change.

Before monitoring can begin the objectives and data collection methods have to be nailed down. The purpose of the monitoring program has to be stated; the objectives have to be identified, e.g., the health of the forest. Health will probably be the first objective, followed by production or aesthetics. Health can be "defined" by a set of indicators of health, such as species, or patterns of health, such as stability or productivity. None of the indices that are measured are really adequate to define the health of the forest because they cannot account for the complexity, richness, and cycling that goes on in the forest. For that reason health indices are data that need to be resolved on the ground in person by someone who knows the history of the forest and has a feel for it.

Descriptions of historic landscape disturbance regimes (e.g., fire magnitude and frequency) and the ecosystem component patterns they maintained (e.g., vegetation composition) provide an initial template for descriptions of ecosystem health. The procedural consequences of these facts involve practical changes in the ecosystem's collective patterning of shared needs and governance which can be defined as its health or soundness.

Monitoring facts and connections is important because it references the ultimate intent of ecosystem management: To discover and describe the structural correspondence between a species and its environment. A healthy species not only fits its environment well but also defines or clarifies the collective life it adapts to. We perceive this clarity to be the richness and wholeness of ecosystem structure. An unhealthy or absent species confuses the collective life, and we perceive this as static, fragmented or degenerating environment. There is a correspondence between the holistic behavior of a species and how it can be seen to relate to other elements of the ecosystem.

The trees and other plants and animals evolve into a community of thousands of different species. The "checks and balances" of a complex number of predators, prey, and decomposers tends to dampen any one species from getting out of control (and becoming a pest). This is not to say that everyone lives in a disneyesque fantasy of good will. Organisms survive by defending themselves or attacking others. But, the defensive and attack strategies "coevolve" (Ehrlich and Raven's term) over time. Organisms specialize to avoid competing. Relationships become more intimate, as organisms cooperate for survival advantage and strive for efficiency.

Efficiency has real social and ecological limits. Mechanical efficiency is a small subset of the kinds of efficiency that need to be cultivated. Noss and others have noted that maintaining a healthy forest ecosystem is more efficient in the long-run than having to duplicate the forest functions to keep it healthy. Furthermore, it is dangerous to take maximum yields out of a system unless all factors are known—a virtual impossibility. Therefore, we must aim for optimum yields and calculate (and hope) that the forest has sufficient flexibility to recover from our "take."

Data sorts itself into description by patterns and rules—this is the essence of fuzzy systems. More data results in more patterns and more rules—up to a point, then the rules sort of "max out." A washing machine may only need 30 rules to deal with load size, water cleanliness, but rules for forest management may number 500; maybe 300 or 8000, we will not know until we try. It would require monitoring thousands of parameters and then creating rules to deal with them. Parameters could include canopy cover, stream flow, number of fungus species, number of food webs, or age. The rules would then dictate how much to cut or not cut or whether to cut at all. Cutting would be adjusted based on canopy, water retention (date of snow melt), floor temperature, humidity, and other factors. The rules could also track and evaluate the health of the forest. We could relate patterns in general using a fuzzy model.

Redefinition: Health as Harmony and Wholeness
Health is the coherence of the pattern of living in other words. If the pattern is disrupted, the local entity dies. Many local patterns flow together through time interdependently, sharing materials. The death of one pattern sometimes leads to the death of other patterns. The body of a human or forest or any entity is a dynamic pattern supported by dynamic processes that include other entities.

In Chinese medical tradition, the highest good is harmony, especially social harmony, or good relations. A good person is one who creates and maintains harmony. Perhaps this is the best working definition of health.

Furthermore harmony is related to wholeness. The word "whole" comes from the Indo-European root *kailo*, which is also the root for the words health and holy. The concept of the whole forest. A forest that has very complete complement of interacting beings. A whole forest can renew itself without replanting and pesticides.

David Bohm, in his theory of the implicate universe, proposes that health is a result of a harmonious interaction of all the analyzable parts that comprise the extricate order—cells, tissues, organs, the body—with the surrounding larger environment. Health is a quality that is grounded in the total order of the environment (or implicate order). Health is a dynamic quality of the entire movement of the environment (holoverse) as it flows. As organisms sometimes interfere with others or with the flow of change, the harmony breaks down—we call that disease. Health is the dance of bodies that interpenetrate (in Paul Shepard's image).

None of the bodies are completely independent or completely bounded; they are interdependent and open systems. A body is only maintained by a flow of energy and materials from its environment—much of this flow is in the form of other entities, usually much smaller, such as prey, insects, bacteria, viruses.

Now we can address health in forests, which are larger entities made up of other beings—large patterns made up of smaller ones. The forest is a constant where every component changes, disappears and appears. It is a pattern like a whirlpool. The pattern forms from a torrent of light, energy, molecules, air, water, and even bigger things.

Comprehending patterns is necessary to protect the scale of the forests that is too large to see (watersheds, except by satellite), the parts of the forest that are too small to see (fungi and viruses), the parts that are too-long-lived for us to observe (long successional changes or evolutions), and the parts that we are ignorant about. Without special effort, we are aware only of what we see working in the forest during a very short time. We trust that our plans will ensure that the forest will remain as a healthy entity for a very long time so that many generations of us can gather our needs from it.

Summary

At a Society of American Foresters meeting last month in California, the body of foresters indicated that the regional group should pursue developing a definition of a healthy forest, which they could use to evaluate certification programs.

Historically, we have used forests without regard to their continuity or to their health. Partial knowledge and technology has allowed us to exploit our environment beyond what is desirable for us or for other species. While continued, moderate exploitation is necessary to live, too much exploitation is unwise. A wise use of resources would not make the world less habitable. We are part of the system and must protect its health as a whole.

More than being just a crisis science, ecoforestry is a medical discipline, aimed at restoring forests to health. As with any medicine, the patient actually does most of the work to become healthy, although the doctor gets the credit and the payment. This would also lead to more respect for the practitioners, but also to more responsibility and more rules. The first rule, which we might take to be basic, is identical to the first vow of the Hippocratic oath, "Do no harm" (see Ecoforester's Oath). Noninterference is a basic ecoforestry principle—do not interfere with the health and stability of the forest—the health, diversity, and stability of the forest are a first consideration.

Ecoforestry is a maturing stage that bases forestry in a community context and limits the use of the forest to that which the forest can afford to provide and remain healthy. Ecoforestry undertakes the responsibility to preserve the healthy functioning of the forests under its domain. Ecoforestry has a responsibility for the ecological production of goods from a forest. The health of ecosystems and human institutions should be measured with a holistic index. We have not developed qualitative indicators of ecological health or quantitative measures of social health, much less an ecocentric view that would value preserves of nature for themselves.

To address the health of forests by ecoforestry would be a temporary medicine, not a constant intervention or even a continuous diet. We have already tried to gain complete control over forests through scientific methods and technological applications. We regard medicine as a foolproof system that tried to eliminate weakness, disease, and mistakes.

Rather than telling the forests what to do, rather than controlling their growth, we need to watch forests to see what they do (this used to be the function of natural history), and we need to let them do it (this requires patience and temperance). Abraham Maslow regards this attitude as "taoistic," and the way to forest health is letting the forest do most of the choosing and working.

Our response to the forest, being concerned with its health (as forest doctors or nurses perhaps), is not benign neglect or complete anticipatory stewardship, it is participation in the process of the forest as a harmonious system, with mutually restrained conflicts and constrained influences. The goodness of our lives reflects an imperfect balance of love and selfishness, reason and passion, sensuous materiality and spirituality. We have the responsibility to be healthy, to contribute to the health of our community, and to contribute to the health of natural forest communities.

Chapter 26

Forest Practices Related to Forest Ecosystem Productivity

Summary
Allowable cuts are based on gross forest productivity (or on imaginary numbers provided by economics). Most calculations are concerned with finding a maximum number for planning. Yet, as we know, maximums are rarely stable. Human use is limited by the biological constraints of ecosystems, by biogeochemical cycles, by our knowledge of these systems, and, possibly, by human psychological and cultural limits. Therefore, an optimum goal is calculated, using a deductive, synthetic, conceptual model based on data generated from research on net primary (NPP) and net community (NCP) productivity.

Industrial Forestry as a Tool of Economics
The myths of the mutant modern economics have tremendous impacts on how forests are treated. The old analogy of the economy as a machine leads to dangerous assumptions about forests:

- Everything is a resource (and its corollary, everything has a price [and its corollary, everything that does not have a price is worthless]): Every forest can be cut to provide wood for human needs. The essence of a resource is that its existence acquires meaning only as it is necessary for human needs and wants. Furthermore, if everything is a resource then everything can be used—the forest, not just the trees, can be used.
- Resources are unlimited: If forests are unlimited then we can keep cutting them (and even if they are not unlimited, by the tenets of modern economics, being a scarce resource makes them even more valuable—because the costs and unused equipment can be written off against other profits). Although many economists admit that forests may not be exactly unlimited, the assumption surfaces again in advertisements about the forest industry where the industry credits itself with planting billions of trees.
- The economy has to keep growing to survive: Growth means that the use of the forests will have to keep increasing. If there are not enough old-growth or good timber trees then all trees will have to be used for pulp and fiber (for paper board and fiberboard). If there are not enough trees then we will have to grow wood in vats . Or, we will have to substitute, for example, steel wall studs for wood ones.
- Any resource can be extended through a substitute: Therefore forests are not unique or intrinsically valuable, because they can be substituted with tree plantations or genetically engineered industrial wood cell production or steel or plastic. In economic equations, human-made capital is considered a perfect substitute for natural capital (although interestingly the equation is never reversed).
- Mass production is most efficient: Clearcutting (the incarnation of mass production in forestry) a forest is the most efficient way to extract good wood; high-value native trees are cut without undue expense,

while the site is automatically prepared to host a plantation containing only desirable market species, such as Douglas fir. Furthermore, as Roy Keene has pointed out, the planning, field work, and accounting are also simplified. Waste, or the wood not used, is minimized by the process—of course, all noncommercial species are considered a waste of potential growing conditions.

- Obsolescence is necessary for successful growth: Obsolescent products made from wood are burned, buried in landfills, or stored in concrete hallways, removing them from the cycle of renewal of forests.
- Quality does not matter very much: Good materials, good wood from old-growth trees, are not necessary to make high quality goods. The quality resides in the perception of the consumer, educated by helpful advertising.
- The future is less valuable than the present (discounting): A forest that may have some value now in a few useful species is worthless beyond a certain time (two years, maybe 20). Modern economics has enshrined this one form of selfish behavior and pretends that it is rational and optimal. Unfortunately, as Costanza and Daly have pointed out, short-term self-interest is usually inconsistent with the long-term best interests of the individual or society. Economic discounting is not in the best interests of keeping forests complete and healthy.
- Economists can control the economy: By simplifying the forest and raising one species on a tree plantation, economists think that fertilizers and pesticides can control all foreseeable circumstances— changes in and shifts of the system state can be controlled.

These dangerous assumptions are translated into unhealthy and unsustainable practices that also generate problems, from which the forest economy is suffering. Overgrowth is one problem—too many trees are cut because of demands for wood. The complexity of the system and the number of costs increase; this is only a surprise when total cost accounting is not used. Economic and social instability result from mechanization and the quest for a narrow mechanical efficiency; powerless families are relocated or dislocated as jobs are eliminated by ruthlessly efficient and sophisticated hardware. Forests experience environmental wobble and ecological instability as they are cut; atmospheric and hydrological cycles are simplified and disrupted. Finally, misdirected effort on ill-conceived products, e.g., upscale firewood or throw-away chopsticks, wastes precious wood. The economy cannot adapt to the time scale of the forests or escape its assumptions, so it turns to planning to solve the problems.

Planning
Planning in general means deciding on goals to be achieved in specific situations. For central planning (by a state or province or federal government), the goals are usually small and not comprehensive, such as a cutting level or a single species preservation, and usually end up being a compromise in cost-benefit analysis. For forestry it is the "ideal timber goal," which is usually stated in terms of consumption, production, growth, or stock, each of which tries to satisfy the needs of the goal that precedes it, for

example, attainment of the stock goal permits attainment of the growth goal.

Most plans address problems, such as building roads in forests. Everything else, from employment to pests, is also considered as a problem, and not a direct effect of the cultural implementation of some technology. Most plans seem to be adequate at compiling area data, from topographic to climactic. Most plans are also development plans that are comprehensive in the sense of seeking to meet all needs of the public, agriculture, and industry. But, they also fall prey to all the assumptions of the industrial culture. They tend to be multipurpose with the aim of providing maximum net benefits through the management of forests and wildlife. Both uses of "multipurpose" and "maximum benefits" are based on misunderstandings, however. Multipurpose in practice means human use—perhaps even just one of those, logging; and maximum benefits have proven to be dangerous. Modern resource management strives for maximum sustainable yield, based on partial knowledge of species and great ignorance of ecosystems.

Formal development is more concerned with an assembly-line model—simple, isolated, efficient, and easy to maintain. Planning tends to neglect or dismiss the distribution of negative, uncertain, or nonmonetary effects. Furthermore, we seem to have no mechanism for developing long-range plans. Certainly, there seems to be no way to deal with long-term, slow catastrophes, such as deforestation. Behind our glass wall of television (life in vitro?), we become remote from, and indifferent to, the system that supports us. We acquire unrealistic images of the world and harmful values and then make bad decisions based upon them. We have not developed qualitative indicators on ecological health or quantitative measures of social health, much less an ecocentric view that would value preserves of nature for themselves. Still, we look to science and agricultural technology to save us.

Science and the Agricultural Model

Since the first agriculture, 10,000 years ago, forests have been cut and burned to create fields for food crops. Technological innovation, combined with accelerating population growth, has lead to clearing of many forests for agriculture.

In the 1700s, forests came to be studied as ecosystems and then resources. Germans forestry in the 1800s was relatively holistic and comprehensive as a science. Enderlin extended silviculture from observation to a more scientific foundation. Heyer analyzed a site factor in a forest and systematically developed a theory of tolerance (capacity to endure shade). Others, including Mayr, Duesberg, and Dittmar, applied biology, chemistry, and physics to silvicultural theory. Ebermayer recommended establishing forest experiment stations in 1861.

As German forests came under management, however, vast plantations replaced the dark woods famous in myth and story, until by the 1930s, most of the forest was managed plantations. Plantations were orderly and neat, with no windfall or "trash" allowed on the ground.

American forestry was founded on this model and has followed it. Forest reserves were set aside starting in 1891 to protect water supplies and to ensure adequate timber. Gifford Pinchot headed the U.S. Forest Service based on the conservation ideas of wise use and sustained yield. Pinchot established his brand of forestry as sustainable, as did Leopold, as do the new foresters. In fact, Pinchot's sustainability was based on the German management concept of sustained-yield from the 18th-century.

Industrial forestry was quick to take advantage of the scientific understanding of forests and trees, starting in the mid-1800s. By the 1980s, scientific research had resulted in technological applications for growing, genetically "improving," harvesting, and processing woody material on an unprecedented, vast, and unconsidered scale. Industrial companies established tree breeding and fertilization programs, whose goal was usually fast growth. Pine plantations in the southern United States were rated as a success and served as a model for plantations elsewhere; Australia, New Zealand, Chile, and South Africa replaced much of their native forests with exotic pine plantations, using species like Monterey pine. Other countries, such as Brazil, India, Spain, and Madagascar, established plantations with exotic species, such as eucalyptus.

A lot of research conducted by industry labs, faculty, or agencies, such as the Western Forests Products Laboratory or the US Forest Products Laboratory, has concentrated on the efficient processing of more kinds and smaller sizes of timber. Technology here has been used to improve processing of wood products, using computer-directed sawing patterns, beam lamination, better adhesives, drying schedules, and preservatives. What this means is that less valuable species and smaller size trees can be used.

Forest managers have adapted newer technologies, including biotechnology, computer modeling, remote sensing, and geographical information systems for monitoring, managing, and cutting. Advances in communication technology permit the rapid exchange of data, information, and opinions. Of course, if the data does not fit the paradigm or belief system, then it is ignored.

Sometimes the data is interpreted for anticipated gains. Because young timber grows faster than old timber, US Forest Service sometimes justifies the removal of two acres of old timber for each acre of young timber in a plantation—ACE, the allowable cut effect. Although that is true of the growth rate, it is not true of total volume. Furthermore, the younger trees in plantations are planned to be part of a short rotation. The 1000-year old trees that were cut were not part of a 1000-year rotation, but part of the inherited capital of the land. If forestry is going to be meaningfully sustainable, then the rotation length should match the forest being cut.

With the plantation system, forestry came to be treated as a special form of agriculture. Agricultural systems are distinctive types of human-modified ecosystems that are laid out to increase production per unit (hectare or acre), which it does by duplicating the highly productive pioneer stages of succession. In order to be successful, however, and increase food output per unit agriculture has had to become industrial: With a large capital outlay for equipment, new technology, from combines to biotechnology, fertilizers, and pesticides, and market supports to hide a few of the real costs—industrial agriculture is very energy-intensive.

Like agriculture, forestry uses soil to produce a crop for the purpose of increasing wealth (or perhaps just revenue). Like agriculture, forestry is renewable (unlike mines or oil extraction). Like agriculture, forestry is based on knowledge of many fields, including botany, soil, and meteorology. Like agriculture, forestry deals with vast areas. Unlike agriculture, forestry deals with wild plants on wild soils. Furthermore, trees are very long-lived, unlike crops of annuals. Unlike agriculture, the crops are not as resilient to being cropped annually. Unlike agriculture, the trees have greater requirements of the forest than grasses do on the grasslands. Trees are directly responsible for soil fertility and tilling.

Beginning in the 1940s, it became evident that the plantation system, with single species even-aged trees, was susceptible to catastrophic change—wind, pollution, insects—in a way that natural forests were not. Some forests began to die as forests—Waldsterben (forest death). The forestry machine was in trouble. (It is basically used for a straight line, once-through pass. As a metaphor for industrial forestry we might think of a high-powered, gas-engined vehicle somewhat like a snow plow but only an accelerator pedal and rearview mirror in the comfortably appointed, fully automated, air-conditioned cabin—no brakes, no steering wheel, no windshield, no reverse, just one forward speed. No wonder it does so much damage. It has no real controls.)

Tragedy and Excess

The success of agriculture in natural grasslands led to its use as a model for agriculture everywhere and then for a model for all organic resources. Humanity has taken over the habitats and ecosystems of other animals and plants, simplified them, and converted them to the production of protein and resources. As ecosystems are degraded, deforestation, desertification, and drawdowns are occurring at the scale of an ice-age or comet impact, but the result is more meaningful to us than these historical events, since we are dependent on the ecosystems we are changing. We take our rapid population growth as a requirement and depend on nonrenewable resources to support our numbers and lifestyles, ignoring long-term deficits, carrying capacities, and limits. Some of these trends—perhaps we should call them gigatrends because of their age and size—have been noticeable for thousands of years, but nothing has been done to halt them. History records that some civilizations tried to manage their resources and failed. These gigatrends might end in tragedy for humanity.

For the Greeks, the operation of tragedy resulted from success taken to great lengths, that is, where successful behavior in one context is applied to all contexts, with the result that the opposite action occurs from the one desired. Humans in moderate numbers were able to take what was needed from natural ecosystems without interfering with the processes. Our dominance, once so successful because of our big brains and tool-using hands, has now become self-destructive. With rapid and intense development, ecosystems collapse or stabilize at a simpler state.

Most of the noncultivated land surface of the earth is being managed; wolves, caribou, salmon, redwoods, irises are managed or else destroyed. Even a modern, balanced exploitation may destroy forests and fisheries.

Currently, many resource managers espouse the ideas of equilibrium maintenance and maximum sustainable yield. These ideas are poor guides to management. By trying to maintain habitats in equilibrium, we often set them up for catastrophic decline, for instance, in fire-climax pine forests, or destroy resident species, e.g., the spotted owl. One possible solution is to relate human use to the natural productivity of ecosystems.

Ecological Productivity of Forest Ecosystems
A forest is an energy/matter system. Needham suggested that living systems are energetic systems competing for materials, but from another perspective, they are energy-organizing material systems. Each perspective emphasizes different aspects of the system: A matter-organizing energy system emphasizes novelty; an energy-organizing matter system emphasizes confirmation. Transmission of information is more closely linked to organization of matter than energy. Furthermore, fitting energy (appropriately low and usable) is what the systems need, not raw, unconverted sunlight. Neither energy or matter is ontologically superior to the other; both are dimensions of a Space/Time/Energy/Mass field (STEM is described more fully in Wittbecker 1976).

The trees in a forest convert sunlight to sugar. The efficiency of conversion hovers about 1 percent, perhaps barely exceeding it on good sites. Over half of the sugar is converted to energy for respiration to keep the tree alive; over half of what remains in leaves and roots is consumed by insects, fungi, and animals dependent on it (and almost a third of what remains is below ground). Most of the massive productivity of forests is not in stem wood. The productivity depends on many other factors besides the interception of sunlight: climate, moisture, nutrients, predation, cycling, fire and others.

The forest is also an ecosystem. Ecosystems result from the interaction of all living and nonliving factors of the environment (refer to Tansley 1935). These systems are profoundly affected by both random and purposive physical and biological factors. As a result, habitats change and organisms adapt. By modifying their habitats in the process of living, organisms change the characteristics of the system and force further adaptation. More important, organisms are limited by the productivity of the system in varying degrees. Human populations (homo sapiens sapiens) inhabit specific ecosystems and are adapted to and limited by the productivity of ecosystems (to what extent remains to be seen).

The structure of a forest is based on material and energetic exchanges. The matter present is biomass (B); the material output is primary productivity (P). Their relation (P/B) is the flow of energy per unit biomass. The total amount of biomass or energy produced by populations of plants (autotrophs) and animals (heterotrophs) through growth and reproduction is the productivity of the system. Primary productivity is the rate at which organic material is created by photosynthesis. Bacterial photosynthesis and chemosynthesis contribute to gross productivity, but much less significantly. Gross Primary Production (GPP) is the rate of energy storage by photosynthesis (equal to the photosynthetic efficiency) in plants.

The maintenance and reproduction of plants is paid for by the energy expenditure of Respiration (R). The amount of energy stored as organic matter after respiration is identified as Net Primary Production (NPP), which equals plant growth efficiency. The calculation of NPP is shown by:

$$NPP = GPP - R$$

To measure net primary productivity, many items must be considered. For example, a field planted in cereal, samples should be taken regularly to determine losses of old leaves, insects, and nonedible material (weeds). At harvest the plants would be dried and weighed. The net productivity would equal the sum of stems, leaves, flowers, fruits, roots, and insect loss, minus the initial seed sown. For a forest the measurement is very difficult. The total amount of organic matter, or standing crop, is the biomass.

NPP accumulates through the history of a system as plant biomass expressed as kilocalories per square meter ($Kcal/m^2$). The kilocalorie is used as a unit of energy flow and production; it is a useful common denominator for these calculations. The problem of confusing production (amounts) with productivity (rates) is avoided here by considering all values per unit area (m^2) over the entire year (m^2/yr). The biomass minus the decomposition in a system is the standing crop biomass of that system.

The energy stored in heterotrophs (consumers such as animals or saprobes) is referred to as secondary production (SP) or assimilation. Secondary productivity is defined as the formation of new protoplasm by heterotroph populations. However, the percentage of net primary productivity eaten by animals is not equal to secondary productivity, because only a small fraction becomes organic matter in the animal. Some food is egested unused; in caterpillars, for instance, this may amount to over 80 percent. Of food that is digested and assimilated, that which is spent on respiration does not become organic matter. The remainder is considered secondary productivity, as growth of the individual or as part of reproduction. Although a mature (non-growing), non-reproducing animal has zero secondary production, a population of animals usually has measurable secondary production since some individuals are usually growing or reproducing.

Secondary productivity corresponds to net productivity in plants. Total assimilation in animals (almost impossible) corresponds to gross productivity in plants. Studies in secondary production are difficult; the species must be measured for population density, age distribution, food consumption and utilization, growth, and reproduction; bacterial, fungal and parasitic populations must also be considered.

Community Production. Biomass accumulation is the increase in community organic matter as the difference between the gross primary productivity and the total community respiration. This difference is also the net ecosystem production or the net community production. It is usually less than the net primary production. The storage of energy or organic matter not used by heterotrophs is Net Community Production (NCP). The relationship between productivities (where Ra=autotroph respiration and Rh=heterotroph respiration) is of the order shown by:

$$GPP = NPP + Ra = Ra + Rh + NCP$$

In a balanced ecosystem, NPP equals respiration; in an accumulating system, NPP usually exceeds respiration by 1 to 10 percent. Although stable ecosystems tend to produce a maximum GPP, species, biomass, and the production to respiration ratio (P/R) continue to change long after the maximum has been achieved. In fact, as the GPP approaches an asymptote, respiration increases. In a mature community, temperate or tropical rainforests for instance, NCP approaches zero, as adapted heterotrophs become more efficient at using production. In accumulating systems, such as grasslands or young forests, NCP can range from 20 to 30 percent. A balanced system is integrated and self-perpetuating, where production (the photosynthetic fixture of carbon) is balanced by respiration (the oxidation of carbon). As a system becomes balanced, the pressure of selection of organisms shifts; the capacity to live in crowded circumstances with limited resources is favored. Populations that depend on rapid individual turnover (r-selection) are not as successful as populations of large, long-lived individuals (K-selection). As an ecosystem ages, pressure is put on populations by other populations. Competition and predation become more complex. Pioneer species, such as Lodgepole pine are replaced by transitional species, such as white fir or by mature species, such as cedar or hemlock. The changes are referred to as succession.

Maturity. Succession decreases the flow of energy per unit biomass—this is Ramon Margalef's concept of maturity. The energy required to maintain an ecosystem is inversely related to complexity. Any ecosystem not subjected to outside disturbance changes in an orderly and directional way: the complexity of structure increases and the energy flow per unit biomass decreases. The physical environment limits the type of change. Homeorhetic (Waddington's term meaning stable flow) mechanisms protect the system from many disruptions. Thus, maturity is self-preserving.

The concept of maturity is important to the understanding of complexity and diversity. Margalef proposes maturity as a quantitative measure of the pattern in which the components of an ecosystem are arranged. The life-form communities and physical elements are related in a definite pattern, which is a real but untouchable property (structure). In general, this structure becomes more complex as time passes, as long as the environment is stable or predictable. The structure acquires a historical character. Maturation, as a function of historical processes, increases the levels of complexity of an ecosystem.

This concept of maturity, as an attribute of a community, is related to structural complexity and organization. Maturity increases with time in an undisturbed community. The species diversity, that is, the information content, of a community also increases with maturity, leading to a more complex spatial structure. Diversity incorporates species richness (how many different kinds are present) as well as a measure of abundance—how many of each, as individuals or biomass. Other aspects of diversity, such as life cycles, are less often considered. The energy in a mature system goes to the maintenance of order and less for the production of new materials. In general, diversity is higher, and life cycles are more complex; symbiosis between species increases, and nutrients are conserved. More mature systems have a richer structure and a lower productivity per unit biomass; there are more steps in the trophic pyramid, higher efficiency in every relation, less energy loss ,and less energy is needed to maintain the system. Complexity and diversity offer advantages for living forms. Complexity allows increases in size, which allows the colonization of harsh environments. Diversity allows more effective behavior through specialization; for example, a specialized organelle may digest less common molecules.

Margalef states that biomass and primary production increase during succession; but the ratio of productivity to total biomass drops. According to E.C. Pielou, species diversity decreases and pattern diversity increases during succession. There is also an increase in the proportion of inert matter, and an increase in structures like paths and burrows. Odum has noted trends in ecological succession:
- the community production decreases;
- individual lives are longer and more complex;
- diversity is high and well-organized;
- there is a closed slow exchange rate, where detritus is important;
- symbiosis has developed, and conservation and stability are good;
- and there is a high biomass in a weblike food chain.

The increase in diversity is related to a multiplication of niches; this process goes with longer food chains and stricter specialization. Animals on top of food chains and those with more special habits show a higher efficiency. This results in gain in efficiency in advanced stages of succession. The more mature systems are found in regions of high temperature: tropical rain forests and coral reefs. Stability should be more important than temperature; there are stable and mature communities in deep ocean, caves, cold areas.

Natural succession may operate in a similar way to a hologram—in the sense that the subparts of the system can maintain a potentiality for all possible behaviors that could flow from any part of the system. Native Hawaiian biota appears to be rejuvenated by volcanic eruption; it is better equipped to reinvade areas than exotic species. Successions are stopped by fluctuations, by volcanoes or storms, Margalef calls the process exploitation; adding that the effect of exploitation in general is rejuvenating. Whatever accelerates change and energy flow in ecosystem reduces potential maturity. In an exploited system, diversity drops and the ratio of primary production to biomass increases. Mature systems can regress to earlier forms when exploited. Ecosystems are constantly evolving under the influence of physicochemical processes poorly understood and so far more powerful than those that result from human activities. The reality is more complex than just systematic succession and maturity.

Ecology to Ecoforestry

Disturbances in a forest are regular but unpredictable events. Many of them kill trees. Mortality is a normal part of the life cycle. Mortality in forests usually occurs from a combination of factors. By trying to prevent one kind of mortality, industrial forestry merely sets up another kind. Ecological forestry accepts a typical percentage of death as the normal condition, necessary for the renewal of the forest. The rate of death per year in an old forest is remarkably consistent at about 1-2 percent, even with wind storms, fires, disease outbreaks, and animal damage. In spite of Boise Cascade's recent advertisement about our public forests ("Let our public forests rot or burn again? What a waste!"), rotting and burning do not produce waste and are an integral part of the cycle of life and death in the forest.

Figure 8. A controlled burn at the Mountain Grove Forest in Oregon (managed by the author, 1997-2000).

As a forest ages, it thins itself naturally. The number of trees decrease as the stand ages; the remaining trees are typically bigger (and more desirable by timber companies). As the frequency of disturbance increases, the forest becomes adapted to the disturbance—even pine plantations in the southeastern US that are managed with controlled burns are less damaged by wildfires; after long periods without disturbance, a catastrophic disturbance is more likely—where wind and fire are absent, the probability of insect and disease outbreaks increases. Yet, even catastrophic disturbances like hurricanes rarely damage more than 5 percent of a forest (forests have not had time to adapt to human disturbance; furthermore, human disturbance can influence up to 100 percent of a forest). More than being agents of mortality, insects, diseases, and animals are native components of complex food webs in ecosystems that contribute to the selection of certain kinds (including healthy) and ages of trees (that determines the composition of the forest, which changes over time). Mammals and birds disseminate seeds. Insects pollinate some trees and overwhelm others (rarely more than 1 percent of a forest). Diseases remove stressed trees (also probably a low percentage on the order of 1 percent). Their effect on the long-term health of a forest can only be regarded as positive.

The mass production (of low-quality pioneer wood) in a biological system causes the system to be unbalanced and vulnerable to attack by vectors (usually kept in check by natural enemies or limited opportunities). This means that the vectors (of disease, infestation, etc.) must be controlled by biocides or other extreme measures such as cutting. Furthermore, after clearcutting, the site has to be "prepared" for young seedlings selected for rapid growth in a simplified managed environment.

Human intervention into mature systems is usually detrimental. In early successional stages, there may be larger net community production totals, which can be harvested by humans without damage to the system. One reason gross productivity cannot be harvested by humans is that it would destroy the ecosystem; the wealth of the system would not be sustainable.

Forestry can never be sustainable under the assumptions of the Newtonian science (where the nature was a stable predictable order), the industrial world view, and the old economics. Besides being based on more appropriate myths (discussed elsewhere in *IJE*), Ecoforestry assumes an ecological economics that is based on ecological science with new paradigms. While some scientists, even in the 1980s were trying to find the causes of forest death to preserve the plantation system, other scientists and managers, were starting to plan for more natural forests (multiple species, uneven-aged, groundcover, and windfall).

Foresters are usually only concerned with a small part of the productivity of the forest, that is, the Current Annual Increment (CAI) of trees as a measure of the material added to the tree stems in a year. This change, called "growth," has to be determined by two different measurements, each of which is usually an estimate. The growth is a composite number arrived at by adding the Survivor Growth (trees still alive at the end of the year) to the Ingrowth (new trees reaching marketable size at an arbitrary DBH) and subtracting the mortality as well as any volume harvested. The amount of growth that can be taken is the "forest yield." Growth is calculated as a volume (cubic meters), because that is how it is estimated; productivity on the other hand is calculated as a weight (grams per square meter) or energy (Kilocalories), depending on how it is measured.

The net growth (without reference to soundness or defectiveness) is calculated with a simple formula:

$$G = A - M - Y + I = V2 - V1$$

where G is the net increase including ingrowth, A is accretion, V1 is the stand volume at the beginning of the year, V2 is volume at the end of the year, Y is the yield volume, I is ingrowth, and M is the mortality volume.

The "Allowable Annual Cut" (AAC) for a plantation is simply equal to the annual growth, whereas the AAC for old growth forests is the total timber volume divided by the rotation period of a typical plantation (50-100 years). In 1984, the BC Ministry of Forests projected the long-term sustained yield would be about 57 million cubic meters of timber per year; in 1989, the AAC was 72 million cubic meters. 85.2 million cubic meters was cut on public and private forests. By contrast, the "allowable sale quantity" (ASQ) on US Forest Service forests was only 50.6 million cubic meters on an approximately equivalent land base. In 1988-89, 246,876 hectares of BC were clearcut, 91% of the cutting that year. Clearcutting is prescribed as the logging method on 90% of the allowable cut, according to Herb Hammond.

The productivity of the forest (NPP) should be the basis for the AAC. Neither the AAC or the NPP are accurate measures due to the complexity of the subject, the forest, and due to the variance of the annual estimates of production. Remember, any operation with a human dimension has subjective and qualitative elements. Which is why we need a holistic, conceptual plan.

A plan should consider the whole system. Ecological planning considers the health of the system, which is based on intimate knowledge of the system.. Direct observation and traditional knowledge yield far more "information" about the societies of plants and animals than autopsies or mathematical models. A comprehensive plan would proceed in stages:

1. Identify the place within its natural boundaries. Most places exist in a uniquely identifiable ecosystem, with recognizable boundaries and a unique history and character.
2. Calculate the optimum amount of wilderness to preserve the natural cycles indefinitely. If the current wild area is less than the calculations, restore the difference and set it aside as a preserve.
3. In the remaining area, zone areas for appropriate use, including conservation and artificial areas, e.g., roads or buildings.
4. Identify the resources needed for human use, including timber and the productivity of the areas. This productivity can be used to calculate rational exploitation through cutting rates.
5. Conduct the harvest in such a way as to minimize damage and maximize value.

We have to examine the natural and cultural histories of a place, as part of our comprehensive plan, which is actually a deductive, synthetic, conceptual model based on data generated from research on biological productivity, the rates of resource use, cultural valuation, minimum wilderness preservation, air and water quality, genetic minima, nonrenewable resources, appropriate technological innovation, the importance of cultural frameworks, adventure, research, beauty, uniqueness, and other intangible experiences. A deductive approach is necessary because accurate measurements of productivities in most ecosystems are lacking and exactness in values is misleading. A synthetic approach is necessary to integrate quantitative and qualitative data. In combining measures of qualitative and quantitative, it is simpler to set aside the first and then to calculate the second. The model must be conceptual because of the inherent fuzziness of the systems.

Ecoforestry Based on Net Community Productivity
For one simple model, consider the forest as a corporation. After all, politicians and law enforcement officers are impotent to stop the destruction of forest ecosystems. The simplest way to give the forest a voice in its development is to incorporate it following international law. A corporation is just a legal entity with its own rights, privileges, and liabilities. Although a corporation is independent from its founders, it is a human construct and the forest corporation would have to have human representation in the human system (a permanent site forester would probably act in the best interests of her home). The forest corporation would not be really different than most corporations. Like other corporations, its primary purpose would be to maintain its own existence and maximize its wealth. It would optimize its values, which would include tree and fungus values, as well as human ones.

This strategy would solve the problem of cutting too much of the forest—the forest itself would be untouchable capital. Most of the shares would be treasury shares; anything more than par value would go to capital surplus to be distributed as dividends—the dividends would be the net community productivity (NCP).

The NCP is profit. The NCP of forests of varying maturities varies itself from almost 0 percent to over 10 percent. The most NCP one would expect in an old-growth forest would be 0-2 percent. In a young Ponderosa pine forest the percentage may range from 5-10 percent. In early seral stages after some catastrophic change, alder growth may produce 15 percent NCP. Some energy-subsidized pine plantations in Britain exceed 30 percent for a short time.

Humans can take part of the profit, the NCP. For example, in a Ponderosa pine forest of 2200 hectares, with an NCP of 2100 kilocalories per square meter per year, $(2200 \times 2100 \times 10,000 = 4.62 \times 10^{10}$ Kcal per forest), the NCP is equal to $1. \times 10^7$ kilograms of dry weight, which is equal to 4.1 million cubic meters of wood (weight of Ponderosa pine is .41 kilogram per cubic meter).

As long as humans limit their take to the NCP, the forest is truly sustainable. If the human managers take part of the NPP, they compete with other animals in exploiting forest resources—competition and exploitation are healthy, remember, so that may not be too bad; but, the forest may not be sustainable indefinitely, unless the human managers replace some of the same functions as the creatures they are competing with. And if they take all of the NPP and most of the GPP, then they interfere with the operation of the ecosystem (and interference is destructive—see Wittbecker 1995).

Obviously, we could cut any forest at any rate—we have been doing so. But, the rate at which we cut determines what the forest will look like over time. For instance, if we were to cut all forests at a 1 percent rate, then they would all probably develop old-growth after several hundred years, depending on the species, even if they started out as young forests characterized by pioneer species. If we were to cut forests at a rate of 10 percent, the forests would never develop beyond an early stage of maturity. In order to exploit rationally, the cut should never exceed growth, the harvest should never exceed renewal, as it can in industrial forestry.

Differences and Recommendations of Ecoforestry
It is obvious, and will become more so, that the differences between industrial forestry and ecoforestry are fundamental. Ecoforestry cannot be derived from industrial forestry. Industrial forestry is a failing system , a small-scope, once-through, temporary process for transforming forests into deserts by killing the forests (silvicide); ecoforestry is a maturing stage that bases forestry in a community context and limits the use of the forest to that which the forest can afford to provide and remain healthy. If the metaphor for industrial forestry is a high-powered plow with a rear-view mirror and an accelerator, then ecoforestry is a solar-powered device with windows, a brake, steering wheel, reverse gear, and hand-holds as well as precision cutters.

Somehow, through an ecoforestry approach, we need to become intimate enough with the forest to fit our needs into the production of the forest without interfering with it. Ecoforestry would optimize cutting instead of maximizing it, harvest a percentage of the natural interest instead of the ecological capital, on very-long turn-arounds 250-750 years instead of 10-100 year rotations, allow self-ordered renewal of the forest instead of rapid planting of selected strains. Ecoforestry would encourage diverse forests instead of single-species , even-aged plantations.

Ecoforestry can make several recommendations which would preserve diversity and complexity (and avoid numerous extinctions).

- Protect the core of the forest, its soil and water.
- Promote cutting practices that respect the productivity and complexity of many aged forests, leaving large important structures such as snags and logs.
- Work with rotation periods geared to ecological times, such as 200 or 500 years.
- Allow natural processes such as fires, infestations, and regeneration to operate in the forest as much as possible.
- Reduce fragmentation through the design of forested areas, taking into account the genetic diversity of the trees, catastrophic conditions, minimum viable populations, corridors, and edge effects.
- Grant timber leases that are contingent on the maintenance of productivity and diversity of the land.
- Prohibit practices that are destructive such as pesticide applications or clearcutting.

The forest is a web-like system that produces many things that are useful to human beings. Ecoforestry proposes ecologically responsible practices that permit use of the forest within its limits of productivity and stability. Limiting the practices to net community productivity or even a percentage of net primary productivity would help save forests, but it would require better planting and restoration of clearcut areas, ecological planning and management of national, corporate, and private forests, improved use of wood products (through reduction and recycling), and a re-evaluation of human needs. Ecoforestry will progress with the replacement of the industrial paradigm, the development of an ecological ethics and a broadened ecological economics, and the participation of forest workers, students, and managers everywhere.

Chapter 27

Good Forestry: Neutrality Death Sowbugs & Ecological Principles

Often, when working with forest landowners, I am asked if I practice good forestry, and if I can apply it to the land in question. I usually describe the interactions of sowbugs and bluebirds, or canopy closure and root rot, to avoid a direct answer or a philosophical discussion of the meaning of the word "good." Recently, Hugh Williams[1] asked, "What is good forestry?" While agreeing with many of his answers and insights, I found that the article raised far more questions than it answered.

This article discusses some of the general questions in forestry and economics. The question of good is addressed in general and with regards to forestry, intrinsic value, anthropocentrism, and intentionality. These things, however, can only be resolved by elucidating some of the problems that both forestry and economics face, such as logical dualisms, difficulties with scales, and inappropriate paradigms.

In fact, industrial forestry cannot address these concerns properly without a new perspective that uses a broader logic and incorporates changes in economics, ethics, and politics. There are many other kinds of forestry; few of them rest on an explicit philosophical foundation. A recent approach to ecological forestry, ecoforestry, has an explicit philosophical basis and is a proper approach to good forestry. The philosophical foundations of ecological forestry rest as much on the ideas of process and creativity ideas of Whitehead as on the egalitarianism and diversity of Arne Naess.[2] Ecoforestry can be described through a series of principles and standards, from creativity to balance; and this approach differentiates it from traditional or industrial forestry.

Forestry and Economics

There are many kinds of forestry, but most of them can be categorized under three basic divisions: traditional, industrial, and ecological. At the moment, industrial forestry is being applied to most forests around the planet, although efforts are being made to preserve traditional forestry in parts of its historical range, especially in the forests of the Amazon and Southeast Asia, and to use more ecologically sound practices on the remaining developed forests in the northern hemisphere.

Traditional Forestry

The Pacific Northwest Kwakiutl took wood from living cedar trees; the boards were begged from the tree, then notched top and bottom on one side of the tree—yew wood wedges were pounded into the top and sides and a lever used to work the board out from top to bottom. Other native forest peoples, such as the Mbuti pygmy in southern Africa, who address the forest as "mother," do not practice forestry at all as a conscious discipline. They do, however, gather wood and find their food and material goods in the forest. The way they collect what they need does not greatly influence the overall shape or extent of the forest. They fit in the forest, within the limits of the forest. There are a few cultures left on the planet that live this way.

In other tropical forests, shifting cultivation—also called slash-and-burn or Swidden agriculture—is practiced by many peoples. The Barafiri people of the Parima highlands in Brazil believe that it is impossible to plant crops on a site unless it is cleared and burned. Shifting cultivation is a cycle of clearing, burning, and cropping small areas of the forest with very long periods of fallowing, up to 300 years; this was only realized recently as biologists studied the composition of canopies from satellite data in northern Amazonia— areas of slash and burn could be located easily since the dominant trees had not fully emerged through the canopy after hundreds of years. As traditional populations get larger or are invaded by other growing populations, more of the forest is used. The demand on the forest by large corporations threatens the existence of indigenous peoples who have been using their forest home carefully for thousands of years. The practices of the Barafiri, Hmong (SE Asia), Hanunoo (Philippines), Kapauku (Indonesia), Bomagai-Angoiang (New Guinea), Yukpa (Columbia), Dyak (Sarawak) and many others are threatened by the scale of industrial forestry.

Industrial Forestry

In the 1700s, forests came to be studied as ecosystems and then as resources, although German forestry in the 1800s was relatively holistic and comprehensive as a science. As German forests came under management, however, vast plantations replaced the dark woods famous in myth and story, until by the 1930s, most of the forest was managed plantations. Plantations were orderly and neat, with no windfall or "trash" allowed on the ground and offered immediate economic benefits. That is changing now as Germans reintroduce diversity.

Beginning in the 1940s, it became evident that the plantation system, with single-species, even-aged trees, was susceptible to catastrophic change—wind, pollution, insects—in a way that natural forests were not. The forests began to die as forests—the Germans called this disturbing phenomenon Waldsterben (forest death). Scientists are trying to find the causes of forest death to preserve the plantation system; air pollution, operating synergistically with climate change, species simplification, diseases, and soil destruction, is considered the basic cause. Other scientists and managers, however, were starting to return to more natural forests, i.e., multiple species, uneven-aged trees, with snags and ground cover.

Mainstream forestry itself has gradually and continually learned more about forests. In the 1970s, forest scientists, such as Jerry Franklin,[3] David Perry,[4] and Michael Amaranthus,[4] showed concern over the composition of the forest floor and its relation to productivity—echoing Ebermayer's work in Germany in the mid-1800s. The new forestry developed out of concerns about the progressive clearcutting of the 1950s and 1960s, as well as about slash removal and about fragmentation from patterns of clearcutting. The goal of new forestry, according to H. Kimmins,[5] is the sustainable management of forests for old-growth conditions, diversity, and resilience—while permitting significant harvesting of "timber values." But, these structures may not be dense enough to support old growth processes.

Dilution, like fragmentation, causes species loss and the ultimate collapse of the system.

Ecological Forestry

There have always been voices for alternatives in forestry. Ever since Muir and Pinchot, forestry has been divided. Thousands of people have worked out their own individual programs of forestry. In the 1960s and 1970s, ecological ideas began to be applied to forestry again, usually in relatively small areas. Merve Wilkinson started selectively logging his forest, Wildwood, in 1945, limiting the yield, even though it meant that the forest could not provide a full-time income. For Wilkinson, what is left behind is more important than what has been removed.[6]

Orville Camp bought over-cut land in Oregon in 1967 and nursed it back to health with selective thinning and clearing. He developed his own holistic program in forestry based on natural selection.[7] Operating as a forest farm, Orville has taken logs and firewood out of the forest while improving the health of the forest.

Herb Hammond evolved his ideas of wholistic and ecologically responsible forestry in the 1970s while working as a profession forester in British Columbia.[8] The staff at the Mountain Grove Center in SW Oregon, where I work now, started practicing a "radical" ecological forestry in 1982. Michael Pilarski uses Restoration Forestry to cover a movement and discipline that draws on old traditions to heal degraded forests and provide a steady yield of high-value timber.[9] Jerry Franklin has modified his conception of new forestry to extend on a spectrum of "variable retention." Many others have been developing new kinds of forestry, from community forestry to excellent forestry, social forestry, and sustainable forestry. All of these kinds of forestry are part of a general movement away from industrial forestry. Industrial forestry itself is renaming its program as "adaptive," or "new," or "stewardship," or "sustainable"—but without changing its basic methods or goals.

The Ecoforestry Institute arose out of a meeting between Orville Camp and Alan Drengson in 1990—both with their concerns about destructive harvesting. Since then many groups have started practicing ecological forestry; there are independent ecoforestry groups in Siberia, Papua New Guinea, and Nova Scotia, for instance.

Goodness and Forestry

Philosophers have puzzled over the term "good" for centuries, constructing partial theories and contradictory systems. According to Plato, technical knowledge is not of ultimate importance for human beings because it knows nothing of "good itself." Knowledge of good is theoretical knowledge for him.[10] The word "good" has an interesting and long history. The current version is derived from the old English, "god," meaning "suitable" or fitting, similar to the words meaning a "suitable time" and to be "suitable" or "pleasing."[11] Ernst Becker concludes that we find things bad when they are built for utility only.[12] In an organic world, good things are defined by a free interplay of energies. Perhaps, as a working definition, we can just use "harmony." In Chinese medical tradition, the highest good is harmony, especially social harmony or good relations. A good person is one who creates and maintains harmony. Harmony is related to wholeness (indeed, the word "whole" comes from the Indo-European root "*kailo*," which is also the root for the words health and holy).[13]

The use of the word "good" with forestry is problematic. Good means different things to different people. Your standards or codes, personally or culturally, might be different from mine. Therefore the meanings of the words will be different. The search for good is measured by personal criteria, personal judgment, and personal reflection. Furthermore, human beings cannot know, or even think of anything, according to Robert Zajonc,[14] without some involvement of emotion, that is, at least a vague feeling of good or bad. On the other hand, there are questions of what one "ought" to do, that is, morality.

Some instances of all three kinds of forestry practices can be called good. Thus, industrial foresters as well as nonindustrial small land owners, Swiddeners, and ecoforesters can practice good forestry, although ecoforesters are more likely to have sustainable and healthy forests. Williams states that good forestry is nonsense without the recognition of intrinsic value,[15] but good forestry can always be practiced, at least on a small scale, independently of notions of intrinsic value or theoretical reasons. As Chris Maser has pointed out, most interactions in a forest are neutral; nature does not assign values.[16] No one being in a forest, from a virus to a mouse, tree, or fungus is more valuable than any other. Some beings, however, value other beings, much as a dung beetle values dung or as human beings value wood or diversity.

Any example of forestry can be good, if it follows its methods, regardless if it is totally external, human-centered, or simply pragmatic. Any application of any forestry can be good, if the worker unconsciously follows certain codes or accidentally does the right thing for the forest—the right thing being an absence of interference with the dynamic systems that shape and maintain forest processes. Good forestry can happen for the most contradictory or trivial reasons: Self-limitation, true love for the forest, accident, intention, techniques, or shame—this last may be why forest management gets altered with public interest and scrutiny, a point Williams brings up, also.

Dueling Problems
Many problems with forestry (and economics as well) are caused by our modern classical logic. This deductive and bivalued, i.e., categories are mutually exclusive, while substance and identity are permanent. Its binary approach results in an either/or situation, rather than a both/and situation. Similarly, there are problems with scale, locality, and limits. Logically, the argument is tossed back and forth about whether forestry should be anthropocentric or ecocentric (or biocentric), when in fact it should be neither. Williams states that the ecocentric view is criticized because of its tendency to subordinate individuals to the whole.[17] This is not a real criticism due to a misunderstanding of the concept of frame and focus.

Focus and Frame

Focus and frame can be understood metaphorically—and, in fact, metaphor itself can be understood as consisting of two parts, according to Max Black: A focus and the frame.[18] The focus (=figure) designates the figurative term signified through the process, and the frame (=ground) refers to the subject or context. Using this distinction, it can be seen that most of the fuss in forestry has occurred at the focus level. Foresters have so long focused on trees that they forget that the forest is a frame that holds many foci (or points of view). Alan Drengson is fond of saying we do not see important operations of nature because we are looking through the wrong paradigm—a paradigm acts like a pair of glasses, focusing on what we want to see.[19]

The elements of a forest are related psychologically, by foresters, as focus or frame, as contrast or uniformity, as dominant or recessive, or in a number of other pairs. For instance, forests can be considered by scientists as either matter systems or energy systems, but the focus on either frame permits subtle differences and limitations in interpretation. Some ecologists describe organisms as being configured by energy through time. But, organisms are material patterns in space as well.

Williams presents biocentric and ecocentric perspectives as holding that the hierarchy of value is ultimately arbitrary.[20] This is not true for either perspective. No special moral significance, such as sentience, is required in either perspective. Furthermore, it is not quite right to say that ecoforestry is ecocentric, because it is actually concerned with the frame and not the focus, the periphery and not the center (perhaps it should be called "drymoperipheral," meaning forest frame.

Williams makes a good point in contrasting anthropocentric needs versus the needs of the forest itself.[21] This dilemma can be answered by considering the focus/frame character of the situation. Both of the two basic views he mentions are merely aspects of one view of good forestry—that happens also to be the ecoforestry position. Anthropocentric values that focus on commodities can only be considered in the context of the values that are contained in the whole; that is, they are derivative from the frame. Furthermore, it is not necessary to ascribe purpose to the latter view, as Williams does, other than perhaps the purpose of being itself, which is to "be."

The framework of ecoforestry is pluralistic and multi-dimensional. Balanced use, rather than unending growth, is emphasized. Massive disruption often results when a community falls out of balance with its local forest environment (industrial forestry only avoids the penalties for such disruption by trading advantageously with other communities in less powerful areas). Ecoforestry has built-in checks, in the form of an oath, to which practicing ecoforesters subscribe much like doctors are bound to the Hippocratic oath.

Scales, Large and Small

Scale has several meanings in ecological forestry. Basically, as it is used here, it has to do with the level of measurement in a space/time/energy/mass (STEM) context.[22] For instance, the measurements of leaf litter can be made at several scales: Single tree, stand, annual measurements, or stand life measurements. Processes that are unimportant at a small scale might be vital at a large scale. Too much litter in one stand in one year, for instance, might suppress a cycle; too much or too little litter over a century might interfere with several regional or global cycles.

Some patterns in forests are scale-dependent; for instance, hemlock trees may dominate small clusters, but be scattered all across the entire forested landscape. That is, the pattern changes with the scale. This is true of processes in forests as well. Canopies shade the understory annually, but fires in a lodgepole pine forest may increase dramatically the amount of light to forest floors once every 200-300 years.

The scale of the system defined, e.g., forest or ecosystem, depends on the scale of the phenomenon being addressed. Mangrove forests are in phase with the frequency of hurricanes, although hurricanes may not influence the life histories of short-lived organisms as much as daily or seasonal cycles. Microbes, for instance, are affected by short-term cycles of precipitation and temperature.

Problems in forestry arise where applications that work on a small scale are expanded to large scales, without thought for the difference or changes in patterns. For instance, it is well known that Douglas-fir is shade intolerant and grows best in openings that get light. Rather than simply remove single trees or small groups of trees, and release or plant the fir in the openings, industrial forestry applies the treatment to the entire landscape with large clearcuts, which alter the other conditions that fir requires—some shade, water, protection from browsing, associated species. This is the formal operation of Greek tragedy—applying a good idea to another situation where it does not fit.

For the forest in which Douglas-fir grow, increasing the scale of cuts to increase the potential for the fir has the unintended effect of drying out the soil, as well as of destroying the infrastructure of plants, animals, and fungi that the forest requires to continue. Other systems of cutting, such as high-grading or thinning from below, may also have unintended effects with changes in scale. High-grading is a form of group selection, in which the best and largest trees are removed. The practice was not overly destructive on a small scale (with many of the matriarchs and patriarchs remaining), but on a large scale, all old, large heritage trees were removed. Thinning from below may be a good idea in a stand, but like high-grading it is biological selection—the gene pool is altered as the small and suppressed trees, some of which may be genetically superior to those left, are removed.

By the 1980s, scientific research had resulted in technological applications for growing, genetically "improving," harvesting, and processing woody material on an unprecedented, vast, and unconsidered scale. Industrial companies established tree breeding and fertilization programs, whose goal was usually fast growth. Pine plantations in the southern United States were rated as a success and served as a model for plantations elsewhere; Australia, New Zealand, Chile, and South Africa replaced much of their native forests with exotic pine plantations, using species like Monterey pine. Many trees in the same species normally grow in clumps within a forest. Industrial forestry changed that scale, also, planting entire forests of a single species.

For thousands of years people have taken things they valued from forests. Culture and habit have resulted in large-scale trends in forest-use: More forests are used; more of each forest is used; fewer products from each forest are used (this seems contradictory, but it is not inconsistent with the way things are done in industrial cultures). The current economic style is too great, fast and reckless for ecological systems to absorb its impacts. The scale of things is an independent problem that can ruin the best intentions of policy.

An inappropriate time scale, its shortness and urgency, is a cause of many of these problems in industrial forestry. Although forests are considered renewable resources, they are slowly renewable, requiring hundreds or thousands of years to renew from catastrophic disturbance; this time is far longer than any economic plans (and really nonrenewable on a human life scale). This has important implications on sustainability.

The forest ecosystem is a large-scale pattern of millions of minute events. A range of elevations across areas would minimize the effects of climactic change—and the possibility of extreme change is rarely considered in wilderness design or forest management. Soils, drainage, and land-use history and ownership would also receive similar considerations. This would allow management for diversity on different scales.

On the other hand, really large scales, bioregional or global, are not considered enough. We should be measuring at larger scales: Watershed, landscape, or biome. Before satellite data analyzed by GISs, ecologists and foresters did not have tools that could address the scale of landscapes. Regional and global data was hard and expensive to collect. Forests are also larger than stands. Although forestry has incorporated some of the technical tools, such as satellite imaging and GIS, it has neglected the conceptual tools, e.g., the Gaia Hypothesis or global design.

Ecoforestry considers those conceptual tools. Ecoforestry recognizes that forests are part of a process that is unending and imperfect, without a final state, and furthermore, that the attempt to perfect it results in disharmony. Ecoforestry accepts a constructive conflict in scale with the ecosystem. There are chaotic events, plagues and random frenzies in every system; rather than deny this ecoforestry tries to account for these factors. This acceptance leads to understanding and the abandonment of stupid strategies.

The tasks of ecoforestry include the return to the natural history and knowledge of forests as the foundation for taking trees, to place forestry on a proper scale (with regards to time and species), to leave the maximum alone, and to encourage the symbiotic relationship of forest economics and wild forests.

One solution to many of the current problems is a reduction in scale for everything from forest use to management units, with local controls and local use primary. Temporal scales, however, should be expanded with long-term research and management. One principle of ecoforestry is that an ecological forestry must observe the proper scales of forests, especially in terms of size, age, and patterns (or diversity).

Williams discusses the cosmic tendencies of diversity and creativity[23] in forestry. Certainly many cosmic tendencies, such as disorder (entropy) and order (ektropy), as well as change, creativity and temporality, are relevant to forestry. But, Williams and many others tend to emphasize only those tendencies that seem positive, such as creativity, complexity (which I will identify with diversity for the moment), and order, and not those with negative connotations, such as entropy, death, and uniformity. The universe comes with the whole package and we risk serious error by choosing only the tendencies we want.

Global and Local Fields
Concepts of scale apply to physical and biological fields. Local fields or events are those separated in space/time from other fields or events. Both quantum theory and relativity consider that local things are generally noncausally and nonlocally connected (that is, no communication is instantaneous), however, through higher dimensional realities.

Local ecosystems are separated from one another, not only in space, but by differences. For example, Dutch elm disease is a local problem (even if it seems ubiquitous in all forests with elms), produced by local actions and requiring local solutions.

Ecosystems are unique and original. Each patch (locality) supports a segment of the total species population in a unique context, with a particular set of predators, competition, food, or physical habitat. The forest ecosystem is a large-scale pattern of millions of minute events. The environment requires an enormous amount of minuscule local adaptations between the earth and its users. Production of pine cones, for example, is adapted to the local microsites; Loblolly pine planted either 50 miles north or south of the seed source are less vigorous. DNA seems to be a local system.

Local systems, however, can affect global systems. Cutting down local forests may contribute to the discharge of greenhouse gases, such as carbon dioxide, into the local atmosphere. This may have the effect of increasing quantities in the global atmosphere, creating a global change, possibly a runaway increase in atmospheric temperature.

Some systems are truly global: Global warming or global biogeochemical cycles. Lovelock suggests that the interaction between hardwood forests and softwood forests may act as a global regulator of oxygen for the planet.[24] Many things, such as human poverty and species extinctions, only seem global because they are happening in many local systems at the same time (and may effect global cycles).

Development plans for forests tend to be global, to call for the eventual development of all resources in an area; British Columbia's intention to cut all the forests in the province is a good example. A one-world planned economy is an even greater threat, being based on unlimited industrial production, unlimited commodity consumption, increased exploitation of nature, and the free flow of resources and labor across cultural borders. This kind of planning requires the abandonment of local controls on development, trade, or lifestyles. Planning is thus characterized by a utilitarian globalism that denies value to the very systems that support it.

The current civilization is attempting to be global, not local. We even have a global mythology with industrialism. A global mythology cannot afford to teach of unimportant local elsewheres. It must teach of a multiplicity of cosmologies. Particularly in agricultural societies, cosmologies are gauged closely to seasons. They are also tuned to the limits of the local ecology, within their knowledge of interactions (the long-range ecological consequences of drainage, irrigation or overexploitation contribute to the deaths of cultures).

Despite cutting and planting on local scales, a number of global trends are evident. We might call them gigatrends, since they are larger and more involving than the megatrends identified by John Naisbitt over a decade ago.[25] These gigatrends, on a global scale, include: Human populations increase exponentially; the impacts of a small percentage of people (the wealthy) increase exponentially; humanity takes over the habitats and functions of other animals, eliminating them and trying to take over their functions with chemicals (unsustainable for long); and ecosystems are simplified and degraded; deforestation, desertification, and exotic take-overs occur on a scale similar to the ice-age or a comet impact, but more meaningful to humanity, since we depend on the ecosystems we are changing.

We are faced with the choice between global logic or local knowledge. Local planning currently ignores limits and carrying capacity, long-term deficits and problems, other species, and ecosystems (regional and global planning are nonexistent). We need to encourage local autonomy and decentralization (to strengthen local regions and encourage self-sufficiency). The implementation of deep changes requires global action as well as local. Every forest could become the basis of local culture rather than the source of profit or support for a global economy.

The ecological social approach (or a redistributive environmental strategy) to development makes it irrelevant to discuss global limits to growth. Local limits are far more significant to majority of population—regardless of how much food exists, people will starve unless they can get it.

Such an approach would also mean limiting humanity and its technological effects, limiting human use to local impacts, and letting other beings live without interference. It is not necessary to dominate or terraform the forest completely to save it. Ecological forestry weaves people back into the fabric that supports them and in a sense makes them subject to the constraints of ecosystem processes.

Limits and Maxima

Limit is the essence of form and pattern, which are defined by their limits. As Pythagoras is thought to have said: "Limit gives form to the formless." A limit is defined as a boundary or the utmost extent (smallest or largest). The limit may be unknown or invisible.

Mathematically, the concept of a limit is fairly simple: It is something that is approached, but not reached, as when fractions descend to zero. A more precise definition, according to David Berlinski,[26] is that a "sequence ...

has a limit at the number L if, as the sequence is extended, its terms get closer and closer to L." This can be qualified by adding more quantifiers.

A maximum is the largest value less than a limit. In applications the limit is usually a number, e.g., the rate of photosynthesis or the carrying capacity. Every finite system has limits.

In some senses, nature does try to maximize or minimize certain conditions. For instance a bubble minimizes surface tension by assuming a spherical shape that has a maximum volume. Of course, we could also say that the bubble is an optimum form for the least possible surface area for a given interior volume.

In principle, mathematicians reduce the question of maxima and minima to a geometric construct, often a three-dimensional surface with peaks and pits. In trying to apply the idea to a tree, however, the calculations become incredibly complex. A tree appears to create a maximum leaf area to collect radiation and a maximum number of seeds for reproduction, but it also tries to minimize evaporation and energy for its metabolism. A forest is even more complex. In reality, the processes of trees and forests have many local maxima and minima; these can be represented graphically as surface potentials on a catastrophe field, as Rene Thom has done.[27]

Industrial forestry, based on the old agricultural model, is concerned with maximizing profits and products. Highly specialized systems are accelerating the change from 'paleotechnic' to 'neotechnic' agriculture. This change minimizes costs and maximizes profits in the short term, but it also narrows the ecological basis of world food production and decreases human livelihood in the long run.

Most forest plans are also development plans that are comprehensive in the sense of seeking to meet all needs of the public, agriculture, and industry. But, they fall prey to all the assumptions of the industrial culture. They tend to be multipurpose with the aim of providing maximum net benefits through management of forests and wildlife. Both uses of "multipurpose" and "maximum benefits" are based on misunderstandings. Multipurpose in practice means human use—perhaps even just one of those, e.g., logging; and the idea of maximum benefits has proven to be slippery and dangerous. Modern resource management strives for maximum sustainable yield, based on partial knowledge of species and great ignorance of ecosystems.

Even a modern, balanced exploitation, however, may destroy forests and fisheries. Currently, many resource managers espouse the ideas of equilibrium maintenance and maximum sustainable yield. These ideas are poor guides to management, according to C. S. Holling.[28] By trying to maintain habitats in equilibrium, we often set them up for catastrophic decline, for instance, in fire-climax pine forests, or destroy resident species, e.g., the condor.

People are concerned with having maximum freedom or producing maximum values. Williams quotes John B. Cobb, Jr. in support of differences between one ecological vision and another. While agreeing with much of what Cobb says, I am concerned about the idea of unlimited value. I agree that values may be indefinite, but would not agree that they are infinite. Like the principle of limited good, I suggest that there is a principle of limited value. Competition, as Cobb mentions, is not the ultimate principle, but neither is cooperation—both are necessary in ecological interactions. Even aesthetic appreciation requires limits, to avoid having our over-appreciation overwhelm the values evident in nature.

Williams also links the maximization of the moral good of creativity to the maximization of public good,[29] then to forests specifically. I wonder if he is equating forest good with public good here. I wonder if we should even consider maximizing the moral good of creativity, or if it is meaningless. Few mammals even try for an optimum; most strive for a satisficing amount, according to F. Varela.[30] Varela analyzes the evolutionary process as satisficing rather than optimizing; that is, a suboptimal solution is adequate. Selection operates as a survival filter that passes any structure than has sufficient integrity to persist. The focus of analysis is on organic patterns in a life history rather than on individual traits. For the evolutionary process, Varela suggests the metaphor of "bricolage," which is the putting together of parts in complicated arrays because they are possible (rather than being part of an ideal design).

Williams also proposes maximal creativity as an ethical imperative. Maximizing creativity, however, would lead to chaos, both in one's self and in ecosystems. There has to be a balance between creativity and stability, between innovation and habit. We should not being trying to maximize creativity in human beings or forests, but rather seeking stability and relative harmony.

The predominant economic theory in industrial countries holds that the full utilization of resources is necessary to ensure full employment and the maximum social good. This economics depends on economic growth to avoid crisis. The major premises assume that: Population will grow, that social good is related to equitable distribution of material products, and if resources are limited, technology will erase the limits. The economist Kenneth Boulding referred to this as a cowboy economy.

A community is forced to accept an upper limit, beyond which it cannot grow any further. Further growth results in destruction or disruption of itself and nature. This is an ecological law of the maximum. Production could be stabilized in a steady state economy, a mature economy, like a climax system, where processes and cycles are constant. A steady state economy is based on natural laws and ethical principles. Traditional economies have a diverse and healthy diet; deliberate underproduction, usually well below the maximum levels; deliberate control of population growth below maximum levels; and deliberate under-use of resources, resulting in a small ratio of people to resources.

Reed Noss and others have noted that maintaining a healthy forest ecosystem is more efficient in the long-run than having to duplicate the forest functions to keep it healthy.[31] Industrial forestry has, without admitting it, demonstrated it in numerous ways: for instance, by providing shade cards instead of shade, deer guards in place of food diversity, fungi plugs to replace healthy soil, and so on.

Furthermore, it is dangerous to take maximum yields out of a system unless all factors are known—a virtual impossibility. Therefore, we must aim for optimum or satisfactory (to our needs) yields and calculate (and hope or guess) that the forest has sufficient flexibility to recover from our "take."

To survive, an ecosystem depends on the interactions and balance of many variables, most of which are not well understood. In forestry we try to maximize one of those variables. When that happens, the balance or harmony is altered, and although it may take decades or centuries for the consequences to be known, the system is affected. William Ophuls noted that nature abhors a maximum.

While it is meaningful to speak of an optimum diversity, as the result of limits and the interaction of many factors, a maximum diversity may never be reached. As Paul Weiss noted, the patterns of organic nature are a combination of order and diversity; order involves constraint while diversity requires freedom for difference.[32] Maximum order would result in a static universe, where a maximum freedom would create a nonordering chaos.

This is why is it not a tragedy of cosmic proportions, as Williams says it is,[33] when species such as sharks or sow bugs fail to respond to the lure of possibility. These species have fitted themselves to relatively unchanging environments and do not need to change for the sake of change or for the sake of diversity.

Williams considers New Brunswick forests to be an expression of an evolving unity in diversity. I wonder if that unity is at the level of the ecosystem, community, or stand—or at all levels. Williams states that diversity has been organized for the good of the forest. This is a good example of a priori reasoning, as good as, but as wrong as, the anthropocentric hypothesis. Diversity is an emergent property of forests, arising from the activities of multitudinous beings learning to use the productivity of the forests to augment their own flesh. The forest is self-organizing, but diversity is not a goal of the forest.

It is true, as Williams says,[34] that diversity "by itself has no value." It is after all just a characteristic of mature ecosystems. But it is never by itself; to think so is to be guilty of misplaced concreteness, according to Whitehead.[35] In fact, nothing exists by itself, that is, not in relation to other beings.

We may not know what is the minimum, optimum or maximum forest cover for a particular watershed. Science might try to identify minima or maxima but philosophy can aim at optima or satisficia. We are so ignorant of the complexities of ecosystems that it is suicidal to pretend to "maximize" their use for resources. A free market has to be limited by conservative calculations of ecological balance. It is almost impossible to estimate the economic value of natural balance.

In a system of ethics, we might consider maximizing a value, but then we have to decide if it is being maximized for one species or the system, or if it is being maximized for the present or the future, or whether it can be maximized at all. John B. Cobb Jr. suggests that we should act to maximize value in general, at least for every entity with intrinsic value, rather than maximize value for one human or all (the greatest good for the greatest number) in the present or the future.[36] Perhaps though, we should aim for an optimum or satisficium here also.

Ecoforestry

Ecoforestry is often defined by contrast with industrial forestry. Where industrial forestry has a corporate structure and corporate values, e.g., self-preservation and profit maximization, ecoforestry has a community form, with community values, e.g., the promotion of health and diversity. Where industrial forestry is global, growth-conscious, isolated, dominant, violent, external, instrumental, large-scale, acquisitive, simple, and single-valued, ecoforestry is local, developmental, reciprocal, interdependent, peaceful, internal, intrinsic, appropriate-scale, generous, complex, and many-valued.[37] Where industrial forestry is anthropocentric in its approach to the forest, ecoforestry tends to be ecocentric. Industrial forestry is based on Aristotelian binary, polar, predicate logic, which encourages simple contrasts like this. Ecoforestry applies a morphogenetic logic (as described by Maruyama),[38] which can incorporate change, harmony, and heterogeneity, as well as unrepeatable and irreversible processes, as exemplified in Mandenka, Navajo, or Eskimo thought. The binary logic can be characterized as classificational quantitative, competitive, uniform, nonreciprocally causal, and hierarchical. The morphogenetic as relational, qualitative, symbiotic, heterogenistic, reciprocally causal, and interactionist. Reciprocal causal processes operating in forests increase structure, differentiation and complexity. Using a morphogenetic logic, it can be seen that ecoforestry is not so much ecocentric as drymoperipheral—a neologism referring to the fact that the forest has many centers, which can only be approached with a sideways crab-like motion attentive to the whole framework and not to a single focus.

Ecoforestry incorporates many dimensions missing from industrial forestry. Where industrial forestry ignores the history of forests and of forest use, ecoforestry starts there.[39] Where industrial forestry assumes the truncated cosmology of the modern age and an economic system based on inequality and short-term returns, ecoforestry considers the adaptive cosmologies of other cultures and a total ecological economics. Industrial forestry bases decisions on an antiquated physics and partial ecology, and ignores the ethical and social consequences of its actions.

Ecoforestry cannot be derived from industrial forestry, however. Industrial forestry is a failing system, a small-scope, once-through, temporary process for transforming forests into commodities, plantations, or deserts; ecoforestry is a maturing stage that considers the health and continuity of forests first. Although ecoforestry sounds like a qualification of the modern, industrial forestry, it is in fact entirely different—it is based on a different metaphysics, a broader ecology, a more comprehensive economics, and it is sensitive to limits, ethics, aesthetics, and spiritual values. Ecoforestry is comprehensive in its application of ethics, for instance, it tolerates all the inhabitants and structures of a forest. The entire community is considered. Human communities are embedded in forest communities; our cultural and spiritual achievements occur in the larger community, which supports human endeavors. The larger perspective of ecoforestry incorporates industrial forestry as a special case, much like the theory of relativity incorporated Newtonian dynamics as a special case . Under certain rare circumstances, such as plantations of exotic trees, the treatments of industrial forestry, such as clearcutting, may be entirely appropriate.

Ecoforestry has been defined as selection forestry or restoration forestry; this is partially true but incomplete. Ecoforestry has also been defined as a context-based community forestry based on traditional wisdom combined with scientific knowledge. As the management of human use of forests for necessary goods at an appropriate scale while respecting the special characteristics and limits of forests. Ecoforestry can be approached obliquely through multiple overlapping definitions.

Ecoforestry is also a crisis science, like Conservation Biology, from which it takes many ideas. Over the planet, entire forests are being removed or converted, while others are degraded or destroyed; in the U.S. alone, 6 million acres of forest are cleared annually, 5 million acres are degraded, and 3 billion cubic meters of wood are consumed. Rarely do these industrial numbers reflect whole trees, associated animals and plants, or living habitats and forests—that 3 billion cubic meters of wood came from 1-6 million trees, each of which was home to lichen, fungi, beetles, birds, and other beings. Some, such as pileated woodpeckers, are so territorial that many die with their tree, just as in Greek myth hamadryads were thought to scream and die as their trees were cut.

The agenda of ecoforestry can be presented through a number of characteristics, principles, and standards. Characteristics are qualities that distinguish unique individuals, systems, or patterns; Gregory Bateson calls them differences that make a difference.[40] Principles are fundamental rules or laws, based on unique characteristics of forest systems that we can use to create models to meet stated objectives, which are goals towards which our actions are directed, e.g., a healthy forest. Standards are models or examples of quality or value established by authority or consent, that can be repeated as procedures. Good forest design means not violating any basic principles.

For example, one characteristic of a mature forest is its wildness. The corresponding principle is that forest is self-making and self-ordering without human control and management. Our objective for this forest is to allow the foresting process to continue, whether we take resources from the forest or not (forests can be influenced or interfered with by acid rain, pollution, and other industrial effects). We can set local standards that are likely to keep mature forests wild: Limit biomass removal to 2 percent of the total forest; use appropriate techniques, e.g., single tree selection, horse skidding; retain mature structure, e.g., 19 snags per hectare, 23 nurse logs per hectare (in mature Ponderosa pine forests in Eastern Washington for instance); preserve surrounding landscape patterns.

The principles of ecoforestry are based on a number of fundamental philosophical, historical, scientific, and cosmological principles that were first presented in other contexts by thinkers such as A. N. Whitehead, A. Einstein,[41] J. Cobb,[42] E. Odum,[43] and H. Hammond.[44] Principles unify our images. These principles are introduced briefly to show the depth and breadth of forestry. Very few of these principles are absolute or universal; in fact, the further one gets from physical or chemical principles, the more likely there are significant variations or exceptions. Nevertheless, they are essential to the understanding of forests and quite useful in applications in forests. Principles, combined with common sense and good judgment, are necessary as guides in the absence of definite knowledge. They give us a broad predictive ability.

- Change (or Process, Evolution). Individuals change; patterns change, forests change. Neither tree plantations or old growth forests can remain unchanging.
- Organism (or Complementarity, Wholeness). Forests are composed of living organisms; the forest itself changes, lives and dies in ways similar to a living entity or organism. The forest as a whole remains the same, partly this is because the whole is a nested system that turns over at a rate much more slowly than the parts.
- Field (Relation, Patterns, Limits, Locality, Polycentric). The forest acts as a field, containing organisms. David Perry notes that any removal, even a single tree, sends ripples through the forest system; this ripple effect may good or bad for the health of the system, depending on the chain of consequences. Thinning produces a larger ripple effect; because more light reaches the ground level, it stimulates herbs and shrubs, which may compete for moisture and slow tree growth or increase the rate of nutrient cycling and enhance tree growth.
- Historicity (Hysteresis, Irreversibility, Uniqueness, Indeterminacy). History creates unique patterns, especially in forests. Each forest is unique in its parts and structure, in its matter, energy, forms, information, and in its dynamics and history. Forests pass through stages that are never repeated, despite superficial similarities; that is, tree-planting cannot reverse clearcutting (although another old-growth forest may develop in time). The history of land limits or determines its future.
- Novelty (Creativity, Ektropy, Complexity, Heterogeneity, Synergy).[45] As an ordering/disordering process, a forest creates new forms and new patterns. Every forest is unique. Forests decay, as well as become more complex. The process of nature is not merely rhythmic change, it is a creative advance, producing new forms everywhere. Creativity as a fundamental metaphor—the dance of creation; in expressing itself it moves around the floor but has no single direction.
- Intrinsic value. Every species has some value (based on Deep Ecology Platform 1). Being unique, creative, historical patterns, all beings have value. These values are independent of the usefulness of the ultrahuman world for human purposes. Whitehead has stated that existence is the upholding of value intensity; for itself and shared with the universe, from which it cannot be separate. The value (intrinsic worth) each being has for itself is shared by others. Each exists for itself and for others and is a value in itself and for others. Value is achieved through an ongoing process in nature.
- Membership. Human beings participate in natural systems as members and cannot unparticipate by choice. The way we participate forms an ethics. Williams mentions universal moral principles, without being more specific.[46] The only universal moral principles I can think of are:

No incest, no killing within the group, and follow the golden rule—do unto others as you would have them do unto you. Cultures are too diverse to be too simple. How can we measure anything against an ultimate sense—we do not even know if any forestry is not going to destroy most forests in the long-term.

- Landscape Structure. Landscapes are heterogeneous and differ structurally in the distribution of energy, materials, and species among the three basic elements of the landscape: patches, corridors, or matrix . Landscapes thus differ functionally in the flows of energy, materials, and species among the structural elements. Trees, owls, water, nutrients, and other 'ecological objects' are distributed, as well as move or flow, among the landscape elements.

- Protection (after Hammond). Ensure that all plans and activities protect, maintain, and, where necessary, restore biological diversity. Maintenance and, where necessary, restoration of all types of biological diversity is necessary to sustain life in forest ecosystems. Maintaining genetic diversity means ensuring that viable natural gene pools, including the gene pools of trees logged from a site, remain on the site or, in the case of previously degraded forests, are restored to the site following human use. Maintaining species diversity means that viable natural populations of plants, animals, and microorganisms are maintained or restored throughout the various successional phases for each ecosystem type within a forest landscape. Maintaining community diversity means maintaining or restoring the variety of forest ecosystem types that result from natural disturbances at a variety of scales through short and long time frames in a forest landscape. We understand that maintaining natural biological diversity is a requirement to ensure fully functioning forests through time.

- Holistic Practice. Forestry practice should be holistic since it affects the whole forest. You should harvest in the context of planning, measuring, monitoring, protection, and restoration. Manage for preservation, reservation, protection, and use. Create a practical plan, based on the forest (especially riparian areas). Emphasize interconnectedness over separate structures or operations. Sustain the forest before any yield.

An examination of these principles reveals how ecoforestry is a continuation of thought based on other metaphors, sometimes older, sometimes newer. Ecoforestry principles can be described in a full spectrum of categories. These principles serve as guides and constraints on our objectives and standards. For each principle, we have to ask the following questions: How will it affect our objectives for that forest? Will the standards vary? Should forest operations be modified?

Summary
To be a good forester, you almost have to be part of a good culture, and that culture must almost certainly have a good image of the world. Otherwise, the culture may destroy whatever good work you do. What I am saying is that you, as a good forester, cannot just care for the forest. You must also

participate in society, and if it is blinded by greed and bad ideas, you must try to influence it to the good. Your being good may contribute to a social good.

I still have trouble weighing good and bad in practice. I have no doubt that when I do more good, more bad is also created. For instance, when I started trying to restore a forest in Idaho, the best knowledge at the time insisted that I should clean the trees out of the stream and remove flammable brush and woody debris from the forest floor. I started to do this in a beautiful cedar grove, but I spent too much time sitting on the ground looking at the trees—I suppose I could blame Artemis, goddess of forests, solitude, young girls, and the hunt, for possessing me—but looking was more rewarding than the limited "housekeeping." Somewhat later, scientific knowledge advanced and I was advised to drop trees into the stream and leave all the woody debris and brush on the forest floor. Had I been less contemplative (or lazy), I would now have much more work to do. Was what I did in the 1970s good? Is what I am doing now good? Do I need more training to determine what is good or bad? Is the failure to do the good of now, or then, bad? Have I failed from ignorance (conflicting knowledge) or from confusion (conflicting intentions)?

That is one problem with forestry today. Which action is good? Which is bad? Which should we do? Unfortunately, the outcome of our exploitation or interference may not be evident for hundreds of years. Perhaps we should aim for harmony, for the health of forest and human communities.

I think that we should do good for one's harmony, for instance, or for the health of the forest, not because the action is an action or for the sake of doing something good. In doing, we choose between good and bad actions; the judgment makes us human and susceptible to error. Good forestry can do nothing more. Perhaps, as John Fowles suggests,[47] all our judgments of good and bad are meaningless in the long run. All actions, good or bad, interweave so extensively as time passes that their individual goodness or badness disappears. Each becomes lost in the other. Judgments evaporate and landscapes remain. Even so, we must consider our actions and perform them, guided by notions of goodness and harmony.

As good forestry, creating good images and good goals, ecoforestry is concerned with resacralizing landscapes, with restoring them to their extents and grandeurs, by regrounding science in ethics (that is, ways of living together), and by changing our attitudes from utilization and flat efficiency towards awe and appreciation.

That means that we, you, me, have to care for each tree, fungus , jay, sowbug, or worm. Each living being matters. We know so little about the lives of trees or of other beings; we do not know what it is like to live for over a thousand years or to stand in one place and draw everything we need into us. We do not know what it is like to live underground and browse between roots. Our detachment from trees and other beings has to end. Our participation in the life of the forest must begin.

Chapter 28

Design of Northwest Forest Ecosystems
(Forest Landscape Analysis and Design)

I. Introduction: Why Design?

Nature is self-making and self-designing, but we humans now influence every natural system, taking what we need from some ecosystems, enhancing a few, misusing others, and interfering with the rest. We need designs to restore the balance between human needs and natural processes.

Ecological designs focus on whole communities that work in the same self-sustaining and self-limiting ways as nature. By consciously creating meaningful order, we can develop ways of producing widespread community wealth while positioning the community for a long, sustainable future in a healthy environment.

There is no guarantee that nature can provide humans with everything they want. Recognizing the lack of guarantee simply recognizes that nature is wild and we must come to terms with nonhuman beings and processes. It is not enough to arrange trees in rows to maximize future harvests; it is not enough to preserve small areas of old-growth without natural disturbances. We must pay attention to the processes that make up the habitat, for example, the role of herbivores on trimming vegetation (and diversifying it by predation). The design of the forest and its management must ensure that the processes operate to maintain a dynamic state. Furthermore, the context must be conserved. The forest, however, cannot be considered outside of the context of the entire landscape, including human images and institutions.

A number of forests in the Northwest United States are still wild. Many others have been impoverished, but have the capacity for regeneration. As afforestation proceeds to reclaim wasted lands, as it has in England and Europe, more attention will be paid to the shape of the forest. Design principles can guide our decisions. Although design in Europe has primarily been concerned with artificial forests, many of the ideas can be applied to wild forests. Human design, until now, has been primarily visual. It has emphasized aesthetic reaction to a place, but also the uniqueness of a place. It is not enough to create ragged edged forests to satisfy human eyes; it is not enough to leave beauty strips of real forest to fool travelers. Design is needed to create natural spatial patterns and temporal phases across watersheds and entire landscapes. Ecological design considers the whole context.

I.A. Definitions

Landscape is a heterogenous area composed of a mosaic of interacting ecosystems of various sizes; an ecosystem is a community of organisms living in place. A forest is an ecosystem characterized by trees—this definition is broad enough to include tree farms and plantations as special kinds of forests. Design is a human project in which, as Oliver Lucas says "visual and physical parts are assembled in order to achieve a specific end result." Ecological design is the creative modification of ecosystems to repair or enhance their ability at self-organization and maintenance of their complexity and diversity. Diversity (is in biological diversity) means species richness, different age and size classes in a population, and genetic differences in a species, as well as kinds of habitats present in an ecosystem and the kinds of communities occupying the habitats; and the kinds of ecological processes that maintain habitats; and the variety and richness of the planet's genetic heritage. Ecosystems that do so are healthy. A definition of health is the condition of being sound in body or well-being.

I.B. Characteristics of a forest

A forest ecosystem can be characterized by a number of words: productivity, openness, efficiency, maturity, stability (many meanings), durability, self-making, flexibility, diversity, richness, wholeness (matrix), and dynamic. This list is not meant to be exhaustive.

Living communities are self-organizing systems with emergent properties; they maintain themselves in a state of flux; species are always coming and going, and changing proportions. In a living system, nothing keeps growing forever; things die and are reborn in cycles. The continuation of the system depends on these cycles. The cycles are bound by limits. Both individuals and communities are usually bound by one or two specific limits. Ecosystem health occurs within those limits.

Signs of ecosystem health include the homeorhesis of the system (after Waddington), the stability of the system (that is, its resilience after stress, such as floods), the diversity of its components, the continuous recycling of elements, and flourishing. Health is related to stress, both good stress and bad stress. Stress may be related to the rate of change for the system, in addition to loss or gain of components or changes in structure. Health is the overall ability of a system to maintain itself under a normal range of environmental conditions. Obviously, a pioneer community may change the conditions to favor a new level of the system with new components.

I.C. As Applied to Design

Design can imitate nature on many levels: from structure and process to landscapes. We can imitate the structure of mature forests by planting on every level of the forest hierarchy, from canopy to below ground. We can use native species. We can imitate the process of forests by allowing birds, bats, and other animals opportunity to distribute seeds and energy to other areas or prey on "pests". We can create microclimates within the landscape that may shift the landscape in new directions. Planting trees, for instance, allows new species to become established in their protection, form soil, etc.

Ecosystem health is one of the goals of design. The goal, of course, is not an end point that can be reached once, but is rather a continual striving.

The landscape provides its own metaphor for design. The landscape is a unique individual, a community, a dynamic system of interacting patterns—the human pattern is a part of it now and should be preserved as part of the whole pattern, but not necessarily as the only pattern or a completely dominant one. Most products of an ecosystem are produced and consumed and recycled within the ecosystem. Humans need to minimize the external inputs in the form of energy and exotic substances. The community must be restored to health. This means balancing human needs with bird or fish needs in a sustainable pattern. Each element in a pattern relates to others and to the whole.

I.D. Design as Lessons from Nature
Natural processes [building up/breaking down, development, disturbance], animal movement, interelement flows, human interaction, shifting mosaics, operate in forests. Conservation Biology (after Michael Soule) suggests a number of rules for reserve design that are based on natural patterns:
- well distributed species are less at risk
- large blocks of habitat are safer from species extinctions
- blocks close together are better
- contiguous blocks are better than fragmented
- interconnected blocks are better than isolated
- corridors can make large blocks functionally
- roadless blocks are better
- human disturbance similar to natural ones are less threatening

Forest fragmentation, for instance, can be reduced through the design of forested areas, taking into account the genetic diversity of the trees, catastrophic conditions, minimum viable populations, corridors, and edge effects. The survival of organisms usually depends on one of two factors in the web of relations. These factors can be modified by design.

Wild landscapes are affected by climate, soils, interactions, and disturbances. Domestic landscape is affected by land use as well. The greatest changes have been brought about by the destruction and creation of forests.

With the predominance of artificial forests, it is important to consider the qualities of naturalness in the landscape. Forests are expected to meet the needs of society by producing timber, creating wildlife habitats, and providing recreational opportunities for people. But, forests are also expected to look natural.

The English Forestry Commission's guidelines to principles and practical applications of forest design may be of use. They represent an established standard. They do not cover every aspect of landscape design or details of design techniques, however,; nor do all of them apply to wild forests. The guidelines indicate what to look out for, and which situations may need special attention. Forest landscape design is a complex subject, as are forestry and ecology.

The values of the land and forest must be most carefully assessed. The characteristic qualities must be identified and measured for uniqueness. Comprehensive landscape plans should be required when planting or extensive felling is planned on a large scale. The patterns established at these times may persist for many years or centuries. Good design may be able to resolve conflicts between characteristic qualities of the landscape and the changes from use. Also, the design:
- should last as long as possible
- should be self-sustaining indefinitely

II. Design Components of a Forest landscape

According to the Forestry Commission, forest landscape design depends upon an appreciation of six key design principles: shape, scale, diversity, visual force, unity, and 'spirit of the place'. These elements need to be expanded and related for ecological design (the following set draws from the Commission, Forman and Godron, and Mollison, but adds a few new ones).

There are basic geometric elements of any design, from the 3-dimensional (volume) to 2 (plane), 1 (line), and 0 (point) dimensions. These elements can vary in numerous ways, by number, position, direction, size, shape, interval, texture, color, and temporal. Furthermore, the elements can be organized into groups by nearness, similarity, and difference (diversity), into structures by rhythm, tension, balance, and scale, and finally into a whole with sensory force and a spirit of place (genius loci). All of the elements interact in complex and unpredictable ways. The spirit of the place is the most important principle to be conserved or enhanced.

II.A. *Elements*

II.A.1. Point. Points may be anywhere.

II.A.2. Plane (lines) The plane may be horizontal, vertical, or diagonal.

II.A.3. Volume. Volumes are three-dimensional.

II.B. *Variations*

II.B.1. Number. Quantitative change is the source of all qualitative change.

II.B.2. Position. Diagonal lines are most pleasing; lines at right angles to the contour are rarely pleasing because the landscape is broadly horizontal; geometrical shapes look artificial (even when they are natural); natural shapes, perhaps fractal, are considered more natural and interesting. The diagonal lines, as opposed to horizontal and vertical lines, of hills is both dynamic and pleasing to human senses.

II.B.3. Direction. Up is preferred to sideways or down. Diagonal lines give the impression of energy and movement, which is quite true as the hills are still being shaped geologically by erosion and wind. The eyes of travelers are drawn down one slope and up the next, then along the series of slopes. As people respond to one element and then another, the elements are perceived as parts of the whole. The sinuous path of the road through the corridor draws the eye toward the end of the corridor. The curves of the fields react to the shapes of the hills. The skyline is made more interesting by the shape of the hills.

II.B.4. Size. The size of an element determines the size and relation of others.

II.B.5. Shape. A complete inventory of elements in the creek starts with the shapes of the features in the area. The large volumes are rounded and natural hills—even the agricultural evidence is almost natural, that is, from the roadside not the air, the fields appear not to be squares, triangles, or circles; a small number of geometric shapes exist in the buildings by the road, but because of their scale are not too intrusive. Although the road itself has been flattened, it is not perfectly straight and does not conflict badly with the curving planes of the hillsides.

Shape to some extent determines how we see our surroundings. Shapes dominate other design factors, so appropriate shapes are critical. Proper scale or diversity cannot save a design if shape is wrong; the mind can pick up incongruities and artificial geometric qualities. Suitable shapes are vital for the unity of the landscape.

"The perception of shape is influenced by overall proportions, viewing position and direction, and the nature of the external boundary edge," (Forest Authority, p. 2)

II.B.6. Interval. The relation between objects in space or time.

II.B.7. Texture. The graininess of the system. Quality as a result of interval of elements.

II.B.8. Color. An indefinite number in nature. Forests have a smaller palette. As a result of reflection or absorption of light energy.

II.B.9. Time. Temporal patterns can be classed into four groups:
- circular, the eternal return, disintegration, reintegration, as of forests, nothing new
- spiral, a circle under stress, a cycle under change, something new
- linear, the straight line of industry, progress to heaven
- nonlinear, the line combined with chaos, to extinction and creation

II.C. *Groups*
II.C.1. Nearness. The spatial relationships of elements.

II.C.2. Similarity. The identity of elements.

II.C.3. Density (connectivity). The number per spatial unit. Closeness.

II.C.4. Diversity (difference). Diversity is the number of differences in a framework. Geology, climate, disturbance, and stability all produce diversity. Landscape diversity is linked to ecological diversity, which depends on diversity of the substrate.

Different ecosystems introduce diversity into a landscape, but different ecosystems often can look similar. Excessive diversity can lead to confusion in a landscape design.

Increased diversity also has the effect of reducing scale, so adding diversity can be used to do reduce the scale. A high level of diversity is acceptable if one element is clearly dominant or if the differences cannot be recognized from a distance.

Psychologists have recognized the need for diversity for people's quality of life and emotional well-being (Kaplan 1973). Ecological diversity in the forest has been reduced by human activities, such as planting or grazing.

The overall landscape diversity of many forests has not fared as badly, due to the addition of human artifacts, which increase it. An increase in ecological diversity would lead to an increase in diversity of the landscape, however. It would also tend to reduce the scale, but this would not be a problem in large forests.

II.D. *Structure* (Paths & Patches)
II.D.1. Rhythm. The repetition of similar elements into a dynamic whole. The units of time may range from milliseconds to thousands of years, although most will seem to be daily, seasonal, or long (years). Generates interest.

II.D.2. Tension. The interaction of elements without resolution.

II.D.3. Balance. A state of apparent stasis. The stable movement of elements around a center (or attractor) in a dynamic equilibrium.

II.D.5. Scale. Scale is a contrast of relative and absolute size. Scale is also determined to some extent by perception. The scale of a forest should reflect the scale of the landscape. But, since scale depends on human perception, which depends on perspective, it is difficult to keep scale for all perspectives. The scale of a landscape increases with the distance and width of the horizon and with elevation. The scale of a landscape is greater from mountain tops than from valleys. The Forest Authority notes that, "Small shapes may appear to be out of scale when viewed from a distance in a large scale landscape."

As the scale of the landscape changes, the scale of forested land should change accordingly. Change between areas should be gradual. When a landscape seems to be composed of two kinds of texture, the ratio between them should be a golden mean, which is judged as most satisfying. The golden mean produces a pleasant symmetry, unlike an equally divided landscape.

The scale of agriculture and development should be related to the scale of the landscape. As it is, the creek area is diminished even more by the scale of cultivated fields. The buffer zone should be increased up the hill, perhaps to a ratio of 1:3. Of course, seen from the air the ratio would decrease radically, as most of the landscape is farmed. The scale also seems greater on hill tops. Because the highway is in the creek corridor, the scale of the landscape is reduced, and because it is smaller scale finer textures can be discerned and therefore must enter the design. Details are more obvious.

II.D.6. Pattern (Network). Process applied to components yields pattern. Nature is composed of patterns. Organisms have characteristic patterns, such as the branching of trees or the cloud forms of tree crowns. Lichens have lobes, wood grain under stress has spirals. The cracks in tree barks form nets.

Patterns are not still. A circular pattern through time can be recognized as a spiral (the earth's orbit for example).

The pattern should allow for surprises and discontinuities; it can do this if it is flexible. The design of forests is vulnerable to surprises because nature is chaotic (unpredictable) and science itself is uncertain (by definition) about patterns of change in forests.

II.E. *Whole*

II.E.1. Unity / Simplicity / Character. Unity is a fundamental objective of landscape design. Unity is the way the elements, including shape and scale, of a landscape are combined.

Visually, a forest usually dominates the landscape. From a distance, even-aged forests have much the same impact, in terms of color, shape, and scale, as uneven-aged forests. Diversity becomes more important visually at a smaller scale.

Natural forms of the forest are unified with the landscape because the margins are very uneven, and open space in the forest is part of the mosaic caused by birth and death of individual or groups of trees.

II.E.2 Spirit of Place. The spirit of each place is unique (Norberg-Schulz 1980). Place is not just location; it is the total sum of objects in the landscape combined into a unique whole. The identity of place often leads to human identity, thus people call themselves Pullmanites or Moscowans or Appaloosa. The more unique a place the stronger the emotional attachment of the inhabitants. Every place has certain characteristics that enforce the spirit of place, for instance, a strong definition of place or indicators of great age (trees or rocks), or where a place distills the essence of larger landscapes. A sense of wildness and water also contribute greatly to the spirit of place. The spirit of Paradise Creek has changed greatly in the past 200 years, but it is still easily defined with elements of water and wildness. The creek is a more intimate place, less dramatic than Steptoe Butte or Palouse falls. The sprit of the place is the best guide to design. 'The spirit of the place.' Each place expresses a unique combination of elements, including contrasts, dramatic features, and the presence of water. Design can work to be consistent with the recognized spirit of place. If the design recognizes this aspect of the landscape, it may be stimulated by spirit and it may further enhance it—what it should not do is degrade it. Forest design can emphasize some features above others.

Goals of good designs include: Relink people with genius of their places, revivify image and identity with places, and develop and maintain identity of places.

II.E.3. Sensory Force. All the elements of design can be combined in an image. Every organism creates an image of its place from what is meaningful to it. This image is what fits the organism to its place. Suckers and caddisworms have simple images; coyotes and humans have more complex ones. Boulding notes that the image as a cognitive construct of the world has several aspects: spatial, temporal, personal, relational, value, and affectional (emotional) for each individual. Cognition is an active relationship that is creatively shaped by the participants. Participation is not an option by the way—every scientist or inhabitant becomes part of the system of observation. The total sum of individual images is a world. Some of the images we impose on nature result from idealized notions of pastoralism or technological futures. Thus landscapes abound in nostalgic or consumptive trends on many levels of explication—some are iconic, some invisible. We originally perceive the landscape symbolically, but the landscape has other functional dimensions that increase according to use.

Visual force is a psychological interpretation of perceived power in a landscape. As a principle, it is embodied in psychology, art, graphic design and architecture. The human mind responds to visual force in predictable and dynamic ways, for instance, visual forces in landscapes draw the eye down convex slopes and up concave ones—the strength depending on the scale and irregularity of the landform.

The effect of a forest landscape is not completely visual, however. Smell, sound, touch, and even taste play a large part of our appreciation of forests. Crawling (recommended by Gary Snyder), climbing, listening, and tasting (soil, bark, lichen, etc.) can expand our perception of other aspects of the forest. De Tocqueville commented on the ceaseless noises in the forests he encountered in Pennsylvania and Ohio (they kept him awake).

III. *Stages of Design*

III.A. Designs can be applied in five stages:

- First, review the situation, observing patterns of movement, population change, land use, building and development, boundaries, limits, and life. Conduct ecological and functional analyses.
- Then record all of the resources, from physical resources to cultural resources. Survey the area and create base maps, from geological to zoological maps.
- Next, evaluate the interactions in terms of impacts, needs, goals, and limits. Assess the whole system and create a series of plans, from the site plans to value plans.
- Start to design, which is a community process requiring the participation of all people (including the elderly, handicapped, and poor, as well those ultrahuman beings who cannot voice their concerns). Synthesize simulations and models (conceptual, capability, and suitability). Make another series of plans, from landscape plans to policy plans, within a master design.
- Finally, implement the design together and start to maintain it. Use appropriate measures and techniques, emphasizing native species over an adequate time period to ensure the stable processes of transformation. Provide services for continuity and management.

Appraising the Forest Landscape. Forest design must work within the components, structure, and function of the forest. Unless it does, it will not be long-lasting or satisfactory. Because design has to work with a forest, whose aspects are often ambiguous, fuzzy, changing, and general, design has to be able to work with these aspects. Furthermore, the design has to work within the constraints of the forest.

Forest design takes far more time than graphic or automobile design, due to the complexity, size and longevity of its subject. A number of factors have to be carefully assessed before design work starts. Forests require a lot of observation before activities can take place. Forests can be highly reactive to change. Some people value different character of forests than others.

III.B. *Levels of Design*
As design process includes planning of ecosystems, this mean a fourth level of design (notice that after the first sample, all apply to normal applications of design): 4. community (forest, city, corporation), 3. systems (ecosystems, traffic, industry), 2. products (habitat, houses, roads, plant sites), and 1. components (trees, fungus, bats, rooms, cars, land). Many problems occur at level three—more are to be expected at the new level four.

All levels of design need to be addressed, from the conceptual to the political, and are involved in all stages of the process. This involves new challenges for ecological design to:
- Relate a project to its total context (fourth level of design); be concerned as much with cultural survival, justice, and wilderness preservation as with efficiency and aesthetics.
- Consider the whole perspective (ecocentric, perhaps); the proper vision is of the whole community in which we dwell. Apply ecological concepts, such as networks and carrying capacity.
- Make designs are anticipatory, flexible, pluralistic, polyvalent, and polytechnic. Make open guidelines for long-term decisions.
- Essentially, work backwards from values and goals, and from the bottom up and inside out, drawing designs from the genius of place.
- Participate in place, care for all inhabitants, and assume responsibility for the designs.

IV. Characteristics of Forest Design
IV.A. *Frugality.* Avoiding excess. The style of use of things.

IV.B. *Adaptation*. Fitting in the forest, within established cycles and functions.

IV.C. *Plurality*. Allowing many values.

IV.D. *Respect*. Recognizing the "beingness," value and rights of the forest itself.

IV.E. *Playfulness*. Activity because it feels good.

IV.F. *Anticipation*. Thinking about possibilities; expecting surprises.

IV.G. *Responsibility*. Undertaking all aspects of the design, regardless of level of success.

IV.H. *Participation*. Designers should participate in a complete design process, guiding involvement and commitment to the art of living together as a community.

IV.I. *Principles*. Principles must be flexible to mirror the flexibility of open systems; flexibility is provided by diversity in fact. Sample principles:

1. Irregular shapes are more pleasing than regular shapes
2. The eye follows diagonal lines
3. The scale of a forest should reflect the landscape
4. Optimum diversity in a landscape is valued
5. Landscape is unified when scale and diversity are optimum
6. Any one sense, even smell or touch, can dominate a landscape
7. Design should follow sensory force
8. Design should enhance the spirit of place
9. Work with the forest. Succession can be assisted, slowed, or speeded up, but not skipped or ignored.
10. Forest is designed by its limits, time, scale, complexity
11. Only details and exceptions form the forest; only generalities are important—so you have to live with contradictions
12. Make the smallest number of changes
13. Adapt process of change to the site (you fit the forest, not other way)
14. Seek best use for products; everything in the forest is a resource for something; many can be directed to human use.
15. Extend the life of things through cycling, then return to forest; things can be recycled indefinitely, as in an old growth cycle.
16. Preserve the components, structure, and function
17. Understand the patterns and connections

V. Applying Design Principles to Forest Landscapes

All design elements are related psychologically by designers, as focus or frame, as contrast or uniformity, as dominant or recessive, or in a number of other pairs. Good forest design means not violating any of the aforementioned principles and ideas.

Design can improve the results of bad practices. Bad harvesting practices often result in geometric wastelands. Good design can correct reliance on straight lines, parallel lines, right angles, and perfect symmetry. In cutting or planting to improve natural appearance a number of things have to be considered, including the age of the forest, windthrow, width of corridors, and minimum size of the habitat.

V.A. Minimum Sizes. At some point in the reduction of forests, species and associations drop out, extirpated or extinct. Every system has a minimum size. Design has to consider minimum viable populations and minimum ecological areas to avoid destroying what it intends to design.

V.B. All the Pieces. There are many key species, or resources, or patterns in a forest. Since it is so difficult to discover all of these design must be cautious and minimal. Is the centipede more important than an owl? Is the mycorrhizal fungus more critical than the tree?

V.C. Forest Shapes. Successful forest design depends on the creation of large-scale visual natural shapes. The shapes are determined by whole blocks of forest, entire woods, external margins, and open spaces. The form and scale of the shapes should be dictated by the land form. Each shape interacts with its neighbors in a larger context.

According the Forest Authority, forest shapes should have gently curved edges—even ragged edges are more effective if superimposed on a curved shape. The shapes should have a diagonal emphasis, starting at near horizontal in flatter country and becoming more strongly diagonal on steep slopes; related to landform, high points are positioned in main hollows or gullies and low points on or near to prominent outcroppings or ridges;

The shapes should not contradict shapes in the surrounding landscape, e.g. be smoothly rounded in a ragged-edged context. Too much symmetry and regularity should be avoided. The more shapes are interlocked with each other, the more natural they look. Forest shapes should conform to visual forces in landscapes by rising up in hollows and falling on spurs and ridges. Forest shapes that reflect such patterns tend to match expectations of a natural landscape. If the shape of a block of trees or a felling area conflicts with the visual forces in the landscape, it looks disruptive and out of place.

V.D. Diversity. Elements of diversity in a forest include macromorphs, such as the land form, presence of water, exposed rocks, trees, wildlife, and special areas, such as archaeological sites.

V.E. Foreground. The immediate foreground allows for specific elements to emerge. At this range edges are prominent and details , such as individual trees, subcanopies, and shrub and herbaceous layers, are evident. The edge is characterized by edge trees with lower limbs, as well as an increase in site diversity (not necessarily overall diversity), especially if it is an established ecotone. Trees get thinner and smaller. (Mention beauty strips.) The visual elements to avoid are a 'wall' and 'size, age, and species uniformity.'

V.F. Margins (fields). Because of their visual nearness, and smaller scale, lower margins should have more variation. This is the usual curvy line idea. Avoid vertical lines dividing the landscape in half. Stick with diagonals in side margins..

The upper margin is usually the most prominent, due to contrast with the color and texture of the sky. The margin should rise uphill in hollows and fall on ridges. It should reflect the smoothness of the topography and be of sufficient density so that it does not look like icing on a cake.

V.F. Horizon. The skyline should contrast with a large mass of trees or expanse of grassland. Boundaries between contrasting areas should cross the skyline close to low features, and away from summits. Curves and diagonals are important again.

V.G. Edges/ Boundaries (Fences). Edges used to have a good reputation for promoting diversity and providing habitat. Because edges exclude interior species, they can never have as much diversity as the entire forest.

V.H. Paths. Roads, skid trails, animal trails, streams (visually) are considered paths. Utility lines are a special kind of path. Fences. Things to consider: edges, animal movement, obtrusiveness.

Paths should follow the land form and connect with natural open spaces, e.g., outcroppings. Width, shape, and direction should be varied when possible. In general, paths should: have minimum impact; follow contours of land form; vary in gradient and curve; cross ridges at low points, which is often done in the case of roads for economic reasons; avoid following lines-of-sight; avoid sensitive areas. Bundle paths (power lines, utilities) to minimize intrusion

V.I. Water. Streams need an adequate area buffer and adequate vegetative cover, especially on steep slopes, where they tend to straighten. Widen at lower elevations. The Forest Authority suggests that the cover of the stream should aim for 50% in full sun and 50% in dappled shade, with an irregular distribution, of course. Lakes and oceans.

V.J. Openings. Forest open space changes scale and increases visual diversity. Openings provide for different kinds of views: feature view, with one element; focal view, a converging to horizon; canopied view, beneath tree canopy; filtered view, through tree stems; panoramic view, from high ground, diverging to horizon. Keep density and openness in natural balance.

V.K. Interior. Preserve the interior/protect riparian zones.

V.L. Character. Character is partially determined by a distinct pattern of elements in a landscape. The character may be desirable or not to different groups of people. Character develops out of the interactions of the elements over a period of time, usually a long time. Some of the character is derived from human perception and values, from color to balance. Design may enhance or ruin character. Allow for human participation. Encourage responsibility. Conserve continuity.

VI. Design Management

Ecological design is not finished with a forest design. The forest may need to be managed as a result of the design. Noninterference matrix management is proposed as a technique for design.

A forest exists as part of matrix that many interacting elements. Any activity in the matrix can have some effect on these elements. The whole matrix needs to be managed with the forest in mind.

Understanding of the principles of ecology can lead to better management. One critical message of ecology is that if we diminish variety in the natural world, we debase its—and our own—stability and wholeness. Many forest ecosystems have been simplified and degraded. Perhaps we do not have sufficient knowledge to manage a complex landscape because it is too complex to understand scientifically. But we can understand the pattern and drive it in a healthy direction with minimal intervention. We must do all that we can to restore its richness and the natural processes that created the richness.

A noninterference approach to forest management (the essence of a Taoist way) is to let forest take its own course. Therefore, once the temporary constructs were in place, whether planting or cutting or any other manipulation, the forest would be allowed to develop without further interference.

In nature, noninterference means letting be. Noninterference matrix management is not indifference, which is diffuse. It is caring. Noninterference will not lead to chaos, poverty, and stagnation. The technocratic vision strives for "life under control," but the forest is self-managing, productive, efficient, and orderly. We need to practice the rule of noninterference so that all beings can enhance themselves. Noninterference can be derived from nonviolence (or taoistic nondoing). This attitude would entail using what is necessary, exploiting parts of some forest ecosystems, changing a place to fit human aspirations, and killing plants and animals for sustenance. But it would also mean limiting humanity and its technological effects, limiting human use to local impacts, and letting other beings live without interference. It is not necessary to dominate or terraform the forest completely to save it. Noninterference matrix management weaves people back into the fabric that supports them and in a sense makes them subject to the constraints of ecosystem processes.

This management (NIMM) would:
- manage the forest system with minimum subsidies
- manage activities that could upset equilibrium
- manage sustainable conditions
- align human activities with natural processes
- work with system instead of attacking it
- restore context

According to Garrett Hardin, many of the ideas necessary to fitting humanity into the pattern of nature are known but not yet popular. For instance, exponential population growth (or economic growth) cannot be maintained very long. Human communities cannot grow 4 percent per year without disastrous consequences to the infrastructure and the quality of life. Growth cannot be continued because the landscape is limited, in terms of productivity, energy, and resilience. Thus, we need to fit our population into the limits of the landscape (although some limits can be expanded by technology or by lowered expectations). The carrying capacity of the area is not only a function of the limits of the community, it is equal to the number of people multiplied by the level of comfort (quality of life style). Having more energy and space means having fewer people.

Design may be costly. For example, grassland restoration costs about $1500 per acre per year, based on the first two years. Forest restoration costs more.

Design may take a long time—longer than human lifetimes.

The forest may be too complex to design. One of the questions in the Distance Learning Course was how to design a forest using artificial pieces, such as giant sponges or shade cards. Every one gave up quickly—too complex. How do we design a forest, a complex, self-making, self-sustaining wild forest? Management has to recognize the limits of design. Limits of

ecological design include:
- forests are wild, we have no real control
- the scale of forests is too large to manage everything
- the longevity of forests is too long, we will never complete the design in human lifetimes
- the costs may be prohibitive—indeed, we have depended on the free goods of forests for economic advantages
- other human limitations apply to our ability to see and understand the forest.

VII. Summary

People often judge the health or wholeness of forests by how they look. Traditional design has emphasized visual results above all else. Ecological design, however, achieves the same results by paying attention to the structure and function of the forest first. Design has been concerned for centuries with making domesticated landscapes out of wild ones. Now, design must address the opposite problem: how to preserve or provide the conditions for wild forests.

Design must address the common good, that is, the good of the entire ambihuman community; it can do so by: promoting the well-being of all individuals in larger community, deciding what is preferable, attempt to regulate and anticipate all effects, encourage convivial activity, recognize links and dependencies, mediate the relation between technology and community, and alleviate some of the problems of modern industrial society.

Designs provide a framework for natural and artificial process to work in. The patterns in design are echoes of patterns in nature. Good designs learn to embrace error and failure, so necessary in open systems.

Most forest designs will not be restorations, because of the uncertainty about the kinds and associations of native vegetation. Furthermore, humans are now an large part, although not yet an integral part, of the system; therefore it could not be restored to a premodern or prehuman state (and even if it could, which state?). This design is not the biotechnological design of a new ecosystem, either; we cannot accurately control and predict ecological events in most ecosystems. However, we can steer some of the events in a known direction—known because we have historical records of the system, although not complete. We can also reduce those human activities that we know alter the conditions of the forest, such as overcutting and pesticide use.

Although ecological design attempts to restore some kind of balance, the balance does not exclude human activity. Rather, it integrates it into the larger community. A moderate number of human impacts can be absorbed by the system—too many destroy the systems capacity for self-maintenance. The design should be open to evolution and to human technological and social development. The design should be based on a model of ecosystem functions, considering diversity, complexity, and the maintenance of natural process—natural here meaning a self-sustaining system composed of elements now lost through human disturbance.

An ecological design involves designers and people in reshaping and

255

recreating a self-sustaining community. Individual resources are limited. The relationships to strive for here are community relationships. Furthermore, there are limits for human manipulation of other communities. Total control has limits, also. We should not aim to try to control the forest and its habitats. We have to trust that natural processes are self-correcting and organizing.

An ecological design is the creation of a clear vision of the forest that is aesthetic, useful, and self-sustaining. Some of the relationships can be captured by maps and drawings, but not the dynamic four-dimensional qualities of the forest itself, which can only be understood by dwelling there for years. Nevertheless, a simulation of the view from foot or airplane is more compelling than a recital of the statistics.

The goal of ecological design is not to restore, but to revitalize and reinhabit the forest. We do not want to live in the dead bones of a mechanistic failure. We want to live in a healthy environment with aesthetic appeal—aesthetic appeal is a requirement for human health. Every forest has physical, biological, economic, and political characteristics. The design, planning, and management for a forest describes the system in a comprehensive interdisciplinary approach, using dynamic concepts such as feedback and stability, recognizing limits to change and sustainability with different levels and scales of structure and function in an anticipatory, flexible planning approach, recognizing human and nonhuman goals, and incorporating personal and institutional interests.

Ecological forest design is the design of communities. We design places as organic wholes to promote the well-being of individuals and the common good. The immediate goals of design are to reverse degradation and reclaim places for communities, but also to work to increase public awareness of the interdependence of communities, to create environmental quality, and to transform public values by generating new metaphors for living.

Chapter 29

An Ecological Forest Care Plan (Mountain Grove, Oregon)

Executive Summary
Based on a complete description and inventory of the Mountain Grove
Forest, the following general and specific management objectives have been
identified:

- To protect, maintain and restore the Mountain Grove forest as a
  fully functioning forest ecosystem in perpetuity as a multi-species,
  multi-aged forest with old-growth characteristics within a landscape
  perspective.
- To recognize all forest values, including terrestrial and aquatic wildlife
  habitat, timber and other forest goods, e.g., mushrooms or medicinals,
  sources of clear water, fresh air, carbon sink, and forest existence.
- To maintain and restore the natural habitat of Woodford Creek to
  encourage migration of indigenous aquatic species.
- To protect the forest against catastrophic fire or interference
  disturbance.
- To practice, demonstrate and teach ecologically responsible forest use
  and restoration. To establish a Model Forest that can demonstrate to
  the public how to sustain and restore a fully functional forest, while
  harvesting a range of forest goods and products.
- To provide a range of forest uses and forest goods on a long-term
  sustainable basis, and to provide employment for people in the local
  community who are sustained by the forest.
- To manage timber for quality wood for high-end uses, recognizing that
  precommercial and commercial thinning are also necessary in some
  situations to restore forest structure, composition and functioning.
- To establish a conservation easement on MGC land to ensure that MGC
  forestry management objectives will continue to be implemented for
  generations to come.

For each of the management zones identified, a specific set of objectives
has been established. Activities, from forest protection to community
participation, monitoring, and record-keeping, are geared towards
supporting the objectives for the forest. Specific recommendations are made
to guide the activities and promote the objectives.

Mountain Grove Forest Care Plan
I. Forest Description
The Mountain Grove Forest occupies 420 acres of forest and meadows that
occupy much of the valley floor and some of the hillsides in the Woodford
Creek Watershed on the north of Buck Horn Mountain, which divides the
Umpqua and Rogue River Watersheds. Woodford Creek flows into Cow
Creek, then into the South Umpqua which later joins the North Umpqua
and flows on to the Pacific Ocean.

This forest is a meeting place and transition zone between the drier

Mediterranean-type climate to the south, characterized by the Pine/Oak/Incense Cedar ecosystem, and the moister temperate rainforest to the north, characterized by the Douglas Fir/Grand Fir/Western Hemlock ecosystem.

I.A. Layers

I.A.1. Geology

The Mountain Grove Forest is located in the geologically complex and deeply broken Klamath (Siskiyou) mountain physiographic province of southwestern Oregon. The province is composed of four belts of island arc-related volcanic and sedimentary rock, intrusive rock, and ultramarine assemblages; the belts are primarily east-dipping, with older plates in the east thrust over younger plates to the west.

Mountain crests are comprised of steeply folded and faulted pre-Tertiary strata, which vary in elevation from 600 to 1200 meters. The region is set apart from the rest of Southern Oregon by a boundary separating its pre-Tertiary rocks, probably the oldest in Oregon, from rock formations outside the area.

Most of the area is decomposed granite or schist. The basin is a highly erodible landscape, according to Walker and McLeod (1991).

I.A.2. Topography

Mountain Grove's forest is situated primarily in the valley floor at an elevation of about 1500 feet. Of the surrounding lands, BLM land is located primarily along the ridges at an elevation of about 2000 ft. and private lands are just upslope. A road and some private property abut Mountain Grove's lower border.

I.A.3. Climate

The Umpqua interior valley is a relatively warm, dry region with a Mediterranean climate. In the rain shadow of the Klamath mountains, the Umpqua valley has hot, dry summers, and mild, wet winters, although potential evapotranspiration in the summer exceeds moisture buildup in winter. Average annual temperature is 12 degrees C. (Average January, 4.2, average July 20.4). Average annual precipitation is 799 millimeters (US Weather Bureau, 1965).

I.A.4. Water

Precipitation (as rain or snow) is intercepted by the forest canopy. Surface water is held by forest floor structure, including duff and debris, before entering the ground water stream. Some of the water may be held in an aquifer. Much of it is channeled into Woodford Creek, part of the Cow Creek watershed.

I.A.5. Soils

Soils in this Klamath Mountains belong in a widespread great group, Haplohumults (reddish brown lateritic soils). The parent materials include sedimentary and basic igneous rocks. The soils are moderately deep (102 meters to bedrock) and possess a silty loam or silty clay loam A horizon underlain by a silty clay B horizon. Scattered upland areas of peridotite or serpentine bedrock have reddish-colored soils classed as Hapludalfs (gray-brown podzolic) or Xerochrepts (Regosols), which are considered unproductive, with very shallow and stony profiles.

I.A.6. Vegetation

The *Tsuga heterophylla* forest zone is considered to end its southern limit just north of MGC, at the North and South Umpqua river divide; it is regularly dominated by the *Pseudotsuga menziesii* subclimax. The *Quercus* woodland of the Umpqua valley has a well-developed canopy of *Pseudotsuga menziesii, Pinus ponderosa*, and *Libocedrus decurrens*; this woodland ranges from open savannas with grass understories to dense forest stands with an abundance of conifer associates.

The vegetational mosaic can be characterized as "Interior Valley" or typologically as a pine-oak-Douglas-fir zone, according to Franklin and Dyrness. The mosaic includes oak woodlands, coniferous forests, grasslands, chaparral (sclerophyllous shrub communities), and riparian forests—all considered semi-natural rather than mature communities, due to human activities. The "mixed-evergreen" occurs usually above 800 feet elevation. Detling (1968) typified the vegetation of the Umpqua as chaparral, with a peripheral belt of pine-oak forest.

Mixed stands of deciduous oaks, *Quercus garryana* and *Q. kelloggii*, and the evergreen Arbutus menziesii are conspicuous. Douglas-fir, *Pseudotsuga menziesii*, is common in the stands. Important shrubs are *Ceanothus integerrimus, Arctostaphylos viscida*, and *Cercocarpus betuloides*, especially on east and southeast slopes. Northeastern slopes and more mesic sites have open stands of Douglas-fir, Ponderosa pine, and incense-cedar, with a well-developed lower canopy of Oregon white oak. Typical understory species include Pacific poison oak, low dogbane, honeysuckle, balsamroot, fescues, lupines, brodiaea, and ground cone.

In coniferous forests, which tend to be in the uplands, Douglas-fir is considered most common, although Ponderosa pine and incense-cedar are conspicuous. Associated hardwoods include bigleaf maple, madrone, and oaks.

Some grassland or prairie communities may be mature sites on some soils or xeric sites. Others appear to be successional, maintained by fire and human activities. Some interior valley zones, dominated by soft brome, dogtail, and ryegrass, are invaded by sweetbriar rose and poison oak.

Chaparral in interior valleys is dominated by buckbrush and manzanita, with other species present: Deerbrush, poison oak, dogwood, tanoak.

Riparian habitats have a typical hardwood component, such as black cottonwood, willow, bigleaf maple, and alder.

I.A.7. Animal, Insect & Bird life

The Mountain Grove Forest is somewhat depauperate. High animal species diversity results from high forest structural and compositional diversity. That depends, in turn, on species, community and age/size diversity in plants from ground to crown. Wildlife is a prime indicator of a healthy, fully functioning forest in their roles as seed/spore dispersal agents, herbivory checks (from rodents to beetles), the health of soil fauna and flora, stream and aquatic habitat health and much more.

Only four of twelve arboreal mammals remain. Two species of woodrats, which are prime food for the northern spotted owl, are no longer seen. Woodrats abandon their stick houses when understory food plants disappear. Other arboreal mammals, including squirrels, porcupines, red tree voles, martens, fishers, chipmunks, forest deer mice and woodrats, are a keystone species guild. They all depend on trees in some important ways, and they are either the main or supplemental food for forest predators, including most raptors.

Many tree-dependent mammals, with the sole exception of the red tree vole, depend even more on an adequate supply of quality understory plants for continuing survival. The forest is more than the trees. Porcupines depend more on understory herbs and shrubs than the inner bark of trees. The northern flying squirrel eats truffles (fruiting bodies of mycorrhizal fungi) and passes spores in feces. Mycorrhizal fungi form a symbiotic relationship with trees by attaching themselves to the roots and greatly increase the uptake of nutrients and water, and by providing a large measure of resistance to disease. When the forest reaches the stem-exclusion stage, northern flying squirrels, like wood rats and owls, move on.

Of the fifty-eight shade-tolerant plant species at MGC, which remain at some population level, most are decreasing or have stopped flowering and fruiting. Many flying insect pollinators will not visit plants in the shade even if they are tolerant enough to flower.

Only twenty-four bird species have been counted so far. The BLM bird checklist for our region indicates that we should have over one hundred bird species as residents or winter/summer visitors. Few bat species remain of the thirteen species known in southwestern Oregon. Deer and elk are not usually seen because their prime browse plants have been shaded out of the forest. Some species of frogs and salamanders are locally extinct.

Finally, our endangered oak/ash wetland and oak/pine woodland are, along with riparian zones, principal habitats for the largest number of species, as well as the most endangered species, especially birds. At least fifty bird species are associated with mature or old growth pine woodland alone in southwestern Oregon. And more animals live in a mixed hardwood/conifer forest than in one dominated only by conifers.

A list of animals and plants found in the forest are attached to this document. They can be added to over time and used as a benchmark to gauge future biodiversity within the forest. This understanding will, in turn, be help guide human interventions in the forest.

I.A.8. Human artifacts/Roads/Buildings
The 1953-54 selective logging compacted much of the soils of the forest. It appears that the forest was cat logged and that the cats were not restricted to the skid roads. The road and skid-trail system was expanded in 1985 by Orville Camp. Starting in June 1985, Camp cut roads through most of the

new timber areas. By October 1986, the road system (12 miles new cut and 2 miles reconstructed) was complete with the exception of rocking the main loop.

Numerous houses and cabins were built in the 1960s and 1970s. Several have been refurbished for occupancy, while others are used to storage. A large Community building exists at the top of the lower meadow, about a mile from the property entrance.

I.B. Contexts/Patterns
I.B.1. Historical/Cultural/Reference
The valley served historically as the meeting ground of the Umpqua Indians. The forest of the Woodford Creek Watershed was initially high-graded in the early 1900s, when a railroad spur was constructed. MGC was then clearcut and burned between the late 1930s and 50s. The current condition of the forest is largely a consequence of past clearcutting, cattle grazing and fire suppression practices which resulted in an even-aged, simplified, overstocked forest with high fire hazard.

The old growth timber was logged by later settlers in the early 1900s with the railroad taking wood to Glendale Junction. One tree, 66" in diameter, 280 years old, indicates that there were fires every 50-80 years in some parts of the valley. Partial logging occurred until the 1950s.

A 420-acre parcel near Glendale, Oregon in the Woodford Creek Valley was purchased in 1969 by the New Education Foundation, with the intent of setting up a rural community school. Existing buildings at the time of purchase consisted of an old homestead and dairy with 320 acres of timber. The remainder of the property was developed as agricultural land.

The New Education Foundation, on purchasing the property, established the conceptual policy of living with and appreciating the forest. This involved not cutting any trees. During the 1970s a series of groups attempted to develop a school that would teach their eclectic theories and also be economically viable. In 1980, the Board of Trustees of the foundation realized that the timber must be managed. Fire hazard was real, with no access to any are of the forest and the faltering financial situation made management of it's resources a necessity. NEF was reorganized as the Mountain Grove Center for New Education (MGC).

In 1982 residents first agreed to practice what they termed "ecological forestry" or "radical forestry." During the early 80's they began commercial thinning using horses. In 1985-86 a network of narrow roads, contoured with the land, were built. They were designed so that trees could be yarded to the roads by cable without compacting the forest soils.

Management and Harvest History. Community members, with the help of county Extension Service cruised the timber in 1982, using 1/10th acre plots. The cruise resulted in projections of three million board feet (3 mmbf) of merchantable timber, without much detail as to the condition of the rest of the stand. The Extension Service forester wrote a management plan, recommending the clear-cutting of large areas

With the timber prices low in 1982, the board enlisted a forester living on the property to commence a horse-logging operation. He cut 25 mbf

261

before he became discouraged with the project and left.

The board then asked David Parker, a former resident of the foundation's community, to take over the timber management project. Parker owned 40 acres to the north of Mountain Grove and had experience thinning, tree planting, etc. Parker agreed to take over the operation. With his input and the interest of other trustees, the Board approached Orville Camp, a local timber management theorist and author (1992) to discuss alternative approaches to forest management. Mr. Camp's method involved an intensive forest management practice of thinning out dead, diseased and smaller trees, leaving the dominant, healthy trees to reseed the forest naturally. An extensive road system was necessary to access the stand on a long term basis. The use of chemicals and clear-cutting were not practiced.

The Board saw Camp's theories as more in keeping with the foundation's early conceptual policies than the clear-cutting recommended by the county agent, and Camp saw Mountain Grove property as an ideal place for an experiment of his methods. Having come to this point, the Board and Camp embarked on a contractual arrangement and established goals and objectives in 1985.

All right of way timber and selected trees within 150-200' of roads were harvested either by cat or tractor winching. The fallers were Mark Camp and David Parker, who cut as they went, leaving dominants on the first round. By September 1986, 360 mbf was harvested, mostly No. 3 and No. 4 grade, with 20% No. 2 grade and 5-6% select and 5 chip cull. The few old growth trees cut proved valuable lessons in economics: usually they were not worth the cost of getting them down and out to the mill.

Although research was intended to be a priority, the largest mistake was not to set up enough control plots. Plots that were set up will provide some information; however, budget for research was not a priority.

In April 1986, Parker purchased a Kubota tractor with a winch. As the roads were completed in 1986, Parker was falling and yarding small saw timber, while Camp was harvesting mature and long saw timber. By October 1986, the initial entry and road system was completed.

Contractual arrangements with the contractors were insufficient. There was a dispute between the Board and Camp over contract provisions and when to stop cutting. Another difficult lesson.

From March 1987 to September 1987 and April 1988 to November 1988, Parker continued to go through each unit, harvesting a total of 110 mbf and 80 cords of firewood from cleanup and more thinning. This was done on a part time basis, since Parker was working on his own property and since bad drought/fire conditions had made the logging season very short.

Continued firewood from cleanup and some thinning in heavy areas was done by community residents and Parker in 1990. The writing of the management plan continued, with plots being assessed and new data collected. Decisions concerning selection thinning, correct tree spacing and economic necessity were weighed.

By October 1990, the Board of Trustees voted to move ahead with another cut of 100 mbf. As the residency of Mountain Grove Center was still in an unsettled state, additional timber revenues will be needed to retire the remaining debts. Several discussions have centered around ways of increasing the cut and maintaining the look of a healthy forest. Future cutting could be on an increased thinning basis, with the intent of opening up holes in the forest canopy to allow light in for the regeneration of Douglas Fir.

Different microclimates and stand types were identified, permanent plots were established and cruise data collected. Precommercial and commercial thinning continued to be practiced through the late 80's and early 90's, leaving the biggest and best trees. However, within five years following each thinning, forest growth again closed the canopy, leaving the forest in a stem-exclusion stage with tree growth slowing down and fuels continuing to accumulate.

In the fall of 1994, following the first Ecoforestry Design Course, a new inventory and monitoring system was designed for MGC by Jerry and Sharon Becker. The survey, again, found that most stands were largely in a stem-exclusion stage and that fire hazard was high. Fire hazard reduction became a top priority.

I.B.2. Physical Landscape
The Mountain Grove forest is part of a mountainous landscape in southwest Oregon, in the center of the Klamath Mountain geologic province.

I.B.3. Watershed
The Mountain Grove Forest is bisected by Woodford Creek, which drains the Woodford Creek subwatershed, which lies with the Cow Creek Watershed (37,937 acres), in the Umpqua River drainage basin.

I.B.4. Forest Zones
Forest zones on Mountain Groce include a mixed evergreen zone and an interior valley zone. The area is environmentally and floristically diverse, including elements of California, north coast, and eastern Oregon floras with indigenous Klamath species.

The floral and faunal communities that existed 500 years ago have been fragmented, removed, or altered. Many of the new associations are maintained only by human intervention, e.g., fire suppression or logging.

I.B.5. Political/Legal
Political entities are changing, as people become more aware of undesirable changes that result from their activities. Nonprofit groups are forming to protect species and watersheds. Some political committees and groups are becoming nonprofit.

I.B.6. Plans/Changes (Factors influencing species)
Plans generated by many local groups are now focused on saving and maintaining habitat rather than individual species. MGC staff are working with several groups, e.g., ODOT and OF&W, about changes to stream structure to encourage anadromous fish returns.

I.C. Maps

MGC has been reworking its maps, using a layer-cake model. Layers include topography, ownerships, roads and trails, habitat types, conservation areas, management zones, and harvest areas. These maps are being put together with the DesignCAD computer program (see attachments).

II. Inventory
II.A. Forest Types

II.A.1. Structure

As a result of historical processes and disturbances, the forest is characterized by four (4) distinct structural and size/age classes:

1. Old-growth Douglas-fir, Incense Cedar and Sugar Pine which predate the fires in the 1920s and the later clearcutting and burning;
2. Trees which regenerated since the fire are now 69 years old and younger;
3. Trees which regenerated since the 1953-54 selective logging and are now 42 years old and younger;
4. Hardwood groves, particularly Chinquapin.

II.A.2. Habitat types

The Mountain Grove forest currently contains at least four (4) distinct ecosystem types:

- Lower elevation , riparian areas with a predominance of Douglas-fir and Grand Fir, with Yew, Madrone, and Vine and Big Leaf Maples. In the absence of disturbance, i.e., fire, the number of Grand Fir is increasing as the stand becomes more dense.
- Mid-level mixed species forest with a predominance of Douglas-fir, Incense Cedar, Madrone, Chinquapin, and scattered Sugar Pine.
- Mid-level hot SW sites dominated by Chinquapin with nearly 100% canopy closure and little regeneration with some Douglas-fir and Madrone.
- Higher elevation, 1,700' and higher, forest dominated by Douglas-fir, Incense Cedar, Madrone, Chinquapin, and a few Sugar pine. In the absence of disturbance, i.e., fire, the Sugar Pine are being crowded by the other species, particularly by the Douglas-fir, and there is little regeneration of Sugar Pine.

II.A.3 Management Zones

Units were originally set up as harvest units rather than as to type and have since been changed to account for this. There were originally 9 units, but they were expanded in 1985 to 15 ecozones, described briefly below.

II.A.3.a. General Descriptions of Management Zones

Fifteen zones have been identified and given short descriptions.

No. 1 Freeway corner NE. Homogeneous, 35-70 year old mature to small saw timber.

No. 1 A Clear cut in 1983 cut and planted in 8x8 spacing in 1988.

No. 2 Over the ridge from unit No. 1. Scattered mature saw timber, mixed with Ponderosa Pine and Pacific Madrone. 35-100 year old trees.

No. 3 Near lower meadow and along creek. 35-70 year old mature saw timber - Douglas Fir, Ponderosa Pine, Incense Cedar, White Fir, Pacific Madrone, Mountain Ash.

No. 4 Similar to No. 2, but with more Douglas Fir volume. Up to BLM land, 5-100 year old trees, mature saw timber, Douglas Fir, Ponderosa Pine - mostly small saw timber to saplings.

No. 5 Wilderness area (possible control area—Phantom Orchid endangered species). 6 acres near big community house. 20-150 year old timber, mature saw timber Douglas Fir, White Fir - 20+ mbf. .

No. 5A Three acres. Mature saw timber. 30-60 year old timber.

No. 6 West slope surrounded by BLM. Small to mature saw timber (Douglas Fir, Pacific Yew, Big Leaf Maple, Mountain Ash, White Fir, Incense Cedar, Red Cedar). 40-60 year old stand.

No. 7 East slope. Douglas Fir, Ponderosa Pine, Chinquapin, Tan Oak, Pacific Madrone. Douglas Fir and Ponderosa Pine small saw timber. Douglas Fir 50 years old after fire.

No. 8 Mostly Tan Oak and Pacific Madrone. 30 years old.

No. 8A Two acres. Clearcut converted to Ponderosa Pine plantation.

No. 9 Small saw timber. Pacific Madrone and Tan Oak in understory with Willow and Manzanita, 10-40 years old.

No. 10 Small to mature saw timber (Douglas Fir and Ponderosa Pine), similar to No. 9 but with bigger madrone. 10-60 years old.

No. 11 West slope similar to No. 6. Small to mature saw timber (Douglas Fir, White Fir, Ponderosa Pine, Sugar Pine, Tan Oak and Pacific Madrone).

No. 12 Similar to No. 9 and No. 10. Small to mature saw timber. 20-70 year old stand with Pacific Madrone, Incense Cedar, Ponderosa Pine and lots of brush.

No. 13 Over the ridge form No. 12 to Woodford Creek (no roads) residual mature saw timber (mostly Ceanothus, Manzanita and Pacific Madrone).

No. 14 Four acres set aside as stream confluence and wilderness area, with road through some right of the way cut. 60-150 year old stand of mature saw timber (Douglas Fir, White Fir and Ponderosa Pine).

No. 15 30-60 year old small saw timber (Douglas Fir and Incense Cedar).

II.A.3.b. Zone Descriptions

II.A.3.b.i. Structure and size

Each of the zones is described more specifically, regarding soils, roads, timber, and health. For instance, Zone 1, the Freeway Stand, is described as: "Size: 45 acres; Slope: 5-60%; Aspect: N/NW; Soils: No. 13 cornut-dubabella well-drained cobbly clay loam; lower section poorly drained; Vegetation: Douglas-fir, some Ponderosa pine; Special populations: Coral Orchid habitat NE corner; dogwood and currant; Structure: Even-aged, 35-70 years; Diversity: Poor; Snags/wildlife trees: 1.5/acre (below minimum); Dead down wood: Adequate; Disease: None observed, but stressed; Fuel load: Hazardous; Roads: 1.8 miles, insloping; Problems: Crowding stress, slumping roads; Harvest History: 25 mbf horse-logged 1982-3, 50 mbf road cut 1985-88, 25% cut by 1990."

II.A.3.b.ii. Volume & Growth
Due to measurement conventions, these figures represent pretty much only Douglas-fir. Pine, cedar, and others species are estimated in past cruises and calculated in 1997, but not included in these calculations.

Volume in 1982. Using the ten administrative units from November 1982, volumes of Douglas-fir in three size categories (9, 13, and 18 inches), the total for MGC is 2.89 mmbf.

Current volume (1997). The sum from all the zones is 6.2 mmbf.

II.A.3.c. Analysis & Summary
MGC is considered to be located in a high Site 3 area, with a growth rate of 700+ mbf per acre per year. The results of the 1982 cruise indicated that MGC had 2.89 mmbf standing. The expected increase in volume, using the index, was 3.15 mmbf (700 bf/ac/yr * 300 acres * 15 years), resulting in 6.04 mmbf.

The total amount cut at MGC in its harvests since 1982 has been 0.65 mmbf. Divided by 15 years, this means MGC is only cutting 43,333 bf per year.

The results of assessment of permanent plots indicates that MGC has 6.20 mmbf of standing timber. The growth rate is calculated at 733 bf/ac/yr.

The AAC is calculated at approximately 50 percent of the growth rate, to account for old growth that will not be cut. MGC does not consider timber the only product of the forest.

Summary
Standing: 6.20 mmbf
Growth rate: 221 mbf/yr
AAC: 111 mbf
Actual Annual Cut: 43 mbf

II.B. State of Forest Health Assessment
Prior to management by European settlers, fire was the dominant process affecting upslope and riparian vegetation (above the floodplain). Many sites burned, or were burned by tribal groups, as often as every 15 years and usually once every 100 years. A complex fire regime created complex vegetation patterns on the stand and landscape level. Fundamental ecosystem cycles—plant succession, nutrient cycling, individual life cycles— were driven by fire disturbance.

Through fire suppression, modern management has excluded fire as a disturbance process. Timber harvest practices, however, produced intense disturbances throughout much of the landscape, as have human habitation and travel patterns. These altered disturbances have fragmented the landscape; both early and late seral vegetation are often absent. The total species diversity is lower, while tree density is higher.

High tree density contributes to higher mortality among pines, from insect attacks. The accumulation of fuel loads increases wildfire hazards.

II.B.1. Plant Communities/Health

The MGC forest is second growth about fifty to seventy years old and entering the shady, densely-stocked "stem exclusion' style of forest development. Because of complete or nearly complete canopy closure, with heavy shade and moisture retention, forest conditions favor regeneration of the more shade tolerant tree and shrub species—grand fir, incense cedar, western hemlock (we have little so far), bigleaf maple, vine maple, red huckleberry, oceanspray (iron wood), Indian plum, bane berry, hazelnut, western yew, salal, creeping snowberry, baldhip rose, longleaf Oregon grape, whipplevine (yerba de selva) and hairy honeysuckle. Seedlings of some species are also favored but usually begin to die in heavy shade before reaching maturity (Douglas fir, cascara, madrone, canyon liveoak) or remain suppressed for decades until released (tanoak, incense cedar).

There is a whole class of plants, constituting a major and (for wildlife and special forest products) critical part of biodiversity at MGC, which are present but are disappearing because of heavy shade. There are few seedlings in any stand interiors of black oak, white oak, sugar pine, Ponderosa pine, bitter cherry, choke cherry, Chinquapin, and mountain mahogany. Douglas fir in the lowland forest is losing ground to grand fir. Most native grass species exist in fragmented remnant stands on forest-meadow edges. Many other native grasses are locally extinct, but probably exist in the greater region in isolated remnant stands. Numerous flowering forbs are missing or exist here only as widely scattered solitary individuals—preventing genetic exchange which will quickly lead to local extirpation.

Even more shade tolerant plant species (e.g., Douglas-fir, tanoak, canyon live oak, madrone, incense cedar, cascara, Indian plum) may fail to mature or remain stunted for long periods of time or die in densely stocked tree stands. At 100 percent canopy closure, even shade tolerant hazelnut and oceanspray will remain suppressed and gradually die. Most shade tolerant trees, shrubs, subshrubs, grasses and forbs will fail to flower and produce fruit or seeds in dense, shady stands. Examples are tanoak, salal, whipplevine, hairy honeysuckle, trailing blackberry, creeping snowberry, wild strawberry and longleaf Oregon grape. Understory diversity is generally very low in the MGC forest, due to shading, fire suppression, and historic overgrazing.

II.B.2. Wildlife

Nearly all of the sun-requiring or shade tolerant (but not maturing or producing fruit or seeds in heavy shade) plants mentioned above are important food plants for wildlife. Dennis Martinez's survey revealed fewer wildlife signs compared to other parts of Southwest Oregon in which he has done vegetation surveys. Even the woodrats—specialist feeders on hard-to-digest leaves of plants like manzanita and madrone—have abandoned their stick houses and moved away or are now residing in human cabins. Only twenty-four bird species have been counted so far, although more should appear as the seasons change. The Medford District BLM bird checklist notes around 150 birds which should be Summer/Winter visitors or residents in this general area. Conservatively, there should be at least 100 bird species at

MGC over all four seasons.

Deer and elk browse is in very short supply. Deer and elk do pass through here—although not in great numbers—and are rarely seen. There is evidence of an elk mud wallow up Woodford Creek (with scat and tracks). Martinez found only one or two corn lily plants—a favorite elk food—in the survey so far. There is little willow, deerbrush, or wedgeleaf buckbrush—the three most important food species for deer—and few nice, tasty succulent forbs or high protein native grasses (high protein browse, especially high after fires, assists deer in digesting the tannin in acorns and oak leaves, another favorite deer food now endangered).

It should clear by now that the forest is more than just the trees. A case in point is species called arboreal mammals. There should be twelve species at MGC and vicinity: red tree vole, western gray squirrel, northern flying squirrel, Douglas' squirrel, dusky-footed woodrat, bushy-tailed woodrat, Townsends chipmunk, porcupine, marten, fisher, raccoon and possibly forest deer mouse.

Except for the red tree vole which rarely comes down from trees, all of the above species, while depending more or less on trees for nests, etc., depend on quality food plants usually found on the ground in at least partly sunny forest openings. I've mentioned the disappearance of MGC's two species of woodrats. Probably a few are left, I've heard raccoons at night talking only once and so far have seen no tracks. There are some porcupines here, but they eat more herbaceous plants in the forest understory than inner bark (cambium) and dwarf mistletoe (in short supply). So they too will probably move on. There are Douglas' squirrel (eats primarily Douglas fir cone seeds but will switch to hazelnut when fir crop fails), western gray squirrel and Townsends chipmunk here. Western gray squirrel needs lots of oak acorns and other native seeds and nuts which typically grow in sunny forest openings or oak/pine woodland and ash/spirea/slough sedges/tufted hairgrass wetlands. These ecotypes are endangered. Martinez did not record any sign of northern flying squirrel which eats mainly fruiting bodies of mycorrhizal fungi (which are plentiful) and lichens, but depends on understory ericaceous shrubs (e.g. huckleberry, salal, manzanita, which form symbiotic mycorrhizal connections underground with the fungi) for cover while foraging on the ground for truffles.

Of the twelve arboreal mammals which should be at MGC, only four are noticeably and consistently here (five if a woodrat is living in your home, or six if a forest deermouse is also present in your home; however we probably have nearly 100% meadow deer mice here). Arboreal rodents are "keystone species" in forest ecosystems; if they are largely missing, many other species will be impacted. An example is the endangered northern spotted owl which feeds on the smaller arboreal mammals (e.g. woodrats) which in turn require food plants which grow well only in sunnier forest openings. This food chain from avian predators like hawks and owls to arboreal rodents includes sun-requiring forest understory species also as keystone species. This translates into forest structure—gaps and edges between groups of trees—as serving a keystone role in forest ecosystems as well.

At present most flagged game trails lead to and through remnant woodland and the riparian zone because of water needs, but also because those communities are the main places where some browse plants remain at MGC. We have asked scientists to check out reptiles and invertebrates (insects) at MGC.

Wildlife and Snags/Large Down Logs. Most of the arboreal mammals mentioned above are cavity nesters. So are bats (there should be thirteen species in this vicinity; it is recommended that a specialist survey and identify bat species). So are many bird species, as well as black bears.

Bird species include *pileated woodpecker, *redbreasted sapsucker, acorn woodpecker, *turkey vulture, *owls and *raptors, osprey, bald eagle, flycatchers and brown creeper. (An asterisk* indicates a species that has been seen at MGC.) All of these classes of forest species use cavities or loose bark in at least eighteen different ways besides just nesting—everything from lookouts and drumming or singing stations to food sources.

Large down logs over 12′1 in diameter are great reservoirs of water in the dry seasons, nursing tree and shrub seedlings through nutrient and water release and inoculating these plants with fungi which hold out when its dry under or in logs. Reptiles also benefit from the moisture, especially salamanders. Invertebrate (insects) decomposers live there and provide food for birds, bears, bats, raccoons, and other species. Specialists will assist in identification of reptiles and invertebrates.

Endangered species and restoration. The endangered species staff from the Oregon State Office with the U.S. Fish and Wildlife Service or the local Oregon Department of Fish and Wildlife office has agreed to assess the property for endangered species. Salamanders that should be present include western red-backed, Pacific giant, ensatina, clouded, northwestern, and Dunn's; the diurnal rough-skinned newt is very commonly seen; choices frogs are common, but red-legged frog and western toad could be here. Wetland restoration should favor both of these. Reptiles, e.g. rubber boa, racer, ringneck, common kingsnake, and possibly mountain kingsnake, sharptail, gopher and western rattlesnake; in abundance are gopher, racer, northwestern garter, Pacific coast aquatic garter, common garter and western terrestrial garter.

II.C. Potential Resources

There are many resources at MGC that are not being collected commercially. For instance fir boughs and ferns could be sold to flower shops for seasonal markets, such as Christmas or Mother's Day. Mike Barnes and other members of MGC are investigating trees with large burls for that market.

III. Objectives

A number of general objectives has been agreed upon by the residents of and workers at MGC. These include:

- To protect, maintain and restore the Mountain Grove forest as a fully functioning forest ecosystem in perpetuity as a multi-species, multi-aged forest with old-growth characteristics within a landscape perspective.

- To recognize all forest values, including terrestrial and aquatic wildlife habitat, timber and other forest goods (e.g., mushrooms, medicinals, etc.), sources of clear water, fresh air, carbon sink, et al.
- To maintain and restore the natural habitat of Woodford Creek to encourage migration of indigenous aquatic species.
- To protect the forest against catastrophic fire.
- To practice, demonstrate and teach ecologically responsible forest use and restoration. To establish a Model Forest that can demonstrate to the public how to sustain and restore a fully functional forest, while harvesting a range of forest products.
- To provide a range of forest uses and forest goods on a long-term sustainable basis, and to provide employment for people in the local community who are sustained by the forest.
- To manage timber for quality wood for high-end uses, recognizing that precommercial and commercial thinning are also necessary in some situations to restore forest structure, composition and functioning.
- To establish a conservation easement on MGC land to ensure that MGC forestry management objectives will continue to be implemented for generations to come.

III.A. Landscape/Bioregion/Watershed

The bioregion has been altered tremendously by recent human impacts. Neighboring forest properties in the watershed have been overcut; this means that MGC cuts do not take place in a vacuum, but have to consider watershed and landscape processes. For example, if surrounding land has been overcut, it would not be wise for MGC to take as large a cut as it could if the surrounding lands were whole and functioning in a healthy way. So MGC forest objectives take place in a matrix of objectives that are being considered for the landscape.

Fire, Wind and Water are key natural disturbances in the valley. As ecoforesters we seek to intervene in a forest ecosystem in ways which protect, maintain and restore the whole system. By augmenting the abundance of the whole system we augment the well-being of all the beings who live in the watershed, as well as above and below us—and all around us.

Since humans are part of the ecological equation—major players, in fact—it is critical that the Ecoforestry Management Plan and subsequent activities take into account the environmental, social, economic and cultural aspects of reality, as we understand it. The MGC Ecoforestry Management Plan is a component of a broader MGC Business Plan to develop an "Eco-village" of sustainable living, as well as an Educational Center for healing and restoring ourselves and our watersheds.

III.B. Forest Functioning/Health

III.B.1. Mapping corridors/roads

MGC has as an important objective, the establishment of cross-valley corridors for wildlife and vegetation shifts. One possible corridor is in the "waist" of MGC property, where it is bordered by relatively whole BLM properties; this would offer the shortest corridor with almost continuous forest cover. However, animals have traditionally used the wide lower field for transit; these fields connect with forest that has been cut heavily by private owners and offers a more dangerous crossing. The field is also a break in forest cover.

III.B.2. The 15 Zones

MGC has set up 15 management zones. It intends to manage these according to set prescriptions, depending on other objectives, historical use, and best potential use. For instance, the six objectives for Zone 6 are:

1. Move towards a multi-age, multi-species forests with at least 3 general age-classes: (1) old-growth (age range to vary according to species); (2) mature second-growth with good canopy position; and (3) regeneration.
2. Reduce fire hazard.
3. Maintain most existing old-growth of every species
4. Increase growth rates by thinning.
5. Emulate fire through management activities, i.e., thinning, some patch cuts, harvesting more heavily of shade-tolerant species, such as Grand fir. Consider use of controlled burns in a few selected small areas.
6. Encourage regeneration of DF with a few Patch Cuts.

III.C. Environmental Protection

III.C.1. Air
- To keep air circulation through diverse structures
- To avoid polluting air with chemicals

III.C.2. Water
- To allow surface water to percolate and cleanse slowly through complex vegetative and soil structures, including pit and mound topography.
- To avoid polluting ground water with chemicals

III.C.3. Soil
- To allow soil to develop stable complex productive deep structures
- To avoid compaction
- To avoid polluting soil with chemicals

III.D. Habitat/Species protection
III.D.1. Riparian
- To provide protection, shade, large woody debris, bed structure, and diverse vegetation along Woodford creek.
- To minimize impacts, crossings, or disruptions around the creek.

III.D.2. Species Guilds
III.D.2.a. Birds
- To maintain habitat, including host species, structures, and food supplies for key old-growth species of birds.
- To provide artificial structures in the interim.
- To minimize disruptions to bird populations.
- To consider impacts for migrating birds.

III.D.2.b. Amphibians, Reptiles, Fish
- To maintain specific moist habitat for amphibians.
- To encourage research on breeding for amphibians.
- To minimize disruptions for amphibian and reptile populations.
- To work with the OR Department of Fish and Wildlife to provide salmon and trout enhancement in Woodford Creek.

III.E. Forest Values/Products
- To identify as many values of the forest as we can, from ecosystem support values to special products.
- To identify specific products that can be safely removed in appropriate quantities without significant impact on the health of the forest.
- To identify cultural or archaeological sites to be protected.

III.E.1. Alternatives
- To identify special products, such as boughs, ferns, shoots, pine cones, lichens.
- To calculate the percentage that can be removed sustainably.

III.E.2. Timber
- To identify the kinds of timber that can be removed.
- To balance the kinds with maintenance and restoration objectives.

III.F. Forest Activities
III.F.1. Forestry and Restoration
- To create old-growth structure in each of the management zones (probably from 20-30%).
- To restore the major plant associations on appropriate sites.
- To maintain productive associations, such as Douglas-fir and incense-cedar, on appropriate sites, e.g., those that have 200-500 year-old trees.
- To allow natural regeneration as much as possible.

III.F.2. Agriculture
- To intercrop agricultural species under the forest canopy, or in places,

where open fields have existed recently.
- To produce many of the agricultural needs of the community through small community gardens, incorporating principles from permaculture and natural farming.

III.F.3. Facilities Construction
Objectives here include:
- Build kiosk for BLM partnership.
- Finish remodeling the Community Center for conferences and workshops.
- Build a shed for the fire truck.
- Rehabilitate cabins for visitors and participants.

III.F.4. Business Planning, Finances, Administration
- To plan for successful business use of products—including education—from the forest.
- To arrange for financing for novel approaches, preparation, research, or long-term projects.
- To administer the forest according to ecoforestry principles as well as ecological economic ones.

III.F.5. Marketing/Income Generation
- To find or create markets for certified wood.
- To find or create markets for alternative forest products.
- To get renumeration for educational conferences, workshops, and courses.

III.F.6. Work Opportunities
- To create work for people in the local community.
- To create work for participants in ecoforestry and MGC courses.

III.F.7. Recreation
- To offer recreation for MGC residents.
- To offer forest experiences for tourists and participants in MGC courses.

III.G. Sustainable Use
- To try to ensure that the forest can sustain itself.
- To try to ensure that all human activities in the forest are sustainable.
- To try to keep product collection below the natural replacement rates, considering the uncertainty of productivity and disturbances.

IV. Activities
IV.A. Preservation of Landscape/Bioregion/Watershed
Further landscape-level planning needs to be conducted over time and in cooperation with surrounding forest land owners/managers of both private and public forest lands. The owners and managers of the Mountain Grove forest will cooperate with adjoining landowners to plan and carry out forest use at a landscape level where opportunities present themselves. Some current ecosystem types will likely change over time, i.e., the Chinquapin grove is first successional on a hot SW slope, which over time will likely become a more heavily conifer dominated forest stand.

Figure 9. Rocks added to Woodford Creek

IV.B. MGC Forest Preservation
IV.B.1. Reservation of Forest Functioning/Health
In order to protect, maintain and restore the forest as a fully functioning ecosystem, the following age classes shall be maintained:
1. Old Growth. Approximately 30% of the forest shall be maintained in late-successional, old growth seral stage. Existing remnant old growth trees, which are scattered across the forest, need to be protected. Other dominant trees need to be identified as old-growth recruits. "Old growth" is defined differently for different species, based on how long such trees live. Douglas Fir can live to be 1,000 year or longer. Generally they do not reach maturity until it is about 125 years old. Old growth begins at about 150-200 years of age. Since the forest is a growing/dying, dynamic community, today's old growth will become snags and eventually fall to the forest ground and become soil from which the forest of the future will grow. Snags can be considered as part of the 30% old growth structure in the forest. Accordingly old growth recruitment is an on-going process. The old growth structure should be scattered across the forest, as well as in small stands along the creeks and riparian areas.

    Old growth trees may be cut if the spacing between two old growth trees in closer than the proximity of old growth stumps in the area and when there is a need to open the canopy for release of smaller trees or for regeneration in the area. In a few areas, old stumps indicate that the previous old growth trees, speaking mostly of Douglas-fir, were approximately 25 yards apart. Further study needs to be done to determine the distances between old growth stumps of various species in different ecosystem types.

2. A minimum of approximately 30% of the forest shall be maintained in the 60 to 150 year age class.
3. A maximum of approximately 30% of the forest can be maintained in the 1 to 60 year age class.

Furthermore, problems with the state of the forest may require direct intervention. Fuel loads, that have risen due to fire suppression, may need to be reduced through burns or removal.

IV.B.2. Protection of Environmental Patterns

Mountain Grove foresters expect to conduct their activities such that there will be minimal impact on or disruption to the patterns of air, water, and soil generation and structure. Many of these patterns need to be monitored regularly, before any actions are taken.

IV.B.3. Protection of Habitat/Species

Special attention is to be given to the requirements of many species for specific habitats. In the interim, staff may provide temporary structures, such as bat boxes, or create snags by girdling.

IV.B.4. Conservation of Forest Values/Products

Certain parts of the forest may need to be set aside to heal or develop. Workers should set boundaries and monitor those areas.

IV.B.5. Restoration of Minimum Areas

Special attention shall be given to minimum viable areas for vegetation (on the landscape level, areas for minimum viable populations of animals and birds can better be addressed). Where necessary, staff shall restore habitat through planting, cutting, and manipulation of species.

IV.B.6. Guided/Sustainable Use or Extraction or Impact

Workers will limit their intrusions or extractions to those amounts that have been calculated to be acceptable.

IV.B.7. Response to Disturbances

Many kinds of disturbances occur in a forest; many species are adapted to regular disturbances. Disturbances include fire, wind, and erosion (especially with local soils). However, it is difficult to manage disturbances in a human system.

Fire can be managed to some extent with jackpot fires or controlled burns (with the possible goal that natural fires may someday be allowed to occur). Some of the effects of fire can be duplicated by silvicultural techniques, such as thinning.

Windthrow can be minimized by cutting patterns and quantities but not eliminated.

Excess erosion can be avoided by careful cutting, as well as management of wide buffers for the riparian areas.

IV.B.8. Maintenance of Infrastructure (roads, trails, etc.)
In general, staff will use existing roads and trails. In some cases, older or marginal trails and roads can be rehabilitated into the forest.

IV.C. Community Participation
IV.C.1. Mountain Grove Center/Community
IV.C.1.a. MGC Nonprofit Educational Corporation.
The Mountain Grove Center is organized as a Nonprofit Education Corporation with an elected Board of Directors, consisting of both resident and nonresident members. Some of the nonresident members were former residents going back to the beginning of MGC in the early 1970s. The Board meets quarterly to set policy.

IV.C.1.b. MGC Community Council (Residents)
The Mountain Grove Center is also organized as an Intentional Community of residents with decisions made by the Community Council at weekly meetings. New members spend a year living at Mountain Grove before they are accepted or rejected as permanent members.

IV.C.1.c. Forestry Committee.
The Forestry Committee makes day to day decisions about forestry and restoration decisions within the context of policies first decided by the Community Council and, then, by the Board.

IV.C.2. Relations with the Broader Community
IV.C.2.a. Labor pool (Training & Projects)
MGC works with a number of people and businesses from the local community. MGC has contracted with the OR Reforestation Cooperative (Hispanic Forest Workers) to do some logging and slash treatments.

MGC has hired four youths (1 from Mountain Grove and 3 from Glendale) as part of a Summer Youth program, a cost share program with the State of Oregon. One is working in the office. The other three young men are building a trail for the Demonstration Forest project, as well as working with us to do monitoring, seed collection, road work, and construction of a pole barn for the fire truck.

IV.C.2.b. Businesses (Milling etc.)
Some of the businesses that MGC works with include:
- Dan Cline Trucking. MGC works with a local log truck driver, Dan Cline, whose yard is a couple of miles down Barton Rd., to haul logs to various mills.
- Tunnel Creek Mill. MGC works with the Tunnel Creek Mill in Wolf Creek to custom mill logs into lumber for buyers wanting to purchase

"certified" lumber, as well as for use at MGC.

- Top Veneer Mill. MGC works with David Fairbairn of Top Veneer Mill, which is currently under construction in Merlin. We have surveyed most of the hardwoods to assess their quality for veneer, burls and lumber. We have sold some veneer-quality Madrone to him for a test run at the Merlin mill.
- Cascadia Forest Goods. MGC will sell most "certified" logs to Cascadia Forest Goods (CFG), a new business set up by Twila Jacobsen and Mike Barnes, to add value to the logs into a variety of products and to market them. For example, CFG is currently processing lumber for Brett KenCairn, the former Director of the Rogue Institute, for his home.
- Ecoforestry Institute. MGC hosts EI conferences and workshops.

IV.C.2.c. Projects/Partnerships (BLM etc.)

MGC residents and the Ecoforestry Institute (EI) are involved in a number of Projects and Partnerships:

- Woodford Creek Watershed Partnership. MGC and EI have formed a partnership with the BLM, who manage about 1,500 or the 2.500 acres in the Woodford Creek Watershed. We have completed a Landscape Analysis and Design process, a Forest Assessment and Inventory for the BLM lands and most of MGC land, and a Stream Survey and Inventory for Woodford Creek. We are currently planning to conduct a Fire History/Age-Class Study.
- Glendale/Azalea Community Action Response Team (CART). Twila Jacobsen is Treasurer of the local CART and participated in a series of week-end training provided by Rural Development Initiatives (RDI) during the winter.
- Douglas County Soil/Water Conservation District. Twila Jacobsen was the first woman ever elected to the Douglas County Soil/Water Conservation District Board, which has a staff of three (3) and an office in Roseburg. The Board meets monthly.
- Umpqua Basin Watershed Council (UBWC). The Douglas County Board of Commissioners appointed David Parker to represent the Ecoforestry Institute on the newly formed UBWC. Twila is the alternate. This Council will play an increasingly important role in restoring habitat for salmonids and other endangered species.

IV.D. Monitoring

Monitoring at Mountain Grove is the systematic recording of soil, air, water, and vegetation numbers, to identify long-term trends in the forest ecosystem, that is, patterns of:

- primary productivity and tree growth rates
- organic matter accumulation
- inorganic inputs and movements through soils and water
- disturbances
- populations in trophic structure
- soil compaction

Before monitoring is begun, the objectives and data collection methods are decided. The purpose of the monitoring program is stated; the objectives are identified, e.g., the health of the forest. Health is the first objective, followed by production and aesthetics. Health can be "defined" by a set of indicators of health, such as species numbers or presence, or by patterns of health, such as stability or productivity. None of the indices that are measured are really adequate to define the health of the forest because they cannot account for the complexity, richness, and cycling that goes on in the forest. For that reason health indices are data that need to be resolved on the ground in person by someone who knows the history of the forest.

Monitoring at MGC is based on the Monitoring Handbook by Richard Hart (1994), which was designed to lay the foundation for a new approach to assessment and monitoring, based on the observation of patterns. MGC addresses the measurement of the parameters that define patterns that indicate health or change in a forest. MGC considers three different levels of monitoring: environmental, biological, and ecological.

Environmental monitoring is an umbrella for many activities, including climatic variables and geological processes; for example, the systematic recording of soil and air temperatures, humidity, air pressure are measured by meteorological organizations to predict long-term climatic change.

Biological monitoring is the regular, systematic use of organisms to determine environmental quality; that is, the state of the environment can be analyzed by how individuals react to pollutants. For instance, MGC already monitors lichens, which are good indicators of sulfur dioxide pollution because different species vary in their tolerance.

Ecological monitoring is the observation of communities to understand long-term ecological processes, such as succession and maturity. G.F. Peterken and C. Backmeroff propose a set of useful rules for ecological monitoring:

1. Any variable or process that can be readily measured and dated may be valuable in detecting changes in ecosystems.
2. Long-term monitoring must be supported by administrative continuity otherwise the program may simply be overlooked or forgotten.
3. Facilities are required to ensure (i) survival of records and duplicate copies or records, (ii) markers locating the transect or quadrat and (iii) that the program is known to exist.
4. Repetitive recording is obviously necessary and although it may not be necessary at regular intervals, further records should be taken after or prior to any formative events.
5. The monitoring locality should be inspected regularly (annually for a forest) even if information is not collected.
6. Although objectives of the monitoring need to be defined, recording aims should be open-ended. The basic systematic record should be supplemented with casual adjuncts which have a habit of being valuable at a later date. This is because we do not know how the data could be applied in the future.

7. Simple variables and processes well recorded are more valuable than poorly recorded complex variables and processes. It's better to record something rather than nothing.
8. Representative records and replicates should be established if possible but even an unrepresentative sample may be valuable in the future analysis.
9. Regular analysis and preparation of reports, even at early stages in the monitoring help to improve the methods for data collection and help to refine the objective. These reports also serve as a reminder of the program.

MGC uses two approaches to monitoring that are vastly different in scale, precision, and cost:

1. Scientific apparatus, which is accurate but often requires you to know the problem in advance; physical and chemical analysis are expensive and technology and labor-intensive. Although identifying the indicator species may be difficult.
2. Mapping biological indicators, which is a quick inexpensive survey in the field. Basically, this means walking or crawling through the forest with your senses open.

W.T. Hinds suggested, based on his work at Battelle Pacific Northwest Laboratories, that nondestructive data on energy transfer between trophic levels in forest ecosystems is a cost-effective basis for long-term modeling (basically this means measure with your eyes instead of grinding up samples). For example, to detect slow, subtle changes in coniferous forests, he and his colleagues look for valid data that could be easily obtained. Since conifers subjected to stress from pollution tend to lose older needles sooner than unstressed trees, Hinds suggests that needlefall is an indicator of stressful conditions.

Richard Hart, in his work at Headwaters, relies on scientific apparatus.

IV.D.1. Patterns & Objects

Hart's scientific approach is based on the observations that the actual substance of which the forest environment is made consists of patterns rather than things or individual species, that the forest environment is generated by a patterning of ecological ebb and flow of energy, substances, individuals, and species across a suitable landscape, and that the forest environment is constituted of a large set of events that are objectively definable by specific outcomes.

The essential intent of ecosystem management is to initiate and maintain a fabric of events that weave through the common connecting points for all species events within the forest in a 'neutral" or beneficial fashion. The ultimate goal is to collect the information that details the establishment and maintenance of "comfort zones": the climate, nutrient and mobility suitable for species viability.

Hart's handbook provides a general characterization of the general "physiology" of the forest (including the ecology, biology, hydrology, etc.), and suggests a set of primary indicators for and guidelines by which its ecosystem can be inventoried, understood and monitored over time. The processes are easy to learn and relatively inexpensive. But we should not be deluded into thinking we are gathering cheap information.

The information needs to be credible, and gathered strategically and carefully. It needs to be entered into a data management system that is both relational and spatial, not for easy cures or the reduction of the forest into numbers, but as a way to gauge relationships and their patterns. What makes something valuable is not in its own properties alone, but also in its relation to the ecosystem as a whole. The "usefulness" of a thing or species does not rest solely in its appearance as an individual entity, but rather in its existence in a web of relationships. How it "is" essentially involves a complex series of relationships between the various events that are going on inside and outside its boundaries over large and small spatial and temporal scales. To understand the physiology of the forest requires that all view-points be considered together (top-down, bottom-up, or across the landscape). The health of forest ecosystems essentially determines human health—physically, socially, economically, and spiritually.

The approach Hart proposes follows ecosystem theory and the hierarchical organization of ecosystem functions throughout the landscape. This can be accomplished by incorporating the three primary attributes of biodiversity, as described by Jerry Franklin—composition, structure and function—into four levels of organization—province, subprovince, watershed, and site (Franklin 1988). Fifteen indicators incorporating composition, structure and function at the appropriate levels of organization have been identified for the Klamath/Siskiyou and Southern Cascade Provinces.

Indicators are measurable surrogates for determining disturbance to indigenous biodiversity that we assume to be of value to the public (Reed Noss, 1990). The 15 indicators were selected based on the following criteria developed by Tom Atzet (Atzet, 1993):

Relevant: meets identified objectives;
Sensitive: quickly detects change, shows trends, identifies critical areas;
Available: inexpensive, known standard, easily applied;
Measurable: accurately quantifiable with acceptable methods;
Defensible: not subject to individual bias;
Coverage: variable is well distributed spatially and temporally;
Acceptable: methods are socially accepted and understood; and,
Usable: provides feedback for adaptive management.

IV.D.2. Procedures
The 15 indicators of biological diversity in an integrated ecosystem monitoring framework are presented: Landscape morphology, stream morphology, soil condition, stream chemistry, species habitat, fragmentation, historic range, Large woody material, demographics, and economics. Each indicator has unique measurables and special methods. The frequencies of measurement depend on the indicator, although for the most part the human dimension requires only a one-time effort, whereas the physical and biological require a first-time baseline and regular validation. Indicators may overlap between the three components,—

Physical, biological and cultural—emphasizing the linkage among functions within the ecosystem.

Crucial to the implementation of this framework is a coordinated strategy to implement the monitoring framework. As an example, the following steps might be taken to implement biodiversity monitoring at the province level (adapted from Tom Atzet's "Strategy for Filtering Information"):

- Establish goals and objectives, such as protect and enhance the condition of the ecosystem, which includes forest processes as well as forest species;
- Gather and integrate pertinent, existing data;
- Recognize "baseline" conditions from current data;
- Compare data to the historical range of biological conditions;
- Determine the natural disturbance regimes for each biogeographical unit, defining the dynamics and structure;
- Through gap analysis, identify "hot spots" and ecosystems at high risk. These are areas of concentrated species richness and ecosystems at high risk of deterioration due to human activity. Control areas for each identified ecosystem need to also be identified. Monitoring trends should be focused in these areas.
- Determine social and economic values;
- Through field observation, formulate specific monitoring questions to be answered. Questions might include:—Is the ratio of native to exotic range grasses increasing or decreasing?—Is the average patch size of managed forests increasing or decreasing?
- Design and implement a sampling scheme for collecting the key indicator data, such as soil compaction, in conjunction with specific biological events immediately associated with the site condition. As an example, an inventory of the attendant plant communities associated with the soil condition will verify site productivity and status of soil repair.
- Data is collected and transferred to a GIS management system, such as ArcInfo, that assesses all the information relationally and makes the outcome(s) obvious.
- Realize trends and recommend actions.

Evaluation of data must occur in an integrated manner that spans biological and physical scales, watersheds, administrative boundaries, as well as functional areas. To understand how ecological processes are connected we need to relate information across disciplines and agencies, and collectively perceive the effect of our actions on the environment. It will be our challenge to find those events within the ecosystem that give us the most succinct information of what those basic needs are.

Funding must be identified and allocated as part of the planning process. Funding would provide for monitoring design, field work, data-management (GIS), and evaluation of results over the period dictated by the sampling scheme.

Involvement with the community, other agencies, research, industry and institutions must be included. Everyone has a stake in the maintenance and enhancement of forest health.

Key Indicators
The following descriptions of each indicator outline the specific attributes
which can be measured, as well as an explanation of why they contribute
to our understanding of forest health. It is anticipated that as our
understanding of ecological processes increases the number and type of
indicators may change.

Physical/Chemical
The structural integrity of forested ecosystems is vital to their continuance.
Mechanical and chemical disturbances affect major changes in the structure
and function of natural communities. The goal is to gain an objective
appraisal of the patterns and processes, including the causes, which lead to
enhancement or impoverishment.

This is an attempt to define a diagnostic procedure for reading
the changing landscape. This procedure will need continual up-grading
as workers fine-tune their ability to perceive and record the changes of
structure, function, and the potential for response to disturbance.

1. Compaction: The availability of water, air, and nutrients to plant
roots decreases as soil porosity decreases. Reduced soil-water recharge
potential can result in reduced productivity and viable nutrient-producing
soil microbe populations. Reduction in root mass and elongation adversely
affects the ability of the roots to supply nutrients and water. Monitoring both
the degree and spatial distribution of compaction can assess soil physical
conditions and long-term productivity.

2. Stream chemistry: Analysis of stream temperatures provides
an indication of water quality, dissolved oxygen concentration, and the
metabolic rates of aquatic organisms. Measures of dissolved oxygen,
nitrogen, turbidity, and pH will further help us sort out the effects of stress to
the aquatic system and its inhabitants.

3. Landscape morphology: Slope stability monitoring provides
information regarding potential: migration of material into watercourses;
effects on landscape structure; changes in overland flow of materials and
nutrients; and, changes in habitat conditions suitable/unsuitable to resident
species.

Soil erosion causes changes in physical properties of soil, such as
structure, texture, bulk density, infiltration rate, depth for favorable root
development, and available water-holding capacity (Batchelder and Jones,
1972; Frye et al., 1982).

4) Stream morphology: Data gathered within stream corridors provides
information on streamflow, velocity, channel shape, and substrate. The extent
of productivity loss and biotic simplification depends largely on landscape
position, runoff and internal drainage.

5. Atmospheric effects: Photosynthetic fixation of solar energy is the
cornerstone of all ecosystem activity. Chronic depression of photosynthesis,
primarily caused by photochemical oxidants, raises critical questions about
indirect effects of air pollution on ecosystem function. Photochemical oxidant

damage is related to the decline in plant productivity (Heck, 1982). This pollution seems to be interacting with such naturally occurring stresses as drought, insects, disease, fire, and windstorm (Bormann, 1983; Burgess, 1984; Postel, 1986). Ozone and sulfur dioxide are the two principle dry airborne indicators influencing ecosystem health.

Biological
Forests have important biological characteristics that can be measured and monitored.

1. Species habitat viability: The best approach to monitoring habitat conditions conducive to species prosperity is to select indicators that meet the following screens: sufficiently sensitive to give warning of impending change; distributed over a broad geographical area; capable of providing a continuous assessment over a wide range of stress; relatively independent of sample size; easy and cost-effective to measure, collect, assay, and/or calculate; differentiates between natural cycles or trends and those induced by humans; and, relevant to ecologically significant phenomena (Cook, 1976; Munn, 1988; Noss, 1990; Sheehan, 1984).

The objective is to assess viable habitat comprehensively. Species, by and large, continually migrate to habitat suitable to their prosperity. These are the events that are to be observed and recorded within the landscape structure.

2. Plant associations: Measurements of ecological components including; structure, composition, and function of plant associations, as well as seral stage, could provide information regarding the effect of species displacement and habitat alterations. Additionally, this would also provide an indication of habitat availability.

3. Habitat linkages/fragmentation: Biotic communities have progressively become reduced in area as a direct result of human activity (Spellerberg, 1991). During this process of reduction, the wildlife habitats and natural communities have become progressively more and more fragmented, resulting in remnants of habitats and communities being surrounded by different land uses.

Monitoring this indicator will provide information at the province and sub-province levels on the rate, extent and potential effect of fragmentation. In addition to monitoring the rate at which biotic communities are affected and the extent to which they are fragmented, monitoring the levels of success or failure arising from attempts to reduce the impact of fragmentation is also needed.

Buffer zones and corridors are two concepts put forward as ways of reducing the isolation effects of reserve areas. Mapping variations in canopy cover may aid in determining the effectiveness of buffer zones and corridors.

Also, monitoring populations in remnant communities and the ecological effects of fragmentation could provide basic data for the development of management strategies for different kinds of protected areas, including late successional reserves. Dennis Martinez has started to do this at MGC.

4. Historic range and variability of species: Use of ecologically-based

historic occurrence of species via elevation, aspect, habitat and natural variability of species significance/ presence would be used to contrast with present human activities. Descriptions of historic landscape disturbance regimes (e.g., fire magnitude and frequency) and the ecosystem component patterns they maintained (e.g., vegetation composition) provide an initial template for descriptions of ecosystem health. This function would provide a baseline or control to effectively gauge other monitoring items.

5. Standing, down, and submerged large woody material (LWM, also referred to as DWD): Measurements of standing, down and submerged LWM, and present state of decay, provides information related to microhabitats, organic soil material, habitat capability, and species viability. As LWM decomposes, it provides an interface between the living and the dead. Monitoring state of decay will insure a continuum of this resource into the next planning cycle.

6. Soil organics: The litter and duff layers are part of a process that produces soil organics. Soil organics are important to soil moisture-holding capacity, nutrient supply, physical aggregation of the soil and reduction of erosion rates. The natural process of decay results in a reduction in the level of soil organic matter—an effect that can be countered only by the frequent addition of fresh residues to the soil. Monitoring of litter and duff thickness is an indicator of soil productivity and health.

Cultural Human Dimension
Human values ultimately determine what, where, and how we monitor forest management activities. Therefore, it is imperative that cultural values and natural processes be determined and monitored simultaneously. In this way, we can begin to understand which natural processes we use (favor/ discourage) in order to benefit our current value system (re Martinez, 1993). Monitoring and evaluation of the indicators below provides necessary information in understanding the human dimension as it relates to ecosystem management.

Social and economic data is often compiled by standard political boundaries, not on provinces, sub-provinces, landscapes or project sites. Thus, socioeconomic data is not strictly comparable or usable at the sub-province, landscape or project site levels (Williams, 1993). However, the indicators listed below can be either described or mapped onto geographic information systems (GIS), providing an opportunity for integration with other physical/chemical and biological indicators.

1. Demographics: This indicator provides information on where people settle, how many there are, and what they do, which has a profound effect on the environment and its sustain able use and management. Data may include; age distribution, in-migration, percent of population using forest resources directly (including recreation visitor days and employment), where people settle—their distribution, and number including rural interface zoning, roads, type of industry, public services, communication lines, and types of community activities.

2. Cultural influences: Information gathered through archaeological sources and studies provides information on historic uses of the forest

and surrounding landscape, and associated human activity. This data may include ethnographic information and historic use patterns.

3. Economics: Data on local and regional economic health would help to predict long range trend and landscape use patterns. Data may include real unit costs and price changes, income tied to forest activities, profitability of forest management, payments to government, percent of wood products recycled or reused, changes in the distribution of employment and income, employment patterns tied to forest activities, rural interface land values, type of commerce, welfare payments into the community, and the economic base.

4. Human values: Data may include various land-use patterns such as percent of land base set-aside in protected areas, dispersed and concentrated recreation areas, grazing allotments, mining activities, "hang-outs" used by certain age groups, vegetative/animal usage (game management areas, firewood, timber, special forest products, fishing spots, water withdrawals), and community contacts (retirees, business people, school faculties and staff, college students, community clubs, tribes).

Humans are part of the equation too. What makes something valuable is not in its own properties but in its relation to the personal preferences of its perceivers. The usefulness of a thing or species does not rest entirely in its appearance but rather in its existence.

IV.E. Planning
IV.E.1. Ecological
Short-term. The staff at MGC continues to work on plans for the forest. Some are considering ecological planning procedures. Others work on forest plans with other watershed or transportation groups.

Long-term. Although it may not be possible to avoid ecological changes—nature is always shifting and changing—some changes such as the greenhouse effect can be taken into consideration with cutting and planting.

IV.E.1.a. Preservation
MGC intends to preserve old-growth Douglas-fir trees in areas where they are doing well. Because the forest itself has been so heavily altered, individual trees will be preserved. Two areas that are relatively healthy and functioning will be set aside as "wilderness" areas.

IV.E.1.b. Restoration
IV.E.1.b.i. Wetland
MGC is probably the southernmost extension of southern Willamette wetland communities. The wetland and riparian species listed are native to both the southern Willamette Valley and the low elevation valleys of the Umpqua Basin. They include federally listed species like Bradshaw's Lomatium. Nelson's checkerspot, rough popcorn flower, Bensoniella, slender meadow foam, Oregon willow herb and red-root yampah probably occurred in Southern Douglas County, part of the Umpqua Basin, as well as in wetland communities as far south as Glendale and MGC in the Cow Creek drainage. We are considering restoring two wetland communities at MGC: wetland prairie and shrub swamp.

Wetland Prairie.
The historically dominant species—tufted hairgrass in the wetter places; perhaps co-dominant with slough and meadow barley; sedges and red fescue in the drier places—are both nearly locally extinct. Only small remnant patches remain. Small remnant stands of more abundant grasses like wild blue-rye, oat grass, sheep's fescue, rough fescue, California fescue and Columbia brome have been located and flagged as future seed sources. Still other grasses and forbs will need to be collected on public or private lands off-site. Some herbaceous wet prairie plants exist here and there on-site and have been flagged, e.g. grass widows, camas, buttercup, kneeling angelica, owl's clover, and field checker-mallow.

One of these plants, southern Oregon buttercup (*Ranunculus austro-oreganus*), is a BLM special status plant proposed for federal listing. We have flagged and will continue to protect the 25 individuals in the population. Restoration techniques, such as mowing (with a weedeater) herbaceous competition around the rare buttercup, may enable it to spread.

There are numerous benefits from restoring these communities: support species diversity from relatively local seed sources, both on and off site within 50 miles; protect existing remnant wetlands from further degradation; expand wetland remnants through restoration techniques; expanded wetlands with greater native species richness may attract the threatened (although not necessarily listed) bird community associated with native grasslands and wet prairies; large logs placed in wet prairie may offer protection and habitat for some native amphibians and reptiles; federally listed wet prairie species like Bradshaw's Lomatium—if local seed is found and necessary permits obtained—could find a permanent protected refuge at Mt. Grove; exotic invaders like tall fescue will be contained and reduced.

Shrub Swamp and Wooded Wetland Communities
Remnants of these once extensive communities exist at MGC. Present wetland remnants combine indicator species of wet prairie, shrub swamp and wooded wetland communities. Moreover, these communities inter-finger with drier oak/pine woodland, as well as riparian gallery forest. There is at present no neat line between all of these communities.

Vegetation surveys at Mt. Grove have located and flagged the following native and exotic wetland indicator species. The list includes both obligate and facultative wetland species: Natives. Kneeling angelica (*Angelica genuflexa*), swamp lupine (*Lupinus polyphyllus*), meadow clumroot (*Heuchera chlorantha*), nootka rose (*Rosa nutkana*), clustered rose (*Rosa pisocarpa*), Oregon ash (*Fraxinus oregana*), Douglas spiraea (*Spirea douglasii*), black hawthorne (*Crataegus douglasii*), sitka willow (*Schix sitkansis*), Scouler's willow (*Salix scoulariana*), red fescue (*Festuca rubra*), tufted hairgrass (*Deschampsia caespitosa*), annual hairgrass (*Deschingosia danthonoides*), yellow monkeyflower (*Mimulus guttatus*), slough sedge (*Carex obnupta*), hairy owl's clover (*orthocampus hispidulus = Castilleja tenuis*), meadow checker-bloom (*sidalcea campestnis*), cow peromp (*Heracleum lenatum*), narrow-leaf blue-eyed grass (*Sisyrinchium anjustifolium = S. idahoense*), water parsley (*Oenanthe sarmentosa*), cattail (*Typha latifolia*), spikerushes

(*Elcocharis* spp.), bulrushes (*Scarpus* spp.), rushes (*Juncus* spp.), common camas (*Camassia guamash*), coyote thistle species (*Eryngium* spp.), western buttercup (*Ranuncolus occidentalis*), northwest cinquefoil (*Potentilla gracilis*), small-flowered forget-me-not (*Myosotis laxa*), witchgrass (*Panicum capillare*), toad rush (*Juncus bufonius*), service-berry (*Amclandrier alnifolia*), common horsetail (*Equisetum arvense*), reed horsetail (*E. hymale*), skunk cabbage (*Lysichitan americanum*), California false hellebone (*Veratrum californicum*), Siberian candy flower (*Claytonia sibirica*), great betony (*Stachys cooleyae*), rigid betony (*Stachys rigida*), stinging nettle (*Urtica dioica*), western water hemlock (*Cicuta douglasii*), short-scale sedge (*Carex deweyana*), woolly sedge (*Canex lanuginosa*), red-osier dogwood (*Cornus sericea = C. stolonifera*), bush snowberry (*Symphoricanpus albus*), Pacific ninebark (*Physocorpus capitatus*), red alder (*Alnus rubra*), white alder (*Alnus rhombifolia*), black cottonwood (*Populus balsamifera* var *trichocarpa*), Gairdner's yampeh (*Perideridia gairdneri*), parched willowherb (*Epilobium paniculatum*), California oatgrass (*Danthornia californica*), and Watson's willow-herb (*Epilobium ciliatium* spp. *watsonii*).

IV.E.1.b.ii. Oak Woodland Restoration Management Plan
MGC's 420 acres consists of approximately 300 acres second growth Douglas Fir dominated forest (around 70 years old) and 100 acres meadow dominated by exotic tall fescue. Included are Woodford Creek riparian zone and remnant native ash\sporaea\slough sedge wetlands. Basically, the forest designated as oak-pine woodland (28 acres) is in the stem-exclusion phase of development with high stocking rates of younger trees (over 1,000 stems/ acre) and high fuel loading/fuel ladders. Quality wildlife habitat is nearly gone while catastrophic fire is a real threat.

We propose to thin (group and individual selection cuts) on a 5 year reentry schedule) for the next 10 years. Thinning goals are:
- generate revenue for restoring wildlife habitat
- reduce fuel load to alleviate fire hazard
- regenerate wildlife plants and habitat understory native shrubs/ forbs/grasses/sedges; Ponderosa and sugar pine; black and white oak, canyon live oak and tan oak; incense cedar, madrone, Chinquapin, bitter cherry, maple and ash; red and white alder, vine maple, cottonwood, big leaf maple, ash, cascara, and red cedar in the riparian zone.
- manage 3 age-classes of Douglas Fir (with a minor component of Grand Fir) through multiple entry group and individual tree selection cuts on a sustained-yield basis.
- recover at least 50% old growth trees as well as retain at least five large logs (over 12" diameter.) per acre (with general fuel loading down to ground 8 tons per acre) and 3 large conifer snags per acre.
- retain 20% diverse under story shrub groupings.
- initiate regular rotational prescription fine following thinning.
- following fire with native grass/forbs/sedge seeding.
- restore patchy forest structure and species-rich composition.
- restore underground mycorrhizal fungal-plant relationship.
- restore wetlands.

Target wildlife species for consideration are:
- deer and elk
- arboreal mammals
- bats
- passerine and game birds/raptors
- anadromous fish (we currently have only cutthroat trout) and eels/ aquatic snails
- amphibians and reptiles
- invertebrates, especially pollinators and decomposers.

We also intend to reduce or eliminate exotic or weedy native pest flora and fauna, such as:
- spotted knapweed
- tall fescue (convert meadows to native grasses/forbs/sedges with scattered pine, oak and incense cedar/maintain by fire)
- canary reedgrass
- cut leaf blackberry/Himalaya blackberry/sweetbrier rose
- deer mice
- ticks
- mistletoe (contain within limits)
- pine and fir beetle/blackstain fungus

To accomplish these goals we propose the following restoration tasks:
- thinning with individual and group selection cuts/selling merchantable timber
- slash piling for both wildlife and burning; scattering for prescription fire; flagging quality habitat plants for protection/enhancement/seed source
- fire prep: limbing, brushing, raking duff away from leave trees, eliminating fuel ladders and jackpots.
- locating off-site and on-site remnant native plants and enhancing/ collecting seed
- manual grubbing to maintain opens until fire
- prescription fire on rotational basis to maintain opens.
- restoring native grasses/forbs/sedge to carry fire in opens when closed parts of forest are damp
- manual grubbing of exotic plants/mowing, burning, reseeding for control of tall fescue/shading out with native plants.
- monitoring results of vegetation manipulations
- restoring large logs to Woodford Creek
- cutting back senescent quality browse species (of *Ceanothus integerrimus*, willow, bitter cherry) for rejuvenation of quality nutrition.

IV.E.1.b.iii. Fir/Hemlock
MGC is examining the native communities in the mixed evergreen and interior valley zones. Examples of these communities will be saved or restored.

Figure 10. A Restored Section of Woodford Creek, Oregon

IV.E.2. Forestry

IV.E.2.a. Annual Operating Plan

Every year, at the annual meeting the board and residents of MGC will plan the annual operations. Depending on the plan, that may include a harvest. Sites will be addressed by the urgency of intervention, depending on weather conditions, canopy closure, fire hazard, or other criteria.

The annual report will address: location, limited area, species, ages, tree health, context, and desirable end. The report will be reviewed by all members of the forest committee.

The annual report will include the following objectives: Improve health of forest; adjust the AAC; provide adequate kind and amount of disturbance for regeneration; pay for harvest and taxes.

The following steps are taken to implement the plan: Mark take trees, indicating direction of fall (towards road) on plan; reconsider; arrange for sale or milling; cut; clean-up.

IV.E.2.b. Silvicultural Plan

Those areas not designated as preservation, restoration (which may, however, produce timber or alternate materials through the restoration process), or riparian shall be covered under the silvicultural plan.

General Strategy

MGC's 420 acres consists of approximately 300 acres second growth Douglas Fir dominated forest (around 70 years old), 100 acres of riparian and grassland meadows, and 20 acres of house sites, gardens, and roads. Basically, the forest is in the stem-exclusion phase of development with high stocking rates of younger trees (over 1,000 stems/acre) and high fuel loading/fuel ladders. Catastrophic fire is a real threat.

We propose to thin (using group and individual selection cuts) each management area on a five-year reentry schedule for the indefinite future. Thinning goals are to:

- generate revenue for paying the overhead and infrastructural costs of

operating MGC as a community.
- reduce fuel load to alleviate catastrophic fire hazard
- regenerate commercial species for forest health and for future takings; this includes understory native shrubs, forbs, grasses, and sedges; Ponderosa and sugar pine; black and white oak, canyon live oak and tan oak; incense cedar, madrone, Chinquapin, bitter cherry, maple and ash; red alder, vine maple, cottonwood, big leaf maple, ash, cascara, and red cedar in the riparian zone.
- manage 3 age-classes of Douglas Fir (with minor components of Grand Fir, cedar, and hardwoods) through multiple entry group and individual tree selection cuts on a sustained-yield basis.
- recover at least 50% old growth trees as well as retain at least five large logs (over 12" diameter.) per acre (with general fuel loading down to ground 8 tons per acre) and 3 large conifer snags per acre.
- retain 20% diverse under story shrub groupings.
- initiate regular rotational prescription fire following thinning.
- restore patchy forest structure and species-rich composition.
- restore underground mycorrhizal fungal-plant relationship.

Certain major plant species are targeted for reintroduction: western hemlock on a small north slope area in the riparian upland; western redcedar, which was extirpated by cutting, on north slopes and draws. Certain wildlife species or guilds are targeted for reintroduction: arboreal mammals, bats, raptors, reptiles and amphibians, and invertebrates, especially pollinators and decomposers. Certain exotic or weedy native pest flora and fauna are targeted for reduction: cut leaf blackberry, Himalaya blackberry, sweetbrier rose, deer mice, ticks, mistletoe (within limits), and pine and fir beetle/blackstain fungus.

To accomplish these goals, the following restoration tasks have to be undertaken:
- thinning with individual and group selection cuts/selling merchantable timber
- planting missing tree species, such as hemlock and western redcedar (which Franklin and Dyrness point out were integral parts of the system)
- flagging quality habitat plants for protection and enhancement
- fire preparation: limbing, brushing, raking duff away from leave trees, eliminating fuel ladders and jackpots.
- locating off-site and on-site remnant native plants and enhancing/collecting seed.
- monitoring results of vegetation manipulations
- cutting back senescent quality browse species (of *Ceanothus integerrimus,* willow, bitter cherry) for rejuvenation of quality nutrition.

Artificial regeneration would be limited to specific cases in zones that are considered out of balance.

Standards

The agenda of ecoforestry can be presented through a number of characteristics, principles, and standards. Characteristics are qualities that distinguish unique individuals, systems, or patterns. Principles are fundamental rules or laws, based on the characteristics of the forest systems, that we can use to create images or models to meet stated objectives, which are goals towards which our action is directed, e.g., a healthy forest. Standards are models or examples of quality or value, established by authority or mutual consent, that can be repeated as procedures.

For example, one characteristic of a mature forest is its wildness. The corresponding principle is that the forest is self-making and self-ordering without human control and management. Our objective for the forest is to allow the foresting process to continue, whether we take resources from the forest or not (by also considering interference from acid rain, pollution, and other industrial effects). We can set standards that are likely to keep mature forests wild: Limit annual biomass removal to 1 percent of the total forest; use appropriate techniques, e.g., single tree selection, with horse skidding; retain mature structure, e.g., 19 snags per hectare and 23 nurse logs per hectare in mature Ponderosa pine; and preserve surrounding landscape patterns. Standards for SW Oregon forests zones are still being developed.

Sustained Yield Management Plan

The silvicultural prescriptions at MGC are based on the data indicating the ecological state of the forest, as well as on state and federal regulations. For the forest as a whole, as well as for each management zone, traditional measures are applied, as are measures dictated by ecoforestry principles and standards.

MGC accepts the general idea of sustained yield, that is, never cutting faster than new trees can grow, but also recognizes two assumptions on which sustained yield is based, according to Herb Hammond: the forest can offer a constant nondeclining yield of timber; and timber growing has priority over other aspects of the forest. MGC does not intend to "normalize" its forest using AACs. Its primary goal is the health and continuity of the forest. For example, MGC expects to subtract the contribution of old-growth trees from the calculation.

Although recognizing that AACs are not necessarily appropriate where harvests are periodic rather than annual, AACs are calculated for each zone and for MGC as a whole. In the past MGC has cut so conservatively that it was felt that the cutting could not exceed growth over a period of time. However, to ensure compliance with its own objectives, MGC relates cutting levels to AACs established by conservative estimates of growth. AACs and target volumes will be calculated annually with change in knowledge or growth, i.e., adaptive management.

Similarly, a basal area per stand and for MGC as a whole is calculated. These numbers are then related to other measures describing the overall characteristics of the forest. However, target basal areas are not calculated, because MGC does not have basal areas as goals.

IV.E.2.c. Future Use
MGC is planning for the quantity and quality of future harvests in the forests. To ensure the character of the forest, it has developed a protocol for legacy trees and natural regeneration.

Legacy trees. Because MGC leaves the majority of old-growth trees, it does not mark legacy trees as such. Legacy trees, however, come under the MGC policy on old-growth. MGC contains scattered old growth or late successional forest fragments across the forest. Old-growth forests are an essential component of the forest landscape which further the ability to maintain fully functioning forests. Areas of late successional or old growth forests need to be included in a protected landscape/stand network. This is particularly the case because of the lack of old growth trees and forest in the surrounding landscape.

Old growth nodes should be located in ways that are connected to other parts of the protected landscape network, e.g., the riparian corridors. While old growth nodes need to be designated for each ecosystem type found in the landscape, it needs to be remembered that the forest is dynamic over time and what is old-growth today will be food for the forest floor in another 500 years. Accordingly, we design cutting patterns which allow for the occasional harvest of mature, old growth trees for high-end uses based on the following criteria: to thin more shade tolerant species at the interfaces of different ecosystem types, e.g., Grand Fir crowding out Douglas-Fir and, in some cases, conifers crowding hardwoods; where the spacing between mature trees are so close that they are within each other's growing space, while recognizing that trees in wetter sites will naturally grow more closely together (Grand Fir/Douglas-Fir) than in drier sites; and to open the canopy to allow for or aid regeneration of other trees of the same of different species.

Old growth recruitment trees need to be identified for each ecosystem type, so that over long periods of time, natural succession will (hopefully) result in the emergence of composition and structures that define an old growth or late successional forest. Through time, old growth recruitment areas need to be developed in all landscapes in order to replace old growth.

Harvest method—Single-tree and group selection. MGC uses single-tree and groups selection as its harvesting method. Since the mid 1980s individual tree selection has been practiced consistently at MGC. Orville Camp's "natural selection" forestry advocated taking only trees already selected out by natural processes (suppressed subdominants). But this light touch practice favors grand fir over Douglas fir (most tree regeneration at MGC is grand fir) and confines pine, oak and associated hardwoods to a suppressed existence in all stand interiors by shading out.

Individual tree selection, as practiced, failed to reduce fire hazard and halt insect and disease infestations. It failed to reverse ecological degradation, including local extinctions of wildlife and plant species. It has not allowed for the reintroduction of light prescription fire-to which most plants and animals have long genetically adapted. Therefore, more group selection cuts will be used at longer intervals.

Important added benefits of group selection cuts are the recovery. of water—lost in overstocked stands through evapotranspiration—and the faster recovery of old-growth trees since their release from competition will speed up growth in girth and height, therefore achieving the structured characteristics of old-growth earlier.

Foresters and Ecoforesters need an understanding of the genetic dynamics of tree populations in order to manage forests wisely. It is difficult to develop strategies to prevent the depletion of genetic information, especially with intense public and industry demands on forests, without knowledge of the diversity and distribution of genes in the tree population. Although study of gene flow and genetic recombination is needed, a lot of this knowledge can come from careful observation of the forest, the mating systems of trees—e.g., pine species require pollination as a mechanism for outcrossing—and the shape and size of a forest area. MGC is not selecting for phenotypes (the physical expression, the best individuals); nor is it selecting for genotypes, although by judicious cutting of all age and size classes, it hopes to increase the long-term flexibility of the forest to respond to certain catastrophic events, such as global warming. The problem with selecting the best looking trees is that they do not always have the best genes—just the best expression of their genes at one particular place at one time (there is a research article that supports this). Observation is necessary, but it is not enough; genetic work has to be part of the program. Management of genetic "resources" is also critical to reforestation efforts. Because trees are genetically correlated with specific place, we are not concerned with a lack of diversity or genetic erosion due to the absence of inplanting. At this time, MGC is not concerned with dysgenic effects as a result of its cutting practices; it does not thin from below exclusively.

Regeneration. MGC follows Franklin and others in expecting Douglas-fir to regenerate from single-tree and group selection cuts, as well as from shelterwood or clearcuts. Newton and Cole suggest that, if periodically thinned, Douglas-fir can maintain good growth for 200 or more years; therefore, rotation times of 200 years are economically feasible. Lorimer points out that pine and oak forests regenerate adequately after group selection cuts.

MGC expects to regulate regeneration only to the extent of opening some areas to light and moisture. Many variables affect regeneration—soils, humus, canopy opening, episodic seeding, rainfall, animal predation and movements—but MGC only attempts to control canopy. Under certain conditions, e.g., restoration of species or replanting from roads, MGC may plant a limited number of trees.

IV.E.3. Financial Planning
MGC is concerned with the values of the whole forest. Even without financial assessments as a commodity, a forest provides renewal services in the form of water, air, and inspiration, game, wood, shelter, and experience. Even without knowing a complete dollar total of the value of the forest, we have a small idea of the quantity and value of the interest and could tailor

our financial system to that. We price forest resources by their value to us.

MGC does practice silviculture, which is regarded as a rational practice for producing the highest quality and quantity of timber, "thus attaining maximum financial returns while maintaining the soil at its highest productive capacity," according to Toumey.

Financial Planning. Once we know what we want to do, and where, in the forest, as well as how, we need to find out how much these activities will cost and how much revenue we can expect from the goods and services.

A financial analysis and plan ensures that our activities are profitable without degrading the forest. If timber cutting or leasing is determined to be unable to generate adequate revenue to cover costs, then this activity can be abandoned before people are tempted to "cut corners" and degrade the forest in order to achieve short-term economic success. Our economic goals include covering costs, ensuring that sufficient revenues are generated to cover future costs, such as equipment replacement, improving forest health and aesthetics, and guaranteeing that workers are paid fairly for their services. Many of these activities may not be considered profitable under conventional timber management.

Part of the analysis is investigating markets or places to sell our products or services. Hammond states that "If you cannot sell a product or service then the activity is probably not advisable, with the notable exception of an activity that fulfills a personal or group need, like collection of food or medicinal plants."

A business plan provides a practical blueprint against which we can check our day-to-day, or month-to-month financial picture; it compares the costs of activities with the revenues received from them. Based on the plan, we also consider borrowing money for short or long term.

Audit. One of the tools used by MGC is an ecological audit, which is a review of MGC's operations from the perspective of deep ecology; the term audit is used more broadly than financial audit to mean any environmental impact. As part of an ecological audit, MGC has a sanctuary forest program, in which 2 old-growth trees are recruited for ever one used. Part of this process is a critical analysis of ecological sustainability, to examine if the products and services of MGC are ecologically sound (nondestructive). The market goals of MGC are made compatible with ecological requirements. Long-term financial goals can be formulated to incorporate ecological goals with high technology and social requirements.

IV.E.3.a. Future Income

MGC expects to get income from a wide variety of sources, including educational courses, ecotourism (bicycling events in SW Oregon), alternate forest products, such as mushrooms and boughs, and timber. Its strategy is to balance the income from many sources.

IV.E.3.b. Forest product utilization/marketing [ref REII 3.3.vii]

Working with Cascadia Forest Goods, MGC is finding markets for oak, madrone, and white fir.

IV.E.3.c. Protection—Easements

The staff at MGC is planning for a conservation easement so that plans for the forest, which has a lifespan much longer than humans, will continue indefinitely into the future.

IV.F. Forest Use

Although MGC has as its primary goal the health of the forest, it is willing to supply some of its financial and other needs from the forest. No harvesting was done between 1950 and 1982. Since 1982, parts of the forest have been thinned; these harvests have been light.

IV.F.1. Timber Harvest
IV.F.1.a. Past Harvest, 1994 (9/24/94)

From the EI intensive and apprentice program, we saw average volumes at MGC to be 5 to 15 (10 MBF) per acre. Translated to 250 acres, that's 2.5 million board feet.

We're growing, based on standard site indexes, +700 bf/ac/yr or 175MBF per year. If we were to adopt a certain % of cut per year as Merve Wilkinson does, say 50%, that would be 87.5 MBF/yr. These growth projections based on industrial forestry lead to eventual clear cut management that is unsustainable.

With changing weather and growth conditions, a much smaller cut on longer rotations are what I advise. The health of the forest is paramount.

Data collection and monitoring plots must be established. During the apprenticeship program, we touched on these skills but little was set up. Community residents do this. Minimal equipment and labor costs can come from logging proceeds.

Taking into account the forest's health, another need, the financial health of the stewards, needs addressing. Retiring the initial read construction loan +-$6,000, back and yearly taxes +-$5,000 and other indebtedness relief is a priority.

Looking at Superior's log prices as of 9/22/94: 11"-tops $700, 12" $750, select mill $850. All Doug. Fir.
Debts to pay $15,000
Log costs and hauling 40%=$6,000+15,000=$21,000
At average $775/M that's 27 MBF or 10 loads
At 50% =$75000+15000=$22,500 29 MBF required 11 loads

Whiter fir is dying through the ecosystem and pine, which is also dying, need to be thinned in many locations. Both these and cedar are within fir price ranges. Markets are less stable and hauling is higher to distant mills. Cost/profits for these species fluctuate.

Choosing to pay for monitoring plots and longer hauls could put MGC expense up to 50%. At %20/plot(standard exam prices), 15 ecozones with 2 plots that's up to +-$600. That leaves $900 for extra hauling or additional revenue for MGC.

So what's sustainable? Thirty thousand board feet mill scale is about one house. We can do that easily this fall. We can get rid of the burdens of debt, cut responsibly and carefully and then get the data together and wait a few years.

We may need some prescribed burns in heavily fueled area. As we discuss in the intensive, mimicking the old growth patterns is a way to achieve the desired old growth type forests that's healthy and productive. Costs of burning are not high—the insurance is. If done at the right time of year we can be safe and set up some demo plots.

IV.F.1.b. Future Harvest
MGC is planning its next harvest for fall 2000, as circumscribed by its plan. The harvest will also be part of the educational process at MGC.

IV.F.2. Alternate Products
Possibilities for alternate forest products will be scoped out at a summer workshop at MGC. Several residents at MGC have taken courses and workshops in finding and developing alternate products.

IV.F.3. Marking Reserve Trees: Wildlife, Legacy, and Heritage Trees
Mountain Grove plans for and marks reserve trees. These trees may be live, green, healthy, large, or green defective, dying, or dead. These trees are left for a variety of reasons:

- To form the basis of a diverse, healthy forest, to provide seed and genetic diversity
- As wildlife trees, for moss, lichen, insects, birds, or other beings, for food, shelter, structure,
- For aesthetic reasons, being well-formed or oddly shaped

Each of these objectives contributes to the kinds, sizes, characteristics, and locations of the trees.

Reserves trees are spread throughout each area, in groups or as individuals (depending on past harvest or disturbance patterns). These areas are carefully marked, due to the effects of single tree and group selection.

The reserve trees are monitored at the same time as permanent plots, to document changes in forest, tree, or hazard conditions.

Reserve tree types
Mountain Grove keeps a dual classification for reserve trees. For its own purposes, healthy conifer and deciduous trees are classified, especially for possible old growth recruitment. The ODF classification for conifer trees is also used.

C1. Healthy normal conifers, native
C2. Healthy suppressed conifers
C3. Healthy defective conifers
C3. Healthy exotic conifers
D1. Healthy normal deciduous
D2. Healthy suppressed deciduous
D3. Healthy defective deciduous
D4. Healthy exotic deciduous
1. Live green, defective conifers
2. Dead, sound conifers

3. Live or dead conifers with unstable tops, with or without lean
4. Live or dead with unstable roots or trunk, with or without bark

Safety. By stating the reserve tree objectives, marking trees and areas, and finally monitoring continuously, Mountain Grove plans for a safe working environment in the forest. Those with ODF class 3 or 4, next to roads, will be removed; others in reserve areas will be marked.

IV.F.3.a. Wildlife Trees
The full spectrum of wildlife in a forest requires a full spectrum of habitat conditions. Each stage of decay hosts a variety of use, not only different species, but different parts of the life cycle of a species. House wrens for instance prefer small green trees, while woodpeckers use trees in the early phases of insect attack. For these reasons Mountain Grove will keep a full complement of wildlife trees in every stage of life and death.
    Selection. Wildlife trees are selected according to
    • species (for target bird and insect populations)
    • hardness (for longevity, as appropriate)
    • size (to have a full range of sizes)
    • location (to be well-distributed throughout the property)
    • quantity (MGC has an interim target of 5 per acre)
Management. Wildlife trees are managed and monitored as an important resource for the health of the forest. Management addresses:
    • posting (all trees are marked with permanent yellow plastic diamonds and identification numbers
    • recording (all trees are recorded in permanent files, the tallied by area, species, and size)
    • protection (trees are protected before, during and after harvest)
    • replacement (as wildlife trees become downed woody debris, new trees are marked, or selected or modified, as necessary)

IV.F.3.b. Heritage Trees
Old-Growth vestiges of an original, pre-management stand or forest are kept in numbers as heritage trees. Heritage trees provide important ecological functions in a stand, forest or landscape that has been converted to secondary forest. These trees perform a lifeboating function to animal and plant populations, especially forest interior-dwelling species, during the regrowth of the forest to a mature condition.
    Selection. Heritage trees are selected according to
    • species (appropriate to native forest)
    • form (in general, trees with a large bole, thick bark, and flat top)
    • vigor (having continued horizontal growth, good crown
    • size (usually very large for the species, e.g., 42 inches for Douglas-fir)
    • location (to be well-distributed throughout the property)
    • quantity (MGC has an interim target of 7 per acre or an occupancy of 30 percent of the canopy)
Management. Heritage trees are managed and monitored as an important resource for the health of the forest. Management addresses:

- posting (all trees are marked with permanent yellow plastic squares and identification numbers)
- recording (all trees are recorded in permanent files, the tallied by area, species, and size)
- protection (trees are protected before, during and after harvest)
- replacement (as heritage trees become downed woody debris, new trees are marked, or selected or modified, as necessary)

IV.F.3.c. Legacy Trees

Healthy mature trees, either dominants or suppressed, are recruited to become old-growth members in the maturing forest. These trees will probably never be harvested, although depending on recruitment success and the longevity of heritage trees, a few may be cut.

Selection. These trees are determined by a phenomenological inspection that addresses vigor, vitality, context, and species. Legacy trees are selected according to

- species (for a native mature forest)
- vigor (with a good growth rate, needle color and density, and crown fullness)
- form (having a low taper, high branching habit, and few bole defects)
- size (usually larger dominants, but possibly good, suppressed trees)
- location (to be well-distributed throughout the property)
- quantity (MGC has an interim target of 16 per acre, due to uncertainty of choice and perseverance)

Management. Legacy trees are managed and monitored as an important resource for the health of the forest. Management addresses:

- posting (all trees are marked with permanent yellow plastic triangles and identification numbers)
- recording (all trees are recorded in permanent files, the tallied by area, species, and size)
- protection (trees are protected before, during and after harvest)
- replacement (as wildlife trees become downed woody debris, new trees are marked, or selected or modified, as necessary)

IV.F.4. Regeneration and Seed Trees

Mountain Grove expects to rely on natural regeneration for 100 percent of forest replacement. At no time does MGC expect that the stand basal area shall fall below 80 square feet per acre (although should it do so, MGC will weight the stocking of seedlings, saplings, or poles to determine basal area, as in the 1997 Oregon Forest Practices Rules (OAR 629-610-010 section 7).

In restoration areas, absent native species, e.g. western redcedar, will be introduced through planting. Native seedlings will be ordered from seed zones appropriate for the site. The planted species will be calculated to comprise an appropriate percentage of the forest, e.g., western redcedar occurred in draws in the Douglas-fir, grand fir, hemlock zone and rarely comprised more than 30 percent of the stand.

In old field conversion (to forest), seedlings will be transplanted for equivalent sites from roads, other fields, or natural nursery areas. If seedlings are purchased, native seedlings will be ordered from seed zones appropriate for the site. The planted species will be calculated to comprise an appropriate percentage of the forest.

IV.F.5. Recreation
MGC plans to participate in several summer bicycle events that go through Glendale, by providing a place where bicyclists can stay. Funding has been applied for to develop MGC as a more community-oriented facility.

IV.F.6. Education and Research
MGC has hosted several educational programs in the past, including the Ecoforestry Institute and a feminist conference.

A research program is being developed in conjunction with faculty from Oregon State University and the University of Idaho. Funding has been applied for two research projects.

V. Records
The Mountain Grove Forest has records for all of its activities after 1982. These records are kept in folders in metal filing cabinets in the community center office, although many records since 1994 are kept on computer disk.

V.A. Surveys/Inventories
A formal survey of the Mountain Grove forest was done in 1982. Numerous walking surveys were done in the late 1980s and early 1990s. As part of the Ecoforestry workshops and courses starting in 1994, many partial surveys were done. Permanent plots were set up in 1996 and are being monitored annually.

V.B. Analysis/Assessments
The use and health of the Mountain Grove forest are assessed regularly. The data from surveys is being entered in computer programs, such as Organon and Timber, so that the data can be summarized conveniently in charts and tables.

V.C. Management Zones
Each management zone is represented by a series of maps and files that contain data from walk-throughs and permanent plots.

V.D. Forest Management
Originally three of the management zones were designated as preservation of wilderness. However, some incursions are necessary to account for missing natural processes, such as low intensity fires. Furthermore, because recent goals are concerned with creating old-growth structure in all of the zones, only small parts of some of the zones will be treated as control plots, where no intervention will occur.

V.D.1. Preservation
Records are also kept for areas that are of special interest for reservation,

conservation, restoration, and special intercession.

V.D.2. Reservation
Records are also kept for areas that are of special interest for reservation, conservation, restoration, and special intercession.

V.D.3. Conservation
Records are also kept for areas that are of special interest for reservation, conservation, restoration, and special intercession.

V.D.4. Restoration
Records are also kept for areas that are of special interest for reservation, conservation, restoration, and special intercession.

V.D.5. Intercession
Records are also kept for areas that are of special interest for reservation, conservation, restoration, and special intercession.

V.D.5.a. Harvest methods
Harvest methods and site preparations are recorded for stands that are being worked on.

V.D.5.b. Site preparation
Records for site preparation, which is mostly concerned with swamping, raking, and fluffing, are kept with harvest records.

V.D.6. Monitoring
Monitoring is performed as a separate activity. Monitoring records are kept in separate files.

V.D.7. Tracking
V.D.7.a. Post-harvest Protocols
Every site will be inspected within 30 days after intervention in the site. The following post-intervention protocols will be observed.

After harvest the Permanent Plots (previously marked with aluminum tags and recorded in the monitoring books) will be monitored. We will make field notes of what further activities are needed. Roads will be inspected and notes made about what activities are needed.

After the harvest of alternate forest products, locations will be examined for possible injury to remaining populations. Remaining numbers will be censused. If vehicles are used, as for gathering firewood, tracks will be covered or repaired (the "duff will be fluffed").

After harvest of wood products, remaining tree tops and branches will be swamped and laid flat over a wide area (not concentrated or piled). Where necessary we will pile and burn. Roads will be inspected for aging and adequacy. Harvest impacts on future trees and on forest will be assessed.

The records will be kept on file and reviewed at subsequent meetings. Subsequent changes will be made to the management plan.

V.D.7.b. Certified Products Custody and Labeling
Products, both timber and alternate, that are to be sold as certified must have tracking paperwork and signage for chain of custody requirements. MGC labels the products (with a stamp) and keeps separate records.

V.E. Financial
Detailed financial records are kept on the income and expenditures of the Mountain Grove Center.

V.E.1. Partnerships
Records are kept for each partnership or joint venture project. This includes correspondence, agreements, and legally binding documents.

V.E.2. Economic Analysis of Costs and Revenues
Mountain Grove harvested in 1997. The majority of logging activities occurred from January through the end of March. After that time, there was further clean-up, piling of slash, and the production of lumber from approximately 12,000 board feet of logs cut during the first phase.

MGC decided to become the employer of record and obtained Worker's Compensation Insurance, filed tax reports, and performed accounting procedures as employer. The following Table indicates the rates per $100 of payroll for the various categories of work performed at MGC for 1997, as well as the new rates for 1998.

Table 6. MGC Harvest Costs

Category	1997	1998
Reforestation/slash piling/drivers	$25.42	$18.97
Logging/lumbering/drivers/yarding	$31.52	$27.90
Road/trail construction/maint/skidding	$9.84	$27.90
Clerical/Administrative	$0.52	$0.43
Wood preserving/Peeling	$0	$11.10
Annual Earned Premium (1998 Est.)	$4,874.00	$4,116.00

Other activities at MGC, such as carpentry work, is also covered by this premium. Another aspect of administering the logging program at MGC was the use of a committee structure to make management decisions. Weekly forestry project meetings were held; each participant was paid for one hour of time at $10 per hour. Minutes were kept, recorded, and distributed. This accounted for $1,696, or 5.5 percent of total expenses of logging and lumber production. The portion of the overall logging expenses that went towards "committee management" and MGC acting as employer totaled $6,360, or 21 percent of total logging expenses.

A full breakdown of expenses for logging, and also the "Logs to Lumber" (the production of dimensional lumber) program for MGC is listed in Appendix A and Appendix C.

The income from timber receipts, the logs sold to Roseburg Forest Products, came to $72,371.50. The expenses of logging and management, not including the lumber production, was 42 percent of this gross income.

The lumber value of $11,739 reflects both actual and projects income from sales, plus the value of lumber reserved for MGC use. This increases the total revenue to $84,111 and total expenses rise to $35,350. Therefore, net income, including lumber value, is $48,762. This project resulted in a cost of $0.30/bf and a net income value of $0.41/bf.

The last 12,000 bf of logs was used in the "logs to Lumber" program, both to provide dimensional lumber for MGC building projects and to provide lumber for sale in the open market. The additional expenses incurred to truck the logs to Tunnel Creek Mill in Wolf Creek, some of the labor involved at MGC, and then the milling, sorting, and stickering at Tunnel Creek added another $4624 to the costs of $30,800, for a total of $35,350 of overall expenses. This is broken down in Appendix C.

Lumber Production
Appendix C is a summary of the "Logs to Lumber" program. The milling of approximately 12,000 bf Scribner resulted in 26,000 bf milled lumber. The current sales have totaled $6,833. The price has been based on $700/mbf (Scribner scale) plus a 10 percent premium and coverage of milling, sticker, and sorting expenses apportioned to each sale.

The total value of the lumber includes what has been sold, the lumber for MGC use, and projected lumber sales of the approximately 5,000 bf of lumber still stored at the Tunnel Creek site. The total projected revenue from sales equals $9,533, and projected net revenue after all expenses are deducted from those sales equals $4,908. The value of the lumber for MGC should be added to this figure to indicate total value to MGC, which equals $7,115.

The comparison of the return between the value-added "Logs to Lumber" project and the sale to Roseburg Forest Products indicates that adding value to the logs increased the price received to $900/mbf, whereas the price of $700 log delivered, or $658/mbf after trucking expenses are deducted, was received from RFP. Further analysis into securing markets, expenses of milling now that the Tunnel Creek Mill has closed, and potential products from species and sizes harvested, will assist MGC in deciding action to be taken at the time of the next harvest.

Madrone Veneer Logs and Lumber
Along with the cutting of primarily Douglas-Fir, MGC also harvested madrone logs for sale to Top Veneer for veneer production. The harvest resulted in a payment of $1000/mbf, or $2,191. After expenses for trucking, wages for logging, and tractor use, this netted $762.

The fall-down was milled for flooring stock. MGC has 1900 bf of madrone 1x air-dried and stacked, ready for sale or to be further processed into flooring. The cost would be $350 to mill and stack. If the lumber is sold for $1/bf as-is, a net of approximately $2,350 could be gained by MGC.

Cost Comparison with Previous Harvests
Appendix E provides a comparison of costs that were recorded for harvests done in 1982 and 1986. All of the costs associated with the 1997 harvest are not included because there were no records from previous harvests that

correlated with these costs, i.e., monitoring, tree marking, and slash.

The most evident change is in the price received, which increased from an average of \$174/mbf received in 1986 to \$681 in 1997. Please note that these numbers are *not adjusted* for inflation. Therefore the best comparisons can be made between costs per thousand board feet harvested.

Logs were sold at the deck to Ron Reed in 1986, therefore the trucking expenses were included in the price received from Reed. Also in 1986, costs incurred by the hiring of a cat and operator were split between road building and logging/skidding. Records which divided those costs were only available from the end of May through August of 1986. A percentage of total costs was used to approximate total costs for the categories of tractor and cat skidding and wages for the first part of 1986.

With those adjustments made:
- The costs of skidding were the greatest in 1986, at \$50.31/mbf compared to \$31.55/mbf in 1997 and \$32.33/mbf in 1982.
- Wages per 1/mbf were highest in 1982, calculated at \$120/mbf, compared to \$111.46/mbf in 1997 and \$63/mbf in 1986.
- Taxes were much higher on a /mbf basis in 1997, at \$18.55 compared to \$5.71 in 1986 and \$0.68 in 1982.
- Administration and accounting cost \$20.19/mbf in 1997, \$4.83/mbf in 1986 and \$3.93 in 1982.
- Logging expenses as a percent of timber receipts was 39% in 1997, 71% in 1986 and 139% in 1982.

Overall, the costs per 1/mbf were \$240.56 in 1997, \$123.85 in 1986 and \$198.27 in 1982. The change that has made it possible to increase costs and make money is the income/mbf which, net of expenses, was \$371.24 in 1997, \$50.31 in 1986 and negative \$55.30 in 1982.

V.E.3. Direct Sales/Revenue
MGC gets income from the direct sale of logs or cut timber. Other income comes from registration fees for courses and workshops, fees for residents, and sales of some services. These records are kept by the Treasurer and audited annually.

Furthermore, MGC also internalizes all environmental costs of operations. Nonmarket goods and services, such as healthy topography (without mudslides) are roughly quantified. The economic plan lists all goods and services.

The revenue has been sufficient to financially support post-harvest management activities, such as road maintenance, silvicultural treatments, and long-term monitoring of forest health. Income is allocated on a priority schedule.

VI. Recommendations
VI.A. General Management recommendations
There are eight recommendations to be applied to all of the zones. These are:
1. Thinning study: in area No. 1 we "thin from below" some of the repressed understory trees and, in area No. 2 cut some of the larger surrounding trees. The objectives would be study increases in growth

rates of released trees, changes in other regeneration of herbs, shrubs and trees, et al.

2. Regeneration study: "Patch cuts."
3. Reintroduction of fire. We might consider introducing fire in the Patch Cuts areas soon after the logging is done.
4. Harvest some large WF more heavily as part of the emulation of fire. Much of the WF component is mature. Loggers in the area tell us that after about 80 years the WF becomes punky.
5. Hardwoods. Prune the multiple stems back to 1 or 2 of the best formed stems, based on specifications for veneer-quality (few knots; straight, no rot). To keep quality hardwoods in the stand, thin surrounding firs for increased light.
6. Harvesting other species: White fir, arbutus.
7. Old-growth and Heritage trees. Protect all existing Old-growth, with the exception of allowing the harvest of some older Grand fir. Given the absence of fire most of the regeneration is GF, and the forest is moving increasingly towards shade-tolerant species.

    In addition to protecting most old-growth trees, we need to mark other large trees of all species as Heritage trees to become old-growth. We recommend that we mark and protect about 30% of the trees as current OG and heritage trees.

    As a long-term goal, plan to leave 20% of the trees to become OG, never to be cut, but slowly reduce the number of trees from 30% to 20% with selective thinning needed due to growth and with scattered patch cuts to regenerate more shade-intolerant species, i.e. DF.
8. Mortality and Snags. There is a fair volume of fallen understory WF and DF. There are a few large WF. While there are a few old snags, generally, there is a shortage of snags. We need to consider girdling a few large DF, perhaps ones with crooked form which could be advantageous for wildlife.

VI.A.1. Forest Protection
VI.A.1.a. Old Growth
Maintain and restore (where necessary) an old-growth forest that is fully functioning, structurally diverse, and aesthetically-pleasing.

VI.A.1.b. Riparian
Maintain and restore Woodford Creek to a condition where it could support anadromous fish (should down-stream changes be made), with an appropriate sinuosity, gradient, bed, and debris. Keep the canopy large and half-open (for ambient light levels enjoyed by fish and birds).

    Some minimal thinning may occur in riparian areas for restoration purposes. They need to meet the requirements of the Oregon State Forest Practices Act and be in accordance with landscape objectives and plans.

VI.A.2. Restoration
Restore several zones to an oak-pine woodland; that is, those zones that are drier, have many oak and pine individuals, and could support the

association with a minimum of future intervention.

Several zones will be restored to the Douglas Fir/Grand Fir/Western Hemlock association. Large Douglas-firs are already resident in these zones. Restore a small western redcedar zone on a north-slope above the riparian.

VI.A.3. Forest Product Management
VI.A.3.a. Alternate forest products
It is recommended that the forest be managed for quality forest goods. This requires more care and, perhaps, some additional initial investment of time, which will pay off with an substantial increase in return. For example, utility grade Douglas Fir or incense-cedar poles usually pay twice as much as saw logs; veneer-quality Madrone and other hardwood logs make these species quite valuable, whereas otherwise they have little value other than firewood. Alternate products, such as boughs and ferns, for the flower shop market, also offer a higher return on labor investment than timber.

VI.A.3.b. Hardwoods Management
Since hardwoods are an important component of a fully functioning forest, this component of the forests needs to be maintained as part of the mix of the forest. The predominate hardwood species are Madrone and Chinquapin throughout most of the forest. Vine Maple, Big Leaf Maple, and Alder are found in the riparian area. Preliminary inventory indicates that there are only a few California Black Oak, which are usually a component of Pine-dominated ecosystem types.

Madrone. The Madrone are scattered throughout most of the forest. Although there are several large ones, most are relatively young and small, and many of them show fire damage. Because of past logging practices, the Madrone are more abundant than they would have been historically. Up to 30% of the Madrone could be harvested based on site-specific considerations of spacing, canopy openings for regeneration, relationship to other species. The Madrone need to be managed for high-end uses, i.e., veneers and burls. There are few Madrone of veneer-quality, although there are numerous burls. Some pruning of branches in the first 10 feet of younger Madrone would produce valuable veneers in the future.

Chinquapin. In some areas there are Chinquapin groves with most trees growing in clumps of 2-5 trees, indicating a fire history and coppicing from the roots. In one Chinquapin grove, located on the SW side of the main ridge running through the center of the property, there is close to 100% canopy closure and almost no regeneration of other species. There is an existing older Doug-Fir component above the Chinquapin canopy. Mature, quality Chinquapin need to be harvested to about 50% of the basal area. Depending on the response, these stands may need to be thinned again in 20-30 years.

Big Leaf Maple. Generally, the Big Leaf Maple are found only in the riparian zones and riparian zones of influence. They provide important shade for the creeks. They can be harvested, but only for high-end uses, such as burls for musical instruments and other high-end uses, and veneers. Maples do coppice and grow back quickly with multiple stems. The multiple stems need to be pruned back to transfer the growth to fewer stems and those stems need to pruned of lower branches to produce quality veneers in the future.

California Black Oak. Because of the scarcity of these trees, thin other trees within 10 feet of the circumference of the crown to encourage possible regeneration. It may be necessary to open up the canopy further to regenerate the California Black Oak. They should not be cut on the property if and until they become more abundant and, then, only for a high-end use.

VI.A.3.c. Conifer Management
For forest growth rate and level/rate of future timber cutting, it is currently estimated that the forest is growing well. Timber Cutting shall be permissible in most zones.

VI.A.4. Fire Hazard Reduction
To reduce fire hazard the forest could be thinned on the perimeter, along the road, and near the community house, while leaving a more dense in the riparian areas and in the interior of the forest. For fire hazard reduction purposes it is recommended that the design of the cutting patterns would be to cut heavier near the house, main road, on the perimeter of the forest and on the ridge lines, leaving denser forest structure on the steeper slopes and in the riparian areas.

The MGC forest has been walked through and studied by scores of ecological scientists, foresters, forest workers, forestry students and a wide array of others visitors. Most have agreed that fire ladder and load are dangerously high. This danger was particularly high on South slopes where predominantly Pine/Oak ecozones thrive, particularly in Zones 1, 2, 4, and 5 near the freeway and along the new Entrance road.

Accordingly, MGC decided to concentrate its study, funding and work during 1997 on the "Pine Oak Woodland Restoration Project." In 1997 MGC harvested 102 MBF of fir, both Doug and Grand (White) fir, from about 30 acres where the fire danger was considered highest. We plan to conduct controlled burns in this area in the fall of 1998.

However, the fire danger remains high across most of the remainder of the MGC forest. Most of the Douglas-fir-Hemlock ecozones are at 90-100% canopy closure with little regeneration. While the DF component is still at a young age; the Grand fir (White) component (60-70 years) is mature. Because of the absence of fire, the forest is increasingly favoring the growth of shade-tolerant species, like GF. Often the GF has grown up right next to an older DF, usually just to the north of the DF in the shade. They grow incredibly, pumping an enormous volume of water. This is why they are called "piss fir" by loggers: at certain times of year (spring) when you cut into one, it will spit sap out. In the hot summer sun this sap will explode in fires. Being right next to older DF, they will carry the fire into the DF, which otherwise may well have survived the fire below quite well. These stands contain an Old-growth component of DF, Incense Cedar (IC) and Yew, which are likewise in danger of a stand-replacement fire event.

We have a lot of understory GF falling into other trees and onto the forest floor, on the W side (Zones 6 and 11). It creates fire ladder. Even in the riparian area (Zone 3) we need to consider the possible consequences of fires. MGC owns 2 miles of forestland on both sides of Woodford Creek. A stand-replacement fire could devastate the riparian area (Zone 3). We need to consider balancing a "no cut policy" of riparian areas with considerations of selectively pulling GF (with root wads) into the creek, as well as selective harvesting to reduce fire danger.

In 1994 many GF were dying on the W side of Woodford Creek (Zone 3), across from the Blueberry meadow. We decided to study the consequences of thinning some of the GF. We cut 10 MBF along the upper portion of Woodford Creek that could be reached from the W Main road. The road allows easy access, particularly with the use of the Sky-line Yarder, now at Mt. Grove.

Because the MGC forest is largely a second-growth, even-aged forest, with 90-100% canopy cover over much of the forest, the fire danger is high. Our cruise data, Permanent Plots and Photo Points indicate the same. Current data indicates that the Woodford Creek Watershed has experienced stand-replacement fires on an 80 year cycle. We are getting close to that age! Living in this forest, while studying it, we experience first hand the dynamics of this fire cycle.

We need to emulate the characteristics of fire by thinning, creating patch cuts, and using controlled burns through most of the MGC forest, both in the Pine/Oak Woodland ecozone and in the Fir-Hemlock stands, as well as in the interface— excluding the riparian zone which has a separate set of recommendations. At the same time we need to thin second-growth hardwood from multiple stems back to 1 or 2 stems, based on form and quality. This will also reduce fire ladder, allow for more regeneration, while transferring growth to the main stems for greater financial return in the future.

Fire spreads in many ways. Some of our criteria are based on these ways fire spreads. Thin trees so that boughs are not touching. The selection of which trees to cut is a decision of applying ecological thinking on a site specific basis within landscape contexts. As we have indicated elsewhere, the "ecological" includes environmental, social, economic, and cultural. The criteria for selecting which trees to cut, likewise, must consider all aspects of reality.

Ecoforestry needs to manage for quality at all levels. Human criteria for quality are based primarily on the characteristics of the wood, i.e.., Douglas-fir's structural strength, and with end uses, i.e., houses, furniture, et al. For example, Pacific madrone is relatively abundance in the MGC forest and surrounding Watersheds— from the perspective "an historical range of variability" for this region. However, markets for madrone and other West coast hardwoods are largely non-existent or under-valued or distant with resulting high transportation costs. The key to successfully marketing madrone is to manage for quality. The specifications for quality madrone include such characteristics are the following: (1) clear wood (the absence of knots for specific lengths); (2) straightness (a limited amount of curve and twist); (3) solid wood without rot; et al. Only 1 in 50-100 madrone trees of 18" dbh and larger meet specifications for veneer quality because so many madrone have cat-faces and rot from past burns or past logging injuries. However, these trees provide good wildlife habitat (for wood rats) and food (for birds).

We need to manage for quality wood, so that as many of the trees that we do harvest can be used for their highest and best uses. For madrone the highest and best uses includes musical instruments, veneers, fine furniture, flooring, et al. Another example is a Broad-leaf Maple tree with figure. While such a tree may be 1 in a 100 Maples, it is quality wood from a human-use perspective and, accordingly, valuable.

There is no inherent contradiction between managing for ecological responsibility and for quality forests, trees and other species, goods and products. This is why the Ecoforestry Institute's education programs seek to prepare ecoforesters to make these decisions by applying ecological understandings on a site specific basis within a quality management context.

This means, for example, that in thinning in the predominantly Douglas-fir-Grand fir ecozones, we need to favor the harvesting of Grand firs.

VI.A.5. Community Development
VI.A.5.a. Cooperate in Community
It is recommended that MGC continue to cooperate with local agencies and groups in the community; coordinate workshops with EI and BLM; and collaborate with partners in the local watershed to improve the watershed. MGC should continue to share resources with the community; promote a network of ecological foresters; and plan for further community involvement in the health of local forests.

VI.A.5.b. Provide Opportunities
MGC should continue to provide opportunities for local residents and foresters to learn and work in its demonstration forest.

VI.B. Management Zones
There are specific recommendations for the management zones. For instance, Zone 1 is: Restore to oak-pine woodland system. Continue thinning. Remove many fir trees; create snags; remove fuel with jackpot fires; limb trees; add culverts to roads; harvest to old growth characteristics. By contrast, Zone 6 has the most requirements for work. Generally the forest is growing well in Zone 6 (See: Core data). While trees are, by in large, well spaced due to past selective logging, selected thinning of all age classes of second and third growth classes could be beneficial. Currently, while there are generally 3 age-classes, there are 4 growth-classes: (1) old-growth; (2) mature second-growth with good canopy position; (3) suppressed second-growth, and (4) regeneration. How do we treat the suppressed second-growth? It is taking up space and yet growing at 30-40 rings/inch, while good growth is 10 R/in. We can thin-from-below, but most of these trees have little commercial value. Or we can cut larger, commercially valuable, trees near by and release them. As the core sample in 2-A (2-98) indicates, this DF went from 33 R/10th " to 13 R/10th " after being released at about age 48-50. It is said that DF does not

respond to release after about age 75. Specific recommendations:

1. Within Zone 6, establish 2 Study Areas: In area No. 1 we "thin from below" some of the repressed understory trees and, in area No. 2 cut some of the larger surrounding trees. The objectives would be study increases in growth rates of released trees, changes in other regeneration of herbs, shrubs and trees, et al.

2. We also study to regeneration of trees by cutting 2 "patch cuts". The size of these cuts needs to be sufficient to regenerate DF, 200 ft on the N-S axis (the taller trees are about 150 ft.). The E-W axis could vary in width, wider on the S end and narrower on the N end, perhaps almost triangular. We might consider introducing fire in the Patch Cuts areas soon after the logging is done.

3. Harvest some large WF more heavily as part of the emulation of fire. Much of the WF component is mature. Loggers in the area tell us that after about 80 years the WF becomes punky.

4. Hardwoods. Prune the multiple stems back to 1 or 2 of the best formed stems, based on specifications for veneer-quality (few knots; straight, no rot). To keep quality hardwoods in the stand, thin surrounding firs for increased light.

5. A few IC could be harvested, but all OG needs to be protected. WRC could be planted in the gullies.

6. Old-growth and Heritage trees. Protect all existing Old-growth, with the exception of allowing the harvest of some older Grand fir. Given the absence of fire most of the regeneration is GF, and the forest is moving increasingly towards shade-tolerant species.

   In addition to protecting most old-growth trees, we need to mark other large trees of all species as Heritage trees to become old-growth. We recommend that we mark and protect about 30% of the trees as current OG and heritage trees.

   As a long-term goal, plan to leave 20% of the trees to become OG, never to be cut, but slowly reduce the number of trees from 30% to 20% with selective thinning needed due to growth and with scattered patch cuts to regenerate more shade-intolerant species, i.e. DF.

7. Regeneration. There is little regeneration. Mostly there is Grand fir (WF) regeneration. In addition to thinning across the zone, there is a need to create a few Patch cuts large enough to regenerate stands of DF, et al.

8. Mortality and Snags. There is a fair volume of fallen understory WF and DF. There are a few large WF. While there are a few old snags, generally, there is a shortage of snags. We need to consider girdling a few large DF, perhaps ones with crooked form which could be advantageous for wildlife.

Thin a few of the Grand fir and Incense-cedar; create snags and DDW.

VII. *Summary*

Ecologically responsible forest use recognizes the forest, including humans, as a whole system. With regard to human uses, ecologically responsible forest use focuses on managing human activities to serve the long-term interests of fully functioning forest ecosystems, rather than on managing ecosystems to serve short-term human interests. Ecologically responsible forest use supports the development and maintenance of stable human communities and diverse, sustainable human economies. Labor-intensive activities and value-added wood products manufacturing in close proximity to the source of wood are cornerstones of the development of ecologically responsible, community-based economies.

Ecologically responsible timber management is ecosystem-based, which means that the character and condition of ecosystems determine the types of human use that can be carried out, and in what manner. In ecologically responsible timber management, there may be reasons to cut small, young trees (for example, when restoring natural stand structure to areas where fire suppression has occurred), as well as for growing trees to old ages (to provide, for example, late successional or old-growth forests and to provide high quality, mature wood).

Trees are selected for cutting in ecologically responsible timber management by considering a variety of criteria, including: stand condition, successional processes, the need for old-growth forests and old-growth trees, and production of high-valued, mature wood. The volume of timber cut each year under ecologically responsible forest use is restricted by the requirement that cutting must maintain fully functioning forests at all scales through time. Thus, annual timber cuts will vary according to the needs of the forest. Once a forest landscape has been managed for an extensive period of time using ecologically responsible approaches, managers will be able to forecast a reliable range of annual cut, based on the annual growth of the forest and depending on the needs of the forest.

Working to protect, maintain, and restore fully functioning forests requires frequent evaluation of our activities to determine whether these activities are meeting the principles and standards of ecologically responsible forest use described in this document.

Chapter 30

Forestry as Poetic Activity, Unified Field Theory,
Ecosystem Medicine, & Right Practice

I thank all of the speakers at this conference, *Forests for the Future*, as well as the many questioners and listeners, for their inspiration. Being sandwiched between David Perry's opening talk yesterday and Merve Wilkinson's workshop tomorrow is quite an honor, although I feel more like the slightly spoiled mayonnaise than the meat or the cheese.

I would like to put the previous talks in a synthetic framework, relating them directly to ecoforestry using keywords from their talks. Then I will try to add the kind of things that have been overlooked or neglected from our discussions in this conference. Finally, I will make a few suggestions for goals for local, regional, and global actions for ecoforestry. I would like to set up the framework by describing ecoforestry as various things. First, I would like to define ecoforestry as poetic activity.

Ecoforestry as Poetic Activity

Ray Travers noted that vision without a task is merely dreaming. But dreaming is a necessary first step. David Perry made reference to Vaclav Havel, the poet-leader of the Czech Republic. These comments are apropos. Ecoforestry is literally a remaking of the landscape (and in fact the Greek word for "making" is "*poeisis*."). We are making the forests of the future with our actions today. We need to use our aesthetic senses to do that. We need to use the poetic process.

Richard Atleo described how native Americans used metaphors in teachings to know relations and change people. Science itself makes extended use of the metaphorical process to construct its models. For example: "A tree is a machine" according to Smith and others or "The brain is a computer" according to Michael Arbib. This kind of metaphor is used to create our images of nature as resource. Brian Nattrass and Mary Altomare presented forests as "resources," which is another metaphor.

The use of the word "ecology" by Ernst Haeckel implied that the natural world was a place to live, a house, rather than a machine to control. Making the earth into a house is fundamentally a poetic activity, according to Gaston Bachelard. Poetry also is a way of understanding the universe through metaphor, a literary device that transfers the characteristics of one term to another.

Poetry is communicative of the quality of things. Like science, it discriminates the unsuspected in the commonplace. It is not different from science, but more diffuse; not better than science, but more comprehensive. It accepts ontological parity, the equality of beings; aspects of the world are not negated or reduced by one another. As metaphorical knowledge, which may be prerational or metarational, poetry can avail itself still of scientific references. Poetry can measure a whole qualitatively and mimetically, a germ or the cosmos with its imagery. Poetry is a tool for comprehending partially what cannot be known totally. A poetic language could include a view of the interrelatedness of all existence in a sublime ecology.

Poetry does this through metaphor, which can be understood as the connection of a focus and a frame. Focus and frame can be understood together metaphorically. A metaphor can be understood as consisting of two parts, according to Max Black: A focus and the frame. The focus (or figure) designates the figurative term signified through the process, and the frame (or ground) refers to the subject or context. Using this distinction, it can be seen that most of the fuss in forestry has occurred at the focus level. Foresters have so long focused on trees that they forget that the forest is a frame that holds many foci (or points of view).

Professor Perry and Audrey Pearson both considered the importance of patterns in forestry. A pattern can be defined as "process applied to components" where the process is actually prior to the components. The elements of a forest are related psychologically, by foresters, as focus or frame, as contrast or uniformity, as dominant or recessive, or in a number of other pairs. For instance, forests can be considered by scientists as either matter systems or energy systems, but the focus on either frame permits subtle differences and limitations in interpretation. Some ecologists describe organisms as being configured by energy through time. But, organisms are material patterns in space as well.

We have neglected slow patterns in forestry, those that move across landscapes over thousands or millions of years. We have neglected the importance of relationships. Perry suggested the metaphor of sailing for us to consider, that foresters are like sailors. This is an excellent metaphor because it reminds us that we cannot control all the forces of nature, only understand them and move with them to get where we want to go.

Ecoforestry as a Unified Field Theory of Forests
Travers used Einstein to illustrate a new level of thinking. Quite so. In fact, ecoforestry is a general theory that incorporates industrial forestry much like general relativity incorporated Newtonian dynamics. That is, industrial forestry is only a small component of ecoforestry, useful under certain restricted conditions.

Perry, Pearson, Nattrass, and Altomare all mentioned the importance of principles for creating sustainability. The agenda of ecoforestry can be presented through a number of characteristics, principles, and standards. Characteristics are qualities that distinguish unique individuals, systems, or patterns; Gregory Bateson calls them differences that make a difference. Principles are fundamental rules or laws, based on unique characteristics of forest systems that we can use to create models to meet stated objectives, which are goals towards which our actions are directed, e.g., a healthy forest. Dr. Perry mentioned how spirituality could be combined with other principles that lead to harmony in society (and doubtless in forestry). Atleo showed how harmony prevails in stories. Standards are models or examples of quality or value established by authority or consent, that can be repeated as procedures. Jim Smith mentioned total forest design. Good forest design means not violating any basic principles.

For example, one characteristic of a mature forest is its wildness.

The corresponding principle is that forest is self-making and self-ordering without human control and management. Our objective for this forest is to allow the foresting process to continue, whether we take resources from the forest or not (forests can be influenced or interfered with by acid rain, pollution, and other industrial effects). We can set local standards that are likely to keep mature forests wild: Limit biomass removal to 2 percent of the total forest; use appropriate techniques, e.g., single tree selection, horse skidding; retain mature structure, e.g., 19 snags per hectare, 23 nurse logs per hectare (in mature Ponderosa pine forests in Eastern Washington for instance); preserve surrounding landscape patterns.

Other characteristics of forests include death, disturbance, and exploitation. These characteristics contribute to the health and self-ordering of a forest. Pearson described how succession is set back by disturbance in forests. But, disturbance is a normal occurrence. The problem in forests is the scale of disturbance caused by industrial clearcutting—this is actually interference with the process of self-ordering in forests.

As a unified field theory of forests, ecoforestry understands changes in scale and employs appropriate techniques. Ecoforestry understands the crucial difference between local and global operations. Insect infestations are local phenomena; clearcutting all forests is a global problem.

Ecoforestry As Ecosystem Medicine
Perry and Atleo both emphasized the importance of health and harmony to forests. Our activities in forests make them less or more healthy. Ecoforestry considers the health of forests to be paramount. It needs to become a science of health, since there is no science of forest health at this time. Pearson likened conservation biology to medicine. In fact, forestry is about where medicine was in the early 1900s—that is, the patient has a better chance of survival being left to itself.

A lot needs to be done to make ecoforestry into forest ecosystem medicine. For a start, there is no standard forest, for measurements, as there is a standard human for human medicine. Then, forests are far more complex than humans, especially since they include humans as part of the system.

One thing we need is more data, as well as a framework to interpret it, rather than less data, as Smith suggested in his talk. We need far more research into fire, biomass, and species interactions.

Walter Briggs said in his talk that thinning from below improved the genetics of the forest. Alas, it is more complex than that, since the only way to be sure that thinning from below improved the DNA would be to test the DNA of every tree taken as well as every tree remaining.

Our language is confused; we talk about improvement as if we knew what we are doing—we do not know much at all. We talk about numbers as if we are collecting every number of everything in the forest. We are not. For instance we sum the number of acres cut (6 million in the US) or the number of cubic meters cut (3 billion in Canada), but how many trees is that? 4 billion? 5 billion? How many woodpeckers? How many owls or salamanders? We discuss the forest in terms of "resources," "waste," and "pests." Do we really know what these words mean? What is a pest? A bear?

A porcupine? Bark beetles? Hikers? Skiers? Perry quoted Mr. Thomas of the Forest Service that nature may be more complex than we know—this idea has been presented many times, by Barry Commoner, James Jeans, and perhaps first by George Perkins Marsh in 1864.

Perhaps it is time for thought experiments about how we use language. We need to think about what would happen if we recast forestry in medical terms rather than restricted economic ones. Perhaps medical terms are too limited, simply because medicine itself uses the wrong words and images. The goal with attaining health is not to rid ourselves of all pests or symptoms: it is to balance those things that exist anyway with human requirements.

Medicine bridges the scientific study of symptoms and diseases as well as human values. Medicine is a discipline of practical experience arising from clinical interactions, e.g., interhuman events. Medicine has the goal of restoration of health of the body (as psychology addresses the mind and religion the soul). Medicine already has to consider the entire spectrum. In disease, the harmony of the body is disrupted by some event. The disruption is a symptom (not exactly). It becomes objectified, it, the disease. But, the disease is a conceptualization of the disharmony of the patient's world, that is self-image and human image. Medicine has to consider multiple etiologies of disease, from environmental causes, social, genetic, somatic, and psychological. Health, being a complex harmony of the environment, society and individuals, the ethic of medicine must consider individual, social, and environmental good—all are rooted in living bodies. Environmentally-related disease are hard to recognize. There is no classification scheme or surveillance mechanisms for them. The impacts of environmental illnesses are not calculate. Medicine focuses on individuals, and less on populations, the chemistry of toxic agents, government regulations, or environmental effects (or backgrounds).

Less than thirty years ago, the environment was of little concern to most people. Now it is the primary issue for most people. Because of the intricate way that the environment inter works with human health and well-being, not only health-care providers but foresters, miners, traffic engineers need to be knowledgeable about the effects of the environment on their areas, as well as on human and community health. A new category of professional is needed: People who address the health of ecosystem themselves. Human physicians may need to be able to identify critical environmental conditions that affect human health, but others are needed to identify the health of those systems themselves. Human physicians need to know the basic principles of diseases related to environmental change or chemical exposure; others need to know the principles of ecosystem health and how that is related to human.

Ecoforestry as Right Practice
Right action is one step in the Buddhist eight-fold path to purification and liberation (the other steps include right thought, right speech, right livelihood, and the others—that is why I'm a Taoist by the way: Fewer steps). It means that we should act for the good of human and forest communities. In Buddhism, virtue is of great value in practice—it is the foundation for

wisdom. A virtuous lifestyle is the basis for moral practice; the mind is free from hatred, greed, worry, distraction. Formal rules are simply used to decrease selfishness. When virtue becomes spontaneous and natural, the action associated with it is considered powerful. We can strive for that kind of power.

In his talk, Tom Milne mentioned that corporations have certain responsibilities. Yes, they do, but they also have real ecological responsibilities, set in the context of an industrial ecology that can recycle materials almost indefinitely, creating links and relationships with other industries and groups. Chris Bailey stressed the importance of sustainability. Nothing, not forestry or cities or toy consumption, can ever be sustainable unless the entire system is designed so that needs, wants, waste, and use are minimized and balanced. Nattrass considered the effects of population in his talk. Smith discussed design. These things are interlinked.

Perry talked about how the beings in a forest are adaptive. We humans are also adaptive. But, as Rene Dubos pointed out, we may be able to adapt to pollution and habitat destruction—simply because, like rats and bacteria, we can adapt to radical changes, even if other plants and animals that we value cannot adapt and perish.

But, we can also design our systems to minimize radical change, to allow those elements that cannot adapt to radical change to keep to habitats that are not altered so dramatically by logging or cities or tourist activities. We can promote this through education.

Education takes place in communities; it is the means for communities to continue. As Plotinus and Novalis recognized, education has an outward, social and civil, aspect as well as an inward, personal and self-revealing, aspect. Education has at least four ends: 1. the appreciation of the richness of nature; 2. the comprehension of human existence; 3. the understanding of the nature of human society, and; 4. the training for a position in human society.

Education has become more universal, but its goal, the well-rounded individual, has been distorted by its fourth aim, training for the economy. To produce wealth for the state and livelihood for the individual, education has become money obsessed. Ethics, in the second and third aims, has been neglected, since it might limit or contradict its economic obsession. In fact, the first three aims are restrictive to a growing, industrial economy. Education, as practiced by public schools, produces unprovocative individuals, adjusted to an unbalanced society. With its emphasis on play, liberation, and community, a radical education integrates all four ends.

Radical education alters and enlarges perception with the selection and presentation of relevant information and forms an ecological consciousness. The survival of human societies depends on the consciousness of the global system in its complexity and connectedness. Buddhist concept of "right action" recognizes that the individual has to make decisions based on consciousness of the effects of those decisions and actions.

Summary: Ecological Advertising & Ecoforestry
Right practice, like good design, requires new images. A contextual forestry requires new language as well as new images. Getting new images and new language to loggers and landowners, as well as to scientists and industrialists, is problematic. That brings us back to poetic activity again.

Poetry and art are undervalued as forms of communication, not to mention as ways of shaping and making. Business has transformed much of art and poetry into advertising, to match the style and attention span of the people in industrial cultures. Advertising, quite literally from the Wall Street Journal to college textbooks, refers to its activities as ``shaping the American dream." Like art, advertising creates an image of a way of experiencing. Unlike art, it limits its focus for a specific goal—profit. Like art, it mirrors us. Unlike art, it intensifies and glorifies only the positive aspects of culture, ignoring the dark, negative aspects or the complex nonhuman framework.

The simplicity of advertising is irresistible. Our environment deteriorates according to ecologists, but always gets better according to economists. And their pictures are prettier. People want to hear that it is getting better. Advertising tells them it is. People want to act stupid, greedy, and selfish, and spend the inheritance of their children on themselves. Advertising tells them their actions are rewarded. The real issues of life and death, destruction and hope, make people feel helpless and anxious, so advertising draws their consciousness to comfortable trivia.

Despite the ugliness of the dreams of progress and growth, of waste and stylistic frenzy, advertising, using sophisticated techniques and narrowing the focus out of context, makes the dreams desirable and irresistible. People in agricultural and hunting cultures interiorize the abstract industrial vision. African farmers are convinced to buy inorganic fertilizers, even though it degrades the soil; women to buy powdered milk for their children, even if it kills them; tractors replace draft animals in the paddies in the Philippines, even though they are costly and less energy-efficient; French winter fashions are found desirable in tropical Brazil, even if they can only be worn in air-conditioned villas. People in industrial societies are convinced that their children will be ruined without personal computers, even if they become isolated game-players, unconcerned with forests in remote locations.

Yet, advertising may be the most effective means to reshape desires and reform buying habits. Advertising presents the symbols of modern experience, even if they are just the trivial ones. It could present healthy symbols equally well. Advertising does incorporate traditional values, like family, friendship, and love, although to sell beer and cereal and, sometimes, churches and hospitals. And, like art, advertising lies (although Jules Henry thought it was instead a new kind of truth—``pecuniary pseudo-truth"—not intended to be believed, or certainly, proved).

Advertising is beginning to support more informational functions, such as the dangers of drug abuse and smoking. Advertising creates values—fur coats, fast cars, dark beer, slim cigarettes are certainly recent and artificial values—but it could be used to create positive ecological values and new identities that show that our needs for prestige, esteem, and belonging can be met without stylistic waste at mindless speeds. Advertising could promote new attitudes about appropriate technology, the rights of other cultures, and the place of people in nature. Good advertising could be as subversive and conservative as ecology. It could avoid confrontation with people's values; emphasize positive aspects without negative ones. A good ad could capture and carry the most self-indulgent viewer; for the most part, ads don't require effort, literacy, or consciousness, just attention.

Perry mentioned that forestry has focused on high volumes of low-value products, then suggested that this needs to be reversed. Advertising could make fun of first world countries selling their resources to second-world countries.

Advertising has been serving the dream of progress, but progress is leading to catastrophe, a long, slow, global catastrophe. When people experience local, sudden catastrophe, they usually respond immediately, with heroism and sacrifice, aiding the victims of earthquakes or floods, sometimes famine. Advertising could bring to consciousness the slow catastrophes of erosion and the destruction of entire forests, and, perhaps, invoke the same altruistic responses to them.

To work towards this service, ecoforestry groups, with conservation groups, preservationists, sportspersons, politicians, and actors, could define and promote an integrative mythology as the basis for the framework of diverse efforts to protect life and the environment. Ecoforestry organizations could provide a meaningful philosophical foundation, as well as coordination for other humane, social, and conservation programs. But, the approach must be egalitarian: Respect for life cannot neglect human life and suffering. The approach must be eutopian: A new cosmology cannot ignore adaptive cultural traditions that arose in place over centuries. Furthermore, in addition to formal education, they could provide re-education through the most effective means, such as advertising. Wildlife groups could spend money advertising ``humane consciousness,'' moderation, and the joy of living (instead of just consuming or winning). Ecological ads would be unique and compelling, simple and effective. They would advertise not a product, but a way; not for a profit, but for a dream.

What should we ecological foresters be advertising? Our successes. Our goals. The goodness and rightness of our efforts to protect, preserve and restore forests. I urge all of you to continue these efforts.

Chapter 31

Virtue Death & Responsibility

Human activities cause a lot of suffering and destruction every year. 17 million mammals are trapped for fur in the US (1984). 12 million unwanted pets are killed every year. 70 million laboratory animals are destroyed after experiments. 4 billion chickens are killed for food. Almost a million cattle die from injuries related to transportation. Well over half a million animals are shot for sport on wildlife refuges. 50 million children die each year from hunger-related diseases.

Forestry causes a lot of suffering. 6 million acres of forest are cleared annually and 5 million more are degraded. Hundreds of species are extirpated from their homes. 3 billion cubic meters of wood are consumed annually. Assuming each tree contains .5 to 3 cubic meters of wood, that means that *1 to 6 billion trees are killed*. Did you ever notice that the number of trees cut is rarely a published statistic? Probably each tree is home to 1 squirrel, 5 birds, 30 lichens, 1000 beetles and insects, and 100,000 fungi. The red-cockaded woodpecker, in the Southeastern US, is so territorial that fewer than 1 percent can relocate to other trees after nest abandonment (which usually happens after the tree is felled). Killing each tree kills the habitat for some species and part of the habitat for others.

Most people are ignorant of these facts. Even those that know them are mostly indifferent. Unlike a Panda or coyote, a tree does not usually inspire strong emotions. Furthermore, habitats are too abstract to attract sympathy. When a forest is considered to harbor snakes and ticks and wet leaves, the reaction may even be negative.

Despite the complexity of the interactions of beings in a forest—G. P. Marsh once said that nature was too complicated for us to ever know her completely—we have knowledge of our effects, knowledge of the costs in lives for poor forestry techniques. This knowledge, however, is outer knowledge. It does nothing to change our detachment or indifference— Gandhi said that the "hardheartedness" of the well-off was what troubled him the most—or to stop the activities that cause suffering and death. Modern science even argues that seeing emotions in nature is a "pathetic fallacy," although it is more likely that seeing nature dead is an "apathetic fallacy."

The knowledge has to be inner knowledge—the poet Novalis considered that both outer knowledge and inner knowledge is necessary to become a whole person—which can be gained by making yourself part of a place and making that place part of you.

This self-knowledge is a modern virtue, although knowledge is rarely considered a virtue. The four Cardinal (or Natural) Virtues can be related to the ideas of good forestry practices:

1. Justice, which means right action, basically, in the distribution of wealth among members of a society. Ecoforestry strives for a more equitable distribution, less concentrated in the hands of the few. Also give others their due; do not take more than is yours.

2. Prudence, which means discipline by reason or simply good judgment. Prudence is required in cutting in a living forest.

3. Fortitude, which means strength of mind as well as strength. It takes fortitude to run counter to the whims of industrial forestry that has bought or brainwashed most forestry workers. It is the determination to continue your actions against any odds.

4. Temperance, which means moderation. Moderation in behavior, especially social behavior. Temperance, not taking too much. Temperance also means not giving in to any passion too much, even to preserve the forests, least the passion inspire its opposite, destruction.

These virtues are similar to the Buddhist eight-fold path to purification and liberation; although other religious figures, such as Jesus and Mohammed, have preached and practiced following the right path, Buddha has the advantage of being simpler and more secular:

1. Right understanding. Understand the relation of people to forest ecosystems, as well as the events that lead us to our situation, now, where we have to kill forests to sustain our addictions to pretty, yellow tools and toys. Remember, chaos helped shape the forest, and us. It is still a part of the process of creation. Accept it. Go with the flow (also good advice from Heraklitus and Lao Tse). There is a forest order and a human order, forest time and human time. In the forest, it's best to adjust orders and times.

2. Right thought. Think about the beings in a forest; think good thoughts about everyone. Everyone has some virtue and some history that has shaped them.

3. Right speech. Say nothing mean or violent or untruthful. Even mean speech is a form of violence, according to Gandhi.

4. Right action. Act for the good of human and forest communities. According to Henryk Skolimowski, frugality is the cornerstone of ecologically responsible behavior. We simply don't need all those neat things. Chastity means being pure in taste and style—it doesn't mean not ever making love. You can also love the forest and its beings.

5. Right livelihood. Work only as the needs of a forest dictate, regardless of economic incentives to cut badly. Work hard, with the efficiency that comes from experience.

6. Right effort. Put your labor into worthy causes. Be moderate. Moderation classically is considered the flower of old age. We can learn it without growing old. Look at animals in a forest: moderate. Look at the climate in a forest: moderate.

7. Right mindfulness, attention or recollection—be aware of nature and the way of nature. Have humility. Humility comes from the Latin word meaning low, from the Greek word meaning on the ground. It means being low of mind, humble, modest. It shares the root for humus, the organic material in soil. Having humility means paying attention to what is important, the soil for instance.

8. Right concentration, or meditation with patience and discipline. Good work can be a form of meditation.

In Buddhism, virtue is of great value in practice—it is the foundation for wisdom. A virtuous lifestyle is the basis for moral practice; the mind is free from hatred, greed, worry, and distraction. Formal rules are simply used to decrease selfishness. When virtue becomes spontaneous and natural, the action associated with it is considered powerful. We can strive for that kind of power.

I suppose we could expand this list of virtues by adding loving knowledge—a deeper kind acquired through loving attention—or patient practice, but this is a good start, especially since virtue is a neglected subject.

We don't think of forests or animals as having virtue, although we used to. Wolves, for instance, were considered to have no virtue, but ants did, and dogs and cattle were considered virtuous. Forests, like wildernesses, were thought by early Europeans to be wastelands, places empty of human values. By the 1800s forests and especially mountains were thought to be beautiful, and in some ways, virtuous, or at least encouraging of virtues in people.

Working in forests can encourage virtues. By respecting the time of the forest we can learn patience. By being humble we can see more of how the forest works. By being part of the forest, we can develop virtues such as fortitude and prudence. Virtue is really required for us to work in something that is too complex to know everything about, or possibly even know adequately.

Many of you are already virtuous. And, compared to most other students and even teachers, you are also enthusiastic (that word is from the Latin meaning possessed with God). Your attitudes can be characterized by the word devotion, that is, you dedicate yourselves by solemn acts (from the Latin words meaning to vow away). You are devoted to local forests, and to your communities. You have the proper attitude to work in forests, which is typified by the word awe, meaning wonder or fear and reverence (from the Greek word meaning anguish).

These words all seem to be religious terms, but perhaps that is the most appropriate attitude after all. Try to keep this attitude and to work on your virtues by learning from the forest and from each other.

Ecoforestry is concerned with resacralizing forests, with restoring them to their extents and grandeurs, by regrounding science in ethics (ways of living together), and by changing our attitudes from utilization and flat efficiency towards awe and appreciation.

That means that you, each of you, has to care for each tree, fungus, jay, or worm. Each living being matters. We know so little about the lives of trees, or of other beings, in the forest; we do not know what it is like to live for over a thousand years or to stand in one place and draw everything we need into us. Our detachment from trees and other beings has to end. Our participation in the life of the forest has to begin.

You are invited to take this oath. An oath is a pledge by which a person swears that she or he is bound, because of beliefs, to perform an act faithfully and truthfully. Like physicians and lawyers (for the forest), ecoforesters are expected to pledge to conduct themselves according to the principles of their profession. (Long before antibiotics, physicians followed a basic rule: *Primum non nocere*—First do no harm.) This oath combines the 1948 Declaration of Geneva for doctors with the 1993 Ecoforester's Way (coined by Alan Drengson).

Ecoforester's Oath

On being admitted as a member to the profession of forestry:
I solemnly pledge to consecrate my life to the forest, as a sacred duty
and trust;
The health, diversity, and stability of the forest will be my first
consideration;
I will respect the lives of all beings that comprise the forest, and I will
continue to learn from them;
I will not harm the forest by taking more from it than it can allow,
or more than the vital needs of my community, or by using
inappropriate technologies or practices;
I will not permit considerations of economics or politics to interfere
with the life of the forest, and I will use only nonviolent
resistance to protect it;
I will not use my knowledge contrary to the good of the forest.
I will maintain by all the means in my power the honor of my
profession;
My colleagues will be my sisters and brothers, teachers, students, and
friends;
I make these promises solemnly, freely, and upon my honor.

Now, be healthy, work wisely, and do no harm!

Chapter Notes

Chapter 2. Interactions in Nature
(see general bibliography)

Chapter 4. Principles of Ecoforestry (Part I)
(see general bibliography)

Chapter 5. Principles of Ecoforestry (Part II)
(see general bibliography)

Chapter 7. *Waldgendankenexperiment*
Bohm, David. 1980. *Wholeness and the Implicate Order*. London: Routledge and
 Kegan Paul.
Einstein, Albert and Leopold Infeld. 1966. *The Evolution of Physics*. New York:
 Simon and Schuster.
Hardin, Garrett. 1977. *The Limits of Altruism: An Ecologist's View of Survival*.
 Bloomington: Indiana University Press.
Lanner, R. M. 1996. *Made for Each Other: A Symbiosis of Birds and Pines*. Oxford:
 Oxford University Press.

Chapter 8. *The Philosophy of Ecological Forestry through Questioning*
(see general bibliography)

Chapter 20. *Fun with Numbers*
Center for Economic Conversion, 1997, Annual Report: *Transforming the
 economy by valuing earth's natural capital*.
Gretchen Daily, 1997, on putting a price on ecosystem services.
Jerry Franklin et al., 1997, *Forestry for the 21st Century*.
Ernie Niemi and Ed Whitelaw, 1997, *Assessing Economic Tradeoffs in Forest
 Management*.
Norgaard, 1997, *Learning from the Past: Timber and Community Well-being in
 Siskiyou County*.
David Perry, 1994, *Forest Ecosystems*.

Chapter 22. Gigatrends
(see general bibliography)

Chapter 23. Ecological Forestry Research
1. Usually isozymes (a variant in size shape or charge of an enzyme, which is
 a catalytic protein) or DNA,
2. Heterozygosity refers to zygotes that have inherited different alleles at a
 locus or loci; heterozygosity is thus a measure of genetic diversity. A
 zygote is a cell formed by the union of two gametes—a gamete is a cell
 with a set of chromosomes like a sperm or egg.
3. In object-oriented programs, objects, themselves composed of code,
 communicate with other objects through messages. Fractal geometry
 was invented by B. Mandelbrot and can duplicate natural features at

multiple scales.

Chapter 24. The Health of Forests
(see general bibliography)

Chapter 25. Forest Practices and Ecosystem Productivities
Costanza, Robert, Herman Daly, and Joy Bartholomew. 1991. Goals, agenda and policy recommendations for ecological economics. In Robert Costanza, ed., *Ecological Economics*. New York: Columbia University Press.
Needham, J. 1941. *Time, the Refreshing River*. Alan & Unwin, London.
Pielou, E. C. 1974. *Population and Community Ecology: Principles and Methods*. New York: Gordon and Breach.
Tansley, A. G. 1935. "The use and abuse of vegetational concepts and terms," *Ecology* 16:284-307.
Toumey, James W. 1947. *Foundations of Silviculture Upon an Ecological Basis*. Second ed. rev. C. F. Korstian. John Wiley & Sons, New York.
Waddington, C. H. 1969. The theory of evolution today. In A. Koestler and J. R. Smythies, eds., *Beyond Reductionism*. London: Hutchinson.
Wittbecker, A. E. 1976. The Poetic Archaeology of the Flesh. Mozart & Reason Wolfe, Ltd., Wilmington.
_____. 1991. "Incorporating the Earth," *Pan Ecology*, V 6, N. 8.
_____. 1995. "The ethics of ecosystem interference," (being reviewed).

Chapter 26. Good Forestry
1. Williams, Hugh. 1996. What is good forestry? *Env. Eth.* 18(4):391-410.
2. Arne Naess. Personal communication. See also Naess, A. 1972. The shallow and the deep, long-range ecology movement. A summary. *Inquiry*, 16: 95-100. And, 1991. *Ecology, Community and Lifestyle*. London: Cambridge University Press.
3. Franklin, Jerry. 1989. "A kinder, gentler forestry in our future: The rise of alternative forestry," *The Trumpeter* 6(3):99-100.
4. Perry, David. 1995. *Forest Ecosystems*. Baltimore: Johns Hopkins University Press. Perry, D. A., T. Bell, and M. P. Amaranthus. 1992. "Mycorrhizal fungi in mixed species forests and other tales of positive feedback, redundancy, and stability," in M. Cannell, D. Malcom, and P. Robertson, eds. *The Ecology of Mixed-species Stands of Trees*. Oxford: Blackwell Scientific.
5. Kimmins, Hamish.1992. *Balancing Act: Environmental Issues in Forestry*. Vancouver: UBC Press.
6. Wilkinson, Merve and Ruth Loomis. 1990. *Wildwood: A Forest for the Future*. Gabriola, BC: Reflections.
7. Camp, Orville. 1984. *The Forest Farmer's Handbook: A Guide to Natural Selection Management*. Ashland: Sky River.
8. Hammond, Herb. 1991. *Seeing the Forest Among the Trees*. Vancouver: Polestar Press.
9. Pilarski, Michael, ed. 1994. *Restoration Forestry*. Durango: Kivaki Press.
10. Plato. 1961. The Republic, Book 6. In *The Collected Dialogues of Plato*.

E. Hamilton and H. Cairns, eds. trans. L. Cooper et al. New York: Pantheon Books.

11. Partridge, Eric. 1983. *Origins*. New York: Greenwich House.
12. Becker, Ernst. *The Problem of Evil*.
13. Partridge, Eric. 1983. *Origins*. New York: Greenwich House.
14. Zajonc, Robert. 1980. Feeling and thinking: Preferences need no inferences. *American Psychologist* 35:151-175.
15. Williams, p. 392.
16. Maser, Chris. 1988. *The Redesigned Forest*. San Diego: R and E Miles.
17. Williams, p. 401.
18. Wheelwright, Philip. 1962. *Metaphor and Reality*. Bloomington, Indiana University Press.
19. Drengson, Alan. 1989. *Beyond Environmental Crisis: From Technocrat to Planetary Person*. New York: Peter Lang.
20. Williams, p. 401.
21. Williams, p. 393.
22. Wittbecker, A. E. 1976. *The Poetic Archeology of the Flesh*. Wilmington: Mozart & Reason Wolfe.
23. Williams, p. 400.
24. Lovelock, James. Lovelock, J. E. 1979. *Gaia: A New Look at Life on Earth*. Oxford: Oxford University Press.
25. Naisbitt, John. 1984. *Megatrends: Ten New Directions Transforming Our Lives*. New York: Warner Books. Wittbecker, Alan. 1996. Gigatrends in Forestry. *International Journal of Ecoforestry*. Vol. 11, No. 2.
26. Berlinski, David. 1995. *A Tour of the Calculus*. New York: Pantheon Books.
27. Thom, Rene. 1975. *Structural Stability and Morphogenesis*. Reading: Benjamin.
28. Holling, C. S. 1973. Resilience and Stability of Ecological Systems. In: *Annual Review of Ecology and Systematics*, R. F. Johnston et al., eds., Vol. 4: 1-24.
29. Williams, p. 399.
30. Varela, F. et al. 1974. Autopoiesis: The organization of living systems. *Biosystems* 5:187-196.
31. Reed Noss. Personal communication.
32. Weiss, Paul. 1953. Presidential Address, Am. Assoc. Advancement of Sci.
33. Williams, p. 405.
34. Williams, p. 406.
35. Whitehead, A. N. 1933. *Adventures of Ideas*. New York: Macmillan, p. 150.
36. John B. Cobb, Jr. Personal communication.
37. This contrast is reminiscent of those between the hemispheres of the human brain. We think of the left hemisphere as verbal, analytic, reductive, rational, linear, convergent, tree-like, and discontinuous; right as nonverbal, holistic, synthetic, intuitive, nonlinear, divergent, mythic, timeless, net-like, and diffuse.
38. Maruyama, Magorah. 1979. Transepistemological Understanding: Wisdom beyond theories. Maruyama, Magorah, ed. *Cultures of the Future*. The Hague: Mouton.
39. Williams suggests that the positivistic deconstructionists, on the basis of

epistemological skepticism, "can reinforce the ecological concern for conservation by pragmatically supporting the importance of restraint and limitation in our interactions with nature ..." How? They ignore the history of the forests. The only restraint shown seems to be in failing to discuss the reality of forests.

40. Bateson, Gregory. 1987. Men are grass. In W. I. Thompson, ed., *Gaia: A Way of Knowing*. Great Barrington: Lindisfarne Press.
41. Einstein, Albert and Leopold Infeld. 1966. *The Evolution of Physics*. New York: Simon and Schuster.
42. John B. Cobb Jr. Personal communication. See also Cobb, John B. Jr. 1992. *Sustainability: Economics, Ecology, and Justice*. Maryknoll: Orbis Books.
43. Odum, Eugene P. 1971. *Fundamentals of Ecology*. Philadelphia: Saunders.
44. Hammond, Herb. 1997. Draft Forest Standards for the Pacific Certification Council.
45. Williams gives great credit to human creativity, suggesting that it should enhance the creativity of "lower-ordered things." While agreeing that human management should be respectful, I think it is folly to require it for ecological value as Williams seems to do. Furthermore, he bases the intrinsic value of organisms on their creative value, whereas it might be better based on simple 'being' value; fitness is certainly as important as creativity and it means limiting creativity. Sharks are certainly have not been creative (other than procreative), but they have been very successful and they have intrinsic value regardless. Nor should ecosystems be ordered to enhance human creativity.
46. Williams, p. 406.
47. Fowles, John. 1979. Seeing Nature Whole. *Harper's*. 259:49-56.

Chapter 27. Ecological Design of Forest Ecosystems

Forestry Commission, 1994. *Forest Landscape Design*. London: HMSO.
Forman, R. T. T. and Michel Godron. 1986. *Landscape Ecology*. New York: John Wiley.
Lucas, Oliver. 1990. *The Design of Forest Landscapes*. New York: Oxford University Press.
Mollison, Bill. 1988. *Permaculture: A Designers' Manual*. Tyalgum, Aus.: Tagari Pubs.
(And see general bibliography)

Chapter 28. An Ecological Forest Care Plan

Barnes, Michael. Personal Communication, 1997.
Camp, Orville. 1992. *The Forest Farmer's Handbook: A Guide to Natural Selection Forest Management*
Diaz, N. and D. Apostol. 1994. *Forest Landscape Analysis and Design: A process for developing and implementing land management objectives for landscape patterns*. Washington: USDA For Serv PNW Region.
Franklin, J. F. and C. T. Dyrness. 1988. *Natural Vegetation of Oregon and Washington*. Washington: USDA For Serv Gen Tech Rep PNW-GTR-8.
Hart, Richard. 1994. *Handbook for Monitoring*. Ashland: Headwaters.
Martinez. Dennis. Personal Communication, 1998.

Walker, G. W. and N. S. Macleod. 1991. *Geologic map of Oregon*. US Geologic
    Survey.

Chapter 30. Forestry as Ecosystem Medicine
(see general bibliography)

Chapter 31. Virtue.
(see general bibliography)

General Bibliography for All Articles
Bachelard, G. (1969) *Poetics of Space*. trans. M. Jolas. Beacon Press, Boston, pp.
    4-6.
Bacon, F. (1901) *Novum Organum*. J. Devey, ed. P.F. Collier, NY.
Bateson, Gregory. 1987. Men are grass. In W. I. Thompson, ed., *Gaia: A Way of
    Knowing*. Great Barrington: Lindisfarne Press.
Bergstraesser, A. (1962) *Goethe's Image of Man and Society*. Herder, Freiburg.
Bly, R., ed. (1980) *News of the Universe: Poems of the Twofold Consciousness*.
    Sierra Club, San Francisco.
Bormann, B. T. and G. E. Likens. 1979. Catastrophic disturbance and the
    steady state in northern hardwood forests. *Am. Sci.* 67:660-669.
Botkin, Daniel B. 1990. *Discordant Harmonies*: A New Ecology for the Twenty-
    first Century. New York: Oxford University Press.
Boulding, K. (1956) *The Image: Knowledge in Life and Society*. University of
    Michigan Press, Ann Arbor.
Callicott, J. Baird. 1980. Animal liberation: A triangular affair. *Environmental
    Ethics* 2:319-321.
Carroll, George. 1976. Personal communication.
Caswell, M. 1986. In M. Begin et al., eds. *Ecology*: Individual, Population, and
    Community. Sunderland: Sinauer Associates.
Cheng, T.C. (1970) *Symbiosis*. Pegasus, NY.
Cobb, John B. Jr. 1972. *Is it Too Late? A Theology of Ecology*. New York: Bruce.
Cobb, John B. Jr. 1992. *Sustainability: Economics, Ecology, and Justice*.
    Maryknoll: Orbis Books.
Colinvaux, P. (1978) *Why Big Fierce Animals Are Rare*. Princeton University
    Press, Princeton, NJ.
Costanza, Robert, Herman Daly, and Joy Bartholomew. 1991. Goals, agenda
    and policy recommendations for ecological economics. In Robert
    Costanza, ed., *Ecological Economics*. New York: Columbia University
    Press.
Costanza, R., B. G. Norton, and B. D. Haskell, eds. 1992. *Ecosystem Health*:
    New Goals for Environmental Management. Washington: Island Press.
Cousteau, J.-Y. et al. (1981) *The Cousteau Almanac*. Doubleday and Company,
    Garden City, NY.
Daly, Herman and J. B. Cobb, Jr. 1989. *For the Common Good*. Boston: Beacon
    Press.
Darwin, Charles. 1962. *The Origin of the Species by Means of Natural Selection
    or the Preservation of Favoured Races in the Struggle for Life*. New York:
    Collier.

Dasmann, R. (1972) *Environmental Conservation*. 3rd Ed.Wiley, NY.

Defenders of Wildlife (1984) *Defenders*. Washington, DC.

Dubos, Rene. 1965. *Man Adapting*. New Haven: Yale University Press.

——————. 1976. Symbiosis between the earth and humankind. *Science* 193:459-462.

——————. (1980) *The Wooing of Earth*. Charles Scribner's Sons, NY.

Earman, John. 1992. *Bayes or Bust: A Critical Examination of Bayesian Confirmation Theory*. Cambridge: MIT Press.

Eckholm, Eric P. 1976. *Losing Ground: Environmental Stress and World Food Prospects*. New York: W. W. Norton.

Ehrlich, Paul and Peter Raven. 1992. Differentiation of populations. In M. Ereshefsky, ed., *The Units of Evolution*. Cambridge: MIT Press.

Elton, C. 1966. *Animal Ecology*. New York: October House.

Eltringham, S.K. (1979) *The Ecology and Conservation of Large African Mammals*. Macmillan Press, London.

Evernden, N. (1981) *Out of Place*. (in press).

Fowles, J. (1979) Seeing Nature Whole. *Harper's*. 259:49-56.

Franklin, Jerry. 1988. "Structural and functional diversity in temperate forests," Pp. 166-175 in E. O. Wilson, ed., *Biodiversity*. Washington: National Academy Press.

Fowles, J. 1979. Seeing Nature Whole. *Harper's*. 259:49-56.

Fox, M.W. 1978. Personal Communication.

——————. (1980a) *One Earth, One Mind*. Coward, McCann and Geoghehan Inc., NY, pp. 174-234.

——————. (1980b) *Returning to Eden: Animal Rights and Human Responsibility*. Viking Press, NY, pp. 19-141.

Franklin, J. and R. Waring. 1980. Distinctive features of the northwestern coniferous forest: Development , structure and function. In R, Waring, ed., *Forests: Fresh Perspectives from Ecosystem Analysis*. Corvallis: OSU Press.

Franklin, Jerry and C. T. Dyrness. 1988. *Natural Vegetation of Oregon and Washington*. Corvallis: Oregon State University Press.

Fuller, Buckminster. 1976. Personal Communication.

Golley, Frank B., K. Petrusewicz, and L. Ryszkowski, eds. 1975. *Small Mammals: Their Productivity and Population Dynamics*. New York: Cambridge University Press.

Gray, Russell D. 1988. Metaphors and methods. In Mae-Wan Ho and S. W. Fox, eds., *Evolutionary Processes and Metaphors*. New York: Wiley.

Greenberg, Cathryn H. et al. 1995. Vegetation recovery following high-intensity wildfire and silvicultural treatments in sand pine scrub. *The American Midland Naturalist* V. 133 (1)149-164.

Haines, Russell. 1994. *Biotechnology in forest tree improvement with special reference to developing countries*. Rome: FAO/UN.

Hardin, Garrett. 1977. *The Limits of Altruism*. Indiana University Press, Bloomington.

Hardin, Garrett. 1959. *Nature and Man's Fate*. New York: Holt, Rinehart and Winston.

Harris, Larry. 1984. *The Fragmented Forest: Island Biogeography Theory and the*

Preservation of Biotic Diversity. Chicago: UC Press.

Hart, Richard. 1994. Personal Communication.

————. 1994. *Monitoring Handbook.* Ashland: Headwaters.

Hebb, D.O. (1958) Alice in Wonderland or psychology among the biological sciences. In *The Biological and Biochemical Bases of Behavior.* H. Harlow and C. Woolsley, eds. University of Wisconsin Press, Madison.

Heidegger, M. (1960) *Being and Time.* 9th ed. Max Niemeyer Verlag, Tubingen.

Ho, Mae-Wan and S. W. Fox. 1988. Processes and metaphors in evolution. In Mae-Wan Ho and S. W. Fox, eds., *Evolutionary Processes and Metaphors.* New York: Wiley.

Holling, C. S. 1973. Resilience and Stability of Ecological Systems. In: *Annual Review of Ecology and Systematics,* R. F. Johnston et al., eds., Vol. 4: 1-24.

————. 1994. Simplifying the Complex: The paradigms of ecological function and structure. *Futures* Vol 26 (6):598.

Hoyt, J. (1984) Letter. Humane Society of the United States, Washington.

International Union for Conservation of Nature. (1984) *World Conservation Strategy in Action.* Gland, Switzerland.

Johnson, Lionel. 1988. The thermodynamic origin of ecosystems: A tale of broken symmetry. In B. H. Weber et al., eds., *Entropy, Information, and Evolution.* Cambridge: MIT Press.

Jonas, Hans. 1974. *Philosophical Essays* Englewood Cliffs: Prentice-Hall Inc.

Kaplan, R. (1983) The role of nature in the urban context. Irwin Altman and Joachim Wohlwill, eds. *Behavior and the Natural Environment.* Plenum Press, NY, pp. 127-159.

Keeling, C. D. and T. P. Whorg. 1992. Muana Loa: Atmospheric CO_2— modern record. In T. A. Boden et al., eds., *Trends 91: A Compendium of Data on Global Change.* Oak Ridge: Oak Ridge National Laboratory.

Kellert, S.R. (1983) Affective, cognitive, and evaluative perceptions of animals. Irwin Altman and Joachim Wohlwill, eds. *Behavior and the Natural Environment.* Plenum Press, NY, pp. 241-265.

Kimmins, Hamish. 1992. *Balancing Act: Environmental Issues in Forestry.* Vancouver: UBC Press.

Klein, David. 1972. "Toward an ecophilosophy." *Tomte Symposium on Ecology and Land Use,* Steinsgard, Norway.

————. 1983. Personal Communication.

Koestler, A. and J.R. Smythies, eds. 1969. *Beyond Reductionism: New Perspectives in the Life Sciences.* London: Hutchinson.

Kropotkin, P.A. (1972) *Mutual Aid: A Factor in Evolution.* New York University Press, NY.

Krutch, J. (1970) *The Best Nature Writing of Joseph Wood Krutch.* Pocket Books, NY.

Kuhn, T. (1970) *The Structure of Scientific Revolutions.* University of Chicago Press, Chicago.

Lackner, S. (1984) *Peaceable Nature.* Harper and Row, NY, passim.

Lanner, Ronald M. 1996. *Made for Each Other: A Symbiosis of Birds and Pines.* New York: Oxford University Press.

Laszlo, E. (1972) *Introduction to Systems Philosophy: Toward a New Paradigm of Contemporary Thought.* Harper Torch Books, NY.

Lehmann, Scott. 1981. Do wildernesses have rights? *Environmental Ethics* 3:129-146.

Leopold, A. 1949. *A Sand County Almanac. And Sketches of Here and There*. New York: Oxford University Press.

Lieth, Helmut. 1973. Primary Production: Terrestrial Ecosystems. *Human Ecology* 1(4):303-332.

Lertzman, Ken et al. 1996. From ecosystem dynamics to ecosystem management, In P. K. Schoonmaker et al., eds., *The Rain Forests of Home: A Profile of a North American Bioregion*. Washington: Island Press.

Lincicome, D.R. (1969) *The Goodness of Parasitism: A New Hypothesis*. Thomas C. Cheng, ed. Aspects of the Biology of Symbiosis. University Park Press, Baltimore, pp. 199-226.

Lorenz, Konrad. 1952. *King Solomon's Ring: New Light on Animal Ways*. trans. M. K. Wilson. New York: Crowell.

Lovejoy, A.O. (1964) *The Great Chain of Being*: A Study of the History of an Idea. Harvard University Press, Cambridge.

Lovelock, J. E. (1979) *Gaia*: A New Look at Life on Earth. Oxford University Press, Oxford, passim.

Ludwig, Donald, Ray Hilborn, and Carl Walters. 1993. Uncertainty, resource exploitation, and conservation: Lessons from history, *Science* 260:17-36.

Mandelbrot, B. B. 1982. *The Fractal Geometry of Nature*. San Francisco.

Mangold, Robert et al. 1993. *Tree Planting in the United States—1993*. Washington: USDA, Forest Service.

Margalef, R. (1968) *Perspectives in Ecological Theory*. University of Chicago Press, Chicago.

Margulis, Lynn. 1991. Big trouble in biology: Physiological autopoiesis versus mechanistic neo-Darwinism. In John Brockman, ed., *Doing Science*. New York: Prentice Hall Press.

Maruyama, Magorah. 1978. Transepistemological Understanding: Wisdom beyond theories. *Cultures of the Future*. The Hague: Mouton.

——————. 1980. Toward Cultural Symbiosis. *Evolution and Consciousness: Human Systems in Transition*. E. Jantsch and C. H. Waddington, eds. Addison-Wesley Publishing Co, Reading, MA, pp. 198-213.

Maser, Chris. 1994. *Sustainable Forestry*. Delray Beach: St. Lucie Press.

Maslow, A. (1968) *Toward a Psychology of Being*. 2nd ed. D. Van Nostrand, NY.

——————. (1971) *The Farther Reaches of Human Nature*. Viking Press, NY, passim.

Meadows, Dennis. 1982. "Fallacies in resource planning," in Charles Hewett, T. Hamilton, and I. Anderson, eds., *Forests in Demand*. Boston: Auburn House.

Mech, L. David. 1981. *The Wolf*. Second edition. Minneapolis: University of Minnesota Press.

Meeker, Joseph. 1974. *The Comedy of Survival*. New York: Charles Scribner's Sons.

Merleau-Ponty, M. (1962) *The Phenomenology of Perception*. trans. C. Smith. Routledge and Kegan Paul, London.

——————. (1968) *The Visible and the Invisible*. trans. A. Lingis. Northwestern University Press, Evanston, IL.

Mollison, Bill. 1988. *Permaculture: A Designers' Manual*. Tyalgum, Australia: Tagari Publications.

Myers, Norman. (1984) *Gaia: An Atlas of Planet Management*. Doubleday and Company, Garden City, NY.

——————. 1984. *The Primary Source*. New York: Norton.

Naess, Arne. (1972) The shallow and the deep, long-range ecology movement. A summary. *Inquiry*, 16: 95-100.

——————. 1983. Personal communication.

——————. 1990. *Ecology, community, and lifestyle: Outline of an Ecosophy*. Cambridge: Cambridge University Press.

Naisbitt, John. 1984. *Megatrends: Ten New Directions Transforming Our Lives*. New York: Warner Books.

Niven, C. (1967) *History of the Humane Movement*. Johnson, London, p. 27.

O'Laughlin, Jay et al. 1994. Defining and measuring forest health, In: R. N. Sampson and D. L. Adams, eds., *Assessing Forest Ecosystem Health in the Inland West*. New York: Food Products Press.

Odum, Eugene. 1971. *Fundamentals of Ecology*. 3rd ed. Philadelphia: Saunders.

Pielou, E. C. 1974. *Population and Community Ecology: Principles and Methods*. New York: Gordon and Breach.

Perry, David. 1994. *Forest Ecosystems*. Baltimore: Johns Hopkins.

Ponge, Francis. 1972. *The Voice of Things*. trans. B. Archer. McGraw-Hill Book Company, NY.

Popper, K. (1982) The place of mind in nature. R.Q. Elvee, ed. *Mind in Nature*. Harper and Row, San Francisco, pp. 31-59.

Portmann, Adolf. 1964. *New Paths in Biology*. Harper and Row, NY.

Rader, Melvin. 1964. *Ethics and the Human Community*. HRW, New York.

Rapport, D.J., et al. 1985. "Ecosystem Behavior Under Stress," *American Naturalist* 125:617-640.

Reinheimer, Herman. 1910. *Evolution by Cooperation: A Study in Bioeconomics*. No city.

Reichel-Dolmatoff, Gerardo. 1971. *Amazonian Cosmos*. Chicago: University of Chicago Press.

Reinheimer, H. 1910. *Evolution by Co-operation: A Study in Bio-economics*. NC: NP.

Rodman, John. 1977. The Liberation of nature? *Inquiry*. 20:83-145.

——————. 1977. Theory and practice in the environmental movement: Notes toward an ecology of experience. *The Search for Absolute Values in a Changing World*. International Cultural Foundation, Tarrytown, NY.

Rorty, Richard. 1982. Mind as ineffable. *Mind in Nature*, ed. R. Elvee. Harper and Row, San Francisco, p. 88.

Roszak, Theodore 1972. *Where the Wasteland Ends*. New York: Harper & Row.

——————. 1979. *Person/Planet*. Harper and Row, NY.

Salk, Jonas, 1973. *Survival of the Wisest*. Harper and Row, NY.

Salthe, Stanley N. 1985. *Evolving Hierarchical Systems*. New York: Columbia U. Press.

Schneider. 1988. Thermodynamics, ecological succession, and natural selection: A common thread. In B. H. Weber et al., eds., *Entropy, Information, and Evolution*. Cambridge: MIT Press.

Schaller, G.B. 1972. *The Serengeti Lion*. University of Chicago Press, Chicago.

Scheler, M.F. 1954. *The Nature of Sympathy*. trans. P. Heath. Routledge and Kegan Paul, London. .

Schweitzer, Albert. 1949. *Out of My Life and Thought*. New York: Henry Holt and Co.

————. 1957. *The Philosophy of Civilization*. Translated by C. T. Campion. New York: Macmillan Co.

Sewell, E. 1960. *The Orphic Voice: Poetry and Natural History*. Harper and Row, NY.

Sharp, Henry. Comparative ethnology of the wolf and the Chipewyan. *Man and Wolf*. NC: NP.

Shepard, Paul. 1974. Animal rights and human rites. *The North American Review Winter*, p. 35.

Shepard, Paul and D. McKinley, eds. 1969. *The Subversive Science*. Boston: Houghton Mifflin.

Shepard, Paul. 1978. *Thinking Animals*. Viking Press, NY, pp. 205-249.

————. 1982. *Nature and Madness*. San Francisco: Sierra Club Books.

Singer, Peter. 1981. *The Expanding Circle: Ethics and Sociobiology*. New York: Farrar, Strauss and Giroux.

Skolimowski, Henryk. 1978. "Ecophilosophy versus the scientific world view." *Ecologist Quarterly* 3 (Autumn): 227-248.

Skolimowski, H. (1981) *Ecophilosophy*. Marion Boyars, Boston.

Snyder, Gary. 1969. *Earth House Hold*. New Directions, NY.

Soleri, Paolo. 1983. *The Food Chain: A Celebration*. Scottsdale, Arizona.

Soleri, Paolo. 1969. *Arcology: The City in the Image of Man*. Cambridge: The MIT Press.

Stanley, S. (1981) *The New Evolutionary Timetable: Fossils, Genes and the Origin of Species*. Basic Books, NY.

Stocek, Karl. 1996. Personal communication.

Stone, Christopher. 1974. *Should Trees Have Standing?* Avon Books, New York.

—. 1987 *Earth and Other Ethics*. New York: Harper & Row.

Tansley, A. G. 1935. "The use and abuse of vegetational concepts and terms," *Ecology* 16:284-307.

Taylor, P. W. 1986. *Respect for Nature*. Princeton: Princeton University Press.

Thomas, L. (1975) *Lives of a Cell*. Bantam Books, Toronto.

Todd, N. J. and J. Todd. 1984. *Bioshelters, Ocean Arks, City Farming: Ecology as the Basis of Design*. San Francisco: Sierra Club Books.

Toumey, James W. 1947. *Foundations of Silviculture Upon an Ecological Basis*. Second ed. rev. C. F. Korstian. John Wiley & Sons, New York.

Tuan, Y.-F. (1974) *Topophilia*. Prentice-Hall Inc., Englewood Cliffs, NJ.

Uexkull, J. von. 1957. A stroll through the world of animals and men. In *Instinctive Behavior*, C. Schiller, ed. New York: International Universities press.

Varela, F. et al. 1974. Autopoiesis: The organization of living systems. *Biosystems* 5:187-196.

Varela, F. 1979. *Principles of Biological Autonomy*. New York: North Holland.

Waddington, C. H., ed. 1969. *Towards a Theoretical Biology*. Chicago: Aldine Publishing Co.

Waddington, C. H. 1969. The theory of evolution today. In A. Koestler and J. R. Smythies, eds., *Beyond Reductionism*. London: Hutchinson.

Waddington, Conrad. 1975.*The Evolution of an Evolutionist* Ithaca: Cornell University Press.

Walters, Carl. (see Ludwig et al.)

Waring, R. H. 1980. Vital signs of forest ecosystems, In: R. H. Waring et al., eds. *Forests: Fresh Perspectives from Ecosystem Analysis. Proceedings, 40th Annual Biology Colloquium*. Corvallis: OSU Press.

Welch, H. (1966) *Taoism: The Parting of the Way*. Revised ed. Beacon Press, Boston.

Weil, S. (1955) *The Need for Roots*. Beacon Press, Boston.

Weiss, Paul A. 1967. "One Plus One Does Not Equal two." In *The Neurosciences: A Study Program*. G.C. Quarton et al., eds. New York: Rockefeller University Press.

Wilson, E. O. 1984. *Biophilia*. Cambridge: Harvard University Press.

Whitehead, Alfred North. 1933. *Adventures of Ideas*. New York: Macmillan.

——————. 1958. *The Function of Reason*. Boston: Beacon Press.

——————. 1967. *Science and the Modern World*. Free Press, NY.

——————. 1968. *Modes of Thought*. New York: Free Press.

——————. 1978. *Process and Reality*. ed. D. R. Griffin and D. W. Sherburne. New York: Free Press.

Whyte, L.L. (1965) *Internal Factors in Evolution*. G. Braziller, NY.

Wilson, E.O. (1984) *Biophilia*. Harvard University Press, Cambridge, MA, pp. 6-11.

Wittbecker, Alan. 1983a. NEP model of an optimum global population. Fargo: ESA paper.

——————. 1995. "Gigatrends in Forestry," *International Journal of Ecoforestry* 11(2/3):69-78.

——————. 1997. "Forest practices related to forest ecosystem productivity," in *Ecoforestry*, A. Drengson and D. Taylor, eds. Gabriola Island: New Society Publishers.

——————. 1997. "What good is philosophy? And other questions about forestry," *International Journal of Ecoforestry* 12 (3/4):7-11.

——————. 1997. "Principles of Ecoforestry. Part I." *International Journal of Ecoforestry* 12 (3/4):35-40.

——————. 1997. "An Ecoforestry Research program," *International Journal of Ecoforestry* 12 (3/4):41-48.

——————. 1998. "The health of forests," *Ecoforestry* 13(1):18-28.

——————. 1999. "Good Forestry," *Ecoforestry* 14(2):4-8.

——————. 1999. "Varieties of Interaction in nature," *The Trumpeter*, Spring (Web Edition: www.athabascau.ca/trumpeter).

——————. 1999. "Forestry Poetic Activity ..." Closing Address, Forests for the Future, Vancouver Island.

——————. 1999. "Global Logic or Local Knowledge," *Ecoforestry* 14(4):4-7.

——————. 2000. "Good Forestry: Neutrality, death, sowbugs, and ecological principles," *Ecoforestry Notes* 15(1):1-18. (Web Edition: www.uidaho.edu/e-journal/ecoforestry)

Index

Author Biography

During a brief career in astrophysics and astronomy at the University of Arizona, where he worked on mathematical models of stars and on spectrometric analysis, Alan Wittbecker spent his daylight hours climbing trees and trying to track mountain lions; he shared his small trailer near the observatory on Mount Lemon with a crow, squirrel and a mouse.

Encouraged by research budget cuts to pursue a different direction, Wittbecker went to graduate school in psychology, anthropology, philosophy, and ecology (his degrees are in these fields). As a graduate student in 1970, he was a cofounder of the G. P. Marsh Institute for Research in Ecology, where he worked for 22 years (including 3 as Director by rotation). When projects were sparse, he worked in other occupations, such as librarian, systems engineer, editor, graphic artist, typesetter, housepainter, television repairman, cook, swimming coach, carpenter, clinical psychologist (drug abuse clinic), auto mechanic (Austins), tree-planter, and instructor.

In 1976, with three partners, Wittbecker cofounded Nieman Ryan Community Designs, specializing in private and urban local landscape design—but, also designing books, posters, journals, packages, landscapes, and buildings. He continued his postgraduate education in landscape ecology, forestry, conservation biology, zoology, and genetics.

In 1992, Wittbecker founded SynGeo ArchiGraph (syngeo.net), a firm specializing in global and regional designs; he created designs for several bioregions, as well as international frameworks. A year later he set up the educational program for the new Ecoforestry Institute (ecoforestry.us), becoming an Instructor in 1994, journal Editor in 1995, and Director in 1997. He has worked on public and private forests from British Columbia to California, and on wildlife projects, from Siberia to Norway. He is the author of three books, including *The Poetic Archaeology of the Flesh: An Investigation into the Phenomenology and Ecology of Being*, and over 100 articles.

A veteran of the US Air Force, Wittbecker is also a returned Peace Corps Volunteer from Bulgaria, where he monitored wolves in the Central Balkan Mountains. When not engaged in preservation activities, he enjoys walking, swimming, reading, and drawing, at the Altazor forest in western Idaho. To discuss any of these essays with him, contact him at "emt@ecoforestry.us". Thank you.

Author's Notes
This book was taken from a much larger, unfinished work-in-progress, which is available as a notebook on our website, www.eutopias.net. The sections in that work are numbered sequentially, and the numbers are kept for this book, even though over eighty percent of them do not appear. There are, alas, a few misspellings, some bad grammar and many unfinished thoughts. Some of the ideas in this work were radical thirty-seven years ago, although some have become acceptable or commonplace. Others are still considered awkward or unpalatable. I trust you will be able to participate in this conversation despite these flaws and shortcomings. Thank you for your consideration.

To make up for the loss of trees and their services, as a result of my use of paper in these books, I have planted over nine thousand trees, during a period of twenty years, at the Altazor Forest in Idaho. More plantings are planned in Oregon forests and Virginia farms.

Colophon

Type: Palatino (designed by Hermann Zapf in 1948 at Stempel AG)
Display Type: Palatino
Book Design: Rian Garcia Calusa Designs
Cover Design: Rian Garcia Calusa
Photographs & Graphics: Alan Wittbecker
Author Photo: Mike Barnes
Editing: J. Garcia B. of Rian Garcia Calusa
Hardware: Macintosh G5
Software: Adobe InDesign & Acrobat
Furious Charge & Entertainment: Pippi Frog
Spiritual & Material Support: Precious Woulfe

www.ingramcontent.com/pod-product-compliance
Lightning Source LLC
Chambersburg PA
CBHW022110210326
41521CB00028B/177